RESEARCHES

ON

MAGNETISM, ELECTRICITY, HEAT, LIGHT, CRYSTALLIZATION, AND CHEMICAL ATTRACTION,

IN THEIR RELATIONS TO

THE VITAL FORCE.

BY

KARL, BARON VON REICHENBACH,

PH. DR.

TRANSLATED AND EDITED, AT THE EXPRESS DESIRE OF THE AUTHOR, WITH A PREFACE, NOTES, AND APPENDIX,

BY

WILLIAM GREGORY, M.D., F.R.S.E.,

PROFESSOR OF CHEMISTRY IN THE UNIVERSITY OF EDINBURGH.

New Forward to 1974 Edition

by LESLIE SHEPARD

With Three Plates and Twenty-three Wood-cuts.

PARTS I. AND II.,

INCLUDING THE SECOND EDITION OF THE FIRST PART, CORRECTED AND IMPROVED.

UNIVERSITY BOOKS SEACAUCUS, NEW JERSEY

Library of Congress Card No. 74-75173

ISBN No. 0-8216-0216-0

Printed in U.S.A. by
NOBLE OFFSET PRINTERS, INC.
New York, N.Y. 10003

NEW FOREWORD TO 1974 EDITION

The solemn and pedantic title of this book conceals one of the great lost masterpieces of scientific research.

This revolutionary work throws light upon the connections between physical science and psychic force. It links together all those studies on the borderland between science and superstition—things like Mesmerism, Animal Magnetism, the Human Aura, Water-witching, Mediumship, and Psychic Healing.

It was based on massive researches extending over many years. Because it ran counter to the mechanistic and doctrinaire orthodoxy of nineteenth century science it was ridiculed, suppressed, and eventually buried in obscurity. A handful of discriminating individuals kept Reichenbach's discoveries alive, but the subject was by-passed by the main line of scientific development. Over a hundred years have passed since publication of this groundbreaking book, and copies have become excessively difficult to find.

Readers who were introduced to the work of Reichenbach by his popular summary *The Odic Force; Letters on Od and Magnetism* (University Books, Inc.) found an exciting and stimulating field for research. This timely revival of interest is now reinforced by reissue of Reichenbach's major work, one of the most complete and scientifically valid investigations of psychic force ever attempted.

BARON KARL VON REICHENBACH (1788-1869) was one of the greatest scientists of the nineteenth century, preeminent expert on meteorites, highly respected as chemist, technologist, and metallurgist, the discoverer of kerosene. He spent over two decades experimenting with the mysterious force which he named "Od" (here translated as *Odyle*), and described its characteristics and effects as manifest through the perception of sensitive persons.

The name *Od* was derived from *Odin,* the Norse deity, and

was intended to indicate a power permeating the whole of nature. Such a concept is an ancient one, and a persistent one. It is the *mana* of primitive societies. It has recurred constantly in European philosophy ever since the *Vis Medicatrix Naturae* of Hippocrates, ca. 400 B.C. It is the *Magnale* of Paracelsus, the *Alcahest* of van Helmont, the *Astral light* of the Kabbalists, the *Azoth* of the Alchemists, the *Spiritus* of Fludd, the *Animal magnetism* of Mesmer, and a score of similar concepts. More recently it was revived in the *Orgone energy* of Dr. Wilhelm Reich. The important thing about Reichenbach's *Od* is that he validated his concept in a purely scientific spirit, divorced from the occultism which had usually surrounded it.

Detection and demonstration of this mysterious force depended upon sensitives—specially gifted individuals. The term is often used in modern times to denote a spiritualist medium, but Reichenbach's senitives were ordinary people from all walks of society. I do not doubt, however, that psychic mediums are rare and talented individuals in whom this type of sensitivity is more highly developed.

Reichenbach's sensitives experienced specific reactions to the proximity of other people—feelings of pleasant coolness and drowsiness or, on the other hand, disagreeable, numbing or exciting feelings.

Reichenbach discovered a special right-hand left-hand polarity in the human body, a polarity to which sensitives were exceptionally responsive. This affected their reactions to other people standing or sitting near to their right or left sides, and particularly to sleeping positions with partners. They were also sympathetic to the color blue, and antipathetic to yellow; they had particular food fads, were sensitive to certain metals, and were unpleasantly affected by mirrors. Incidentally I have a suspicion that some of the elaborate personal rules of some religions, such as the Jewish *Kitzur Shulchan Aruch*, are related to such special sensitivities and polarities.

In a long series of experiments with some two hundred

individuals, Reichenbach established that the sensitive person could see emanations from crystals and magnets in total darkness, and could detect alternations of electric current. Sensitives could also perceive an aura surrounding the human body.

Reichenbach studied the various manifestations of this vital force in its relationship to electricity, magnetism, and chemistry. He showed that it was connected with the phenomena of water-witching, mesmerism, and similar psychic subjects. He demonstrated that the force could move objects without conscious effort, as in the table-turning of the spiritualists. He studied the negative and positive polar effects of *Od* (or *Odyle*) and demonstrated that liquids and solids could be charged with it by gifted individuals.

Reichenbach was not a spiritualist or a mesmerist. His interest was purely scientific, and his hundreds of careful experiments were conducted as precisely as in the other varied subject-areas in which he had earlier become so famous. Unfortunately the unavoidable reliance upon sensitiveness laid Reichenbach open to accusations of suggestion or mesmerism. At that time orthodox scientists were striving to maintain a mechanistic view of life.

The painful story of how Reichenbach's experiments were wrecked by arrogant and unscrupulous dogmatists has been dealt with in detail in the prefatory matter to the reissue of *The Odic Force; Letters on Od and Magnetism*. Reichenbach did not secure a fair hearing. In spite of his great reputation in other fields of science, his researches on Odic phenomena and his numerous publications on the subject were ridiculed, attacked, and suppressed until his death a quarter of a century later. His great work vanished in obscurity and misrepresentation. A conspiracy of silence prevented its proper revival. Reichenbach died at the age of eighty and, as Gustav Fechner, another great scientist, commented: "Up to the last days of his life, he grieved at the thought of having to die without obtaining recognition for his system, and such was the tragic fate that actually befell him."

New Foreword to 1974 Edition

Some thirteen years after Reichenbach's death, there was a belated revival of interest by the Society for Psychical Research in Britain, who formed a Reichenbach Committee which included eminent individuals such as Professor W. F. Barrett, Edmund Gurney, and Frederic W. H. Myers. It was precisely the connection with psychic phenomena that inspired this renewal of interest in a subject pointedly ignored by orthodox science. The Committee made careful investigations but was less fortunate than Reichenbach in obtaining suitable sensitives. Only three out of the forty-five individuals tested possessed the sensitivity postulated by Reichenbach, but these three provided interesting confirmation of Reichenbach's observations.

In the early twentieth century, further reinforcement of Reichenbach's discoveries was claimed by Dr. Bernard Hollander in Britain. He confirmed that non-sensitives—normal healthy individuals without special sensitivity—could also see the visual phenomena of magnets in total darkness *provided that they were in a hypnotic state*. This would indicate that the requisite sensitivity is latent in ordinary individuals, although not immediately accessible as in the case of sensitives. It also suggests that sensitivity might be developed, as indeed many dowsers or water-witchers have often claimed.

In 1908, Dr. Walter J. Kilner, who was familiar with the work of Reichenbach, developed a technique for making the human aura visible. His findings were published in his book *The Human Atmosphere*, first published 1911. The revised edition of 1920 was reissued by University Books, Inc. under the title *The Human Aura* in 1965. Confirmation of Dr. Kilner's findings, with additional original research, was described by Oscar Bagnall in his book *The Origin and Properties of the Human Aura*, London, 1937; a new revised edition of this important book was published by University Books, Inc., 1970. Two other books which have a bearing on the subject of vital radiation are: *Mitogenetic Analysis of the Excitation of the Nervous System* by Alexander Gurwitsch, a Russian cytologist (English translation published Amsterdam 1937), and *Invisible Radiations of Organisms* by Otto Rahn, Professor of Bacteriology, Cornell University (published Berlin, 1936). In 1936 the American physician George Crile reviewed many aspects of cellular electricity in his book *The Phenomena of Life; a Radio-Electric Interpretation.* Crile had also experimented with the measurement of emotion, and the detection of vital currents in vegetable life.

Reichenbach's discovery of a mysterious vital force in plants, animals, and human beings, anticipated the *Orgone energy* of Dr. Wilhelm Reich, and it is sad to record that Reich was also ridiculed by orthodoxy. As a young pupil of Freud, and the first man to place psychoanalysis upon a sound psychsomatic basis, he deserved better treatment. His claimed invention of an apparatus to accumulate *Orgone energy* was the subject of a sustained campaign by the F.D.A. in the U.S. Reich's work was judged and condemned by anonymous investigators. When he refused to defend himself (on the ground that issues of natural science could not be judged in a court of law) he was sentenced for contempt. He was heavily fined, and imprisoned in Lewisburg Penitentiary, Pennsylvania, where he died November 3, 1957. Under an extraordinary court ruling, his books which described *Orgone energy* were burnt by the U.S. authorities on August 23, 1956, in New York. After Reich's tragic death the books have been freely reprinted. It is ironic that Reich's theories of sexual energy (which he related to Orgone) have become symbols of today's permissive society.

Reich's work adds an important new dimension to the researches of Reichenbach, because Reich emphasized the question of sexual energy and its effects in changes of body electric potential. Many of Reich's observations also confirm the treatises of the ancient Hindus, who related the cosmic energy *Kundalini* to specific expressions of sexuality and mystic states through different phases of *prana* or vital currents in the human body. In recent times Pandit Gopi Krishna has validated the reality of the fabulous Kundalini in a book which describes his personal experiences—*Kundalini; the*

Evolutionary Energy in Man (Berkeley, California/London, 1970).

The present reissue of Reichenbach's great work is a timely one, for it supplies a missing chapter in the study of human sensitivity in relation to vitalistic theories of life. A revival of this work offers exciting and rewarding avenues for new research. We do not know enough about the actual nature of vital energies in the body, nor of vital forces in nature. There is a wide range of electro-magnetic phenomena in living things, from the effects of cell radiation investigated by Gurwitsch, Rahn, and Crile, to the changes in skin potential measured by Reich. There is a vast and complex field between subtle phenomena observed by faint luminosity or measurable only in microvolts, and the sensational effects claimed by spiritualist mediums in the materialization of phantom forms and the movement of heavy objects at a distance. And a revival of interest in Kundalini Yoga may lead us nearer to the more metaphysical riddle of vital forces that may be expressed either in procreation or in higher mysticism. It may be some time before we can express a unified field theory or equations to cover such a variety of phenomena, but we can be certain that the work of Karl von Reichenbach is an essential part of this great puzzle.

The present book was first published in Germany in 1850. There were actually two English translations. This is the official translation by Professor William Gregory, who also translated Reichenbach's preliminary observations of 1845 as well as the later *Letters on Od and Magnetism.* In 1851, Dr. John Ashburner, a physician very much concerned with Mesmerism and Phrenology, produced an independent translation under the formidable title *Physico-Physiological Researches on the Dynamics of Magnetism, Electricity, Heat, Light, Crystallization, and Chemism, in their relations to Vital Force*—a faithful but unnecessarily pedantic equivalent of the original German title. There is very little to choose between the Gregory and Ashburner translation, but on balance Gregory's official translation is regarded as the best. There is, however, one important detail to bear in mind. Gregory translated Reichenbach's term *Od* as *Odyle,* perhaps feeling that it would sound more acceptable to scientists.

This, then, is a reissue of the famous *Researches* of Reichenbach, after more than a hundred and twenty years. In conjunction with his *Letters on Od and Magnetism* it comprises a complete presentation of the Odic Force, and should suggest fruitful lines of research for modern scientists.

It is time that Reichenbach had a new hearing. At a period when Witchcraft and Magic are accepted subjects at American universities, perhaps it is not too much to hope for a sympathetic investigation of this long-neglected work of a great scientist.

LESLIE SHEPARD

CONTENTS OF PART I.

CONTENTS OF PART II.

EDITOR'S PREFACE.

THE present publication is the first volume of a work, in which the whole of the observations made, up to this time, by Baron VON REICHENBACH, on the very interesting and important subject on which it treats, are to be permanently placed on record; and, at the request of the Author, I have undertaken to lay these researches, as they shall appear, before the British Public.

The present volume, which includes all that has yet been published in Germany, consists of two Parts.

Part I. is a new and improved edition of that part of the work which was published by the Author in LIEBIG'S Annalen, March and May 1845; but no essential alteration has been made in it. It contains a general historical sketch and summary of the whole investigations, as made up to the latter part of 1844, when it was written; and only enters into such details as appeared necessary to establish the existence of the new Imponderable or Influence, Odyle; and to trace it in the numerous sources from which it is found to flow. These are, magnets, crystals, the human body, the sun, the moon, the stars, heat, electricity, friction, chemical action, and the whole material universe; and the reader will find various interesting and valuable applications of the facts thus ascertained; for example, to the explanation of corpse-lights, the origin of many ghost-stories, and further on, to dietetics, &c. None of these sources of odyle are, in Part I., treated fully or minutely. The experiments detailed in it were made on upwards of twelve sensitive persons, of whom one half, those who exhibited the highest degree of sensitiveness, were females affected with various diseases of the nervous system, while the remainder were strong healthy men. It is necessary here to mention this, because in various passages, written in 1844, the Author expresses the opinion which he then held, that the sensitive state is essentially a morbid one, and that perfectly healthy persons are perhaps never sensitive. His subsequent researches, as detailed in Part II., have proved the fallacy of this opinion; but the original passages remain, and might, without explanation, puzzle the reader.

Of this First Part, or general summary of the investigations down to the end of 1844, I published an Abstract early in 1846. My object in doing so, was simply to direct the attention of my countrymen to those admirable researches; and to render the work more readable and popular, I condensed the translation into about one-half the bulk of the original, without, however, omitting any essential point. The abridged portions were those consisting of minute and often repeated details of the experiments, essentially necessary, no doubt, to the permanent value of the work, as embracing the evidence produced; but not, in all their details, required for the purpose of directing public attention to the subject; especially as I always intended to translate the complete work when it should appear. I was well aware, and mentioned in the preface to that Abstract, that I was not doing full justice to the Author, in omitting any part of that evidence, but I felt convinced, that in the meantime, a more popular sketch would answer better the intended purpose.

The reception of the Abstract in this country, so much more favourable than that accorded to the original on the Continent, has, I think, fully justified this abridgement. I now present to the public Part I. in its full extent, and in a permanent form; and I am persuaded that, on comparison with the Abstract, it will be found that, in every point but that of the full detail of *all* the numerous and similar experiments, the summary of Baron VON REICHENBACH was faithfully represented in that publication.

The very favourable manner in which my Abstract has been received in this country, demands my warmest acknowledgements. Not only was the edition very rapidly sold, but I have been, ever since 1846, favoured with letters of enquiry concerning a new edition, and of high approval of the work, so numerous, that I have found it quite impossible to return answers individually to nearly the whole of them. I beg here to apologise for all omissions, and to explain, that I should, long ere this, have republished the Abstract, had I not been in constant expectation of receiving from the Author the succeeding parts of his great work, which I had undertaken to publish in full. The present volume is, then, the commencement of that publication, delayed for reasons which are given in the Author's Preface. But I may here state, that the chief of these is not there sufficiently brought into view; namely, the fact, that upwards of three years were devoted by the Author, after publishing the summary in 1845, to a laborious and minute study of all the branches of the subject. The results of these three years' researches, (as far as they concern only one of these branches,) are now given in Part II., which, if published much sooner, must necessarily have been far less complete than it now is.

It may here be mentioned, that the Abstract of Part I. was favourably noticed in various scientific and literary journals, as well as in the daily press. Indeed, up to this time, I have not become acquainted with any scientific criticisms, published in this country, on the Author's researches, which require any notice from me in this place. This, as will be seen by the Author's Preface, forms a strong and favourable contrast with the reception given to Part I. by various men of science in Germany. It is pleasing to reflect, that a work so truly scientific in its character, has, in spite of the startling nature of the facts recorded in it, received from the British public that respectful and becoming attention, to which, from the known scientific reputation of its Author, it was justly entitled. It must be gratifying to the numerous English readers of the Abstract to know, and to this I can myself testify, that the lamented BERZELIUS took a very deep interest in the investigation, and expressed, in a letter to the Editor, his conviction, that it could not possibly have been in better hands than those of Baron VON REICHENBACH.

So much in reference to Part I.

Part II. consists of several sections. In the first, the Author enumerates, by name, upwards of fifty new sensitive persons, of whom thirty-five are perfectly healthy and vigorous, yet among whom he has found all degrees of sensitiveness, even to the highest, and whose descriptions and statements exhibit a perfect harmony. It is unnecessary here to enter into any detail. In the second section, the author institutes a minute comparison of the new influence, odyle, with the known influences or imponderables, heat, electricity, and magnetism (proper); and establishes, in a very beautiful and convincing manner, both its great analogy to these, and the striking differences which compel him to distinguish it from them, and which leave no choice, for the present, but that of giving it a distinct place and name, although future discoveries may possibly enable us to refer *all* the imponderables to a common force. But that time is still distant; and, in every event, a name will be required for the new group of distinct phenomena, just as we find it necessary to apply the names of magnetism and electricity to two such groups, the close relation, and possibly common origin, of which must be universally admitted. These sections constitute the Introduction.

The third, and chief section of Part II. is a very full, minute, and able developement of the various forms, characters, and properties of the light given out by magnets of all kinds, and visible, in the dark, to the sensitive. It may here be mentioned, that the Author now finds fully one-third of people in general to be more or less sensitive. This, as far as my own observations extend, I am inclined to consider as not exaggerated. The highest degree of sensitiveness is comparatively rare, but is still common enough, even among the healthy. This, the light *from magnets,* is the only form of odylic light treated of in Part II. The observations, generally, of the fifty new sensitives, are first given. Then come, specially, the *forms* of the light, such as glow, flames, fibrous downy light, scintillations, and luminous nebulæ, vapours, smoke, and clouds; to which is subjoined a most ingenious theory of the spectres and demons of the Walpurgis-nacht on the mountains of the Harz forest. Next, we have the effects produced on it by different media, such as rarefied air, or the vacuum, water, and solid media. The *colours* of the odylic light of magnets are then treated of; and the Author demonstrates in it the presence of all the prismatic colours, very often in the form of a regular Iris; with many other highly interesting details. The next investigates most successfully, and with most beautiful results, the remarkable effects produced on the light and its colours, by the varied positions of the magnet, and by the magnetism of the earth. Lastly, he examines the effects of giving to the poles of magnets differently-shaped terminations, and of employing disc-shaped and spherical magnets. The results are singularly beautiful and interesting.

The last section consists of an application of these most remarkable observations, to the explanation of the Aurora Borealis, which, in fact, the Author had produced, *with all its characteristic phenomena in high intensity and perfection,* and in a form visible to the sensitive, in the last mentioned experiments with spherical magnets. To the consideration of this subject, the Author brings a vast store of valuable information, and all the resources of a mind accomplished in physical science. The result is, a theory or explanation of the aurora, which must, I think, be admitted to have the highest probability; and to be, in the present state of our knowledge, the best within our reach. The reader is referred to this very striking chapter for all details.

Such is a very brief and condensed sketch of the contents of Part II., which, as the reader is aware, now appears for the first time.

I have not felt justified in omitting any part of the Author's Preface, or of his notes, both of which are controversial, and may appear to some rather energetic in their tone. But as they contain the Author's replies to attacks of a very violent and unjustifiable kind, which have been widely circulated in Germany, and may possibly have influenced such as have read them in this country, I thought it but just to let him be heard. It is at all times deeply to be regretted, when scientific discussions assume an angry or acrimonious tone; but in this case, there was nothing in Part I. to justify the style in which it was treated; and consequently, the introduction of unscientific language into the discussion, must be ascribed to those who so unsparingly employed it in their criticisms, if indeed we can truly give that name to such productions, written, as they are, without due knowledge, either of the subject or of the work.

I might here conclude; but although, in this country, no such irrational and unwarrantable attacks have appeared, yet I have occasionally met, in private, with persons disposed, like the German critics, to reject the whole investigation, without even giving it a careful study, or indeed any study at all. I have further observed, to a certain extent, the prevalence of very erroneous views as to the proper objects of scientific enquiry, and especially as to the true nature and value of scientific evidence, in relation to obscure and difficult subjects. I trust, therefore, the reader will excuse me, if I venture to trespass on his patience with some brief remarks on these points.

And first, as to the objects of enquiry. It may be safely laid down, that no well-ascertained fact, or series of facts, can possibly exist in nature, which are not worthy of our earnest study. It is no answer to say, that there are many facts of no practical value; for the fallacy of such an assertion has been too often proved, by the unexpected application of apparently trifling facts to important purposes, theoretical as well as practical. When LIEBIG and SOUBEIRAN first described a peculiar volatile fragrant liquid, produced by the action of bleaching powder on alcohol, this fact was recorded, and stood for many years in our manuals, as a mere curiosity of science. Yet when the time came, this insignificant compound, under the hands of my accomplished colleague, Dr. SIMPSON, started up into the highest practical value, as CHLOROFORM. Had *all* the properties of this body been at first *carefully studied*, we should not have had to wait till 1846 or 1847 for this boon to suffering humanity. It would be easy to fill a volume with parallel cases, from the Steam Engine to Gun Cotton, all proving the great truth, that no natural fact is insignificant or unimportant.

Secondly; as to the question, whether the phenomena studied in this work be *facts* or not; I may safely leave this to the decision of the careful reader of the work; and I shall only express my conviction, that if these facts be not facts, then do no facts exist in any department of science.

Thirdly; as to the mode of investigation adopted by the Author: it is the inductive method, the same to which we owe all the progress of modern science; and in which the conclusions are deduced from carefully observed phenomena, varied by experiments performed with due attention to accuracy. No other method is known, by which natural phenomena, and especially obscure natural phenomena, can be investigated with any chance of success.

This method has been most diligently, laboriously, and successfully applied by the Author to a class of phenomena, which might perhaps be considered at once more obscure and more interesting than those which form the objects of many other departments of science, if it could justly be said that one part of nature is more interesting than another. I feel constrained to say, that in the course of a life devoted to science, I have met with no researches in which the true and universally approved rules of investigation have been more perfectly adhered to and followed out, than in those before us; which, were it necessary, might serve as a model to all experimental inquirers.

The qualifications of the Author for such an inquiry are of the very highest kind. He possesses a thorough scientific education, combined with extensive knowledge. His life has been devoted to science, and to its application to the practical purposes of mankind.

He is known as a distinguished improver of the iron manufacture in his own native country, Austria. He is a thorough practical Chemist; and, by his well known researches on Tar, has acquired a very high position. But in Geology, Physics, and Mineralogy, he has been equally active. In particular, he is the highest living authority on the subject of meteorites or aerolites, of which remarkable bodies he possesses a magnificent collection. Of his knowledge on this subject, good use is made in this work.

But these are the least of his qualifications. He has a turn of mind, observing, minute, accurate, patient, and persevering, in a rare degree. All his previous researches bear testimony to this; and, at the same time, prove that he possesses great ingenuity and skill in devising and performing experiments; great sagacity in reflection on the results; and, more important than all, extreme caution in adopting conclusions; reserve in propounding theories; and conscientiousness in reporting his observations. He has been found fault with for too great minuteness of detail; but this fault, if in such matters it be a fault, arises from his intense love of truth and accuracy; a quality which, when applied to such researches as the present, becomes invaluable, and cannot easily be pushed to excess.

It therefore appears, that BERZELIUS, who well knew the value of the Author's labours, was right in saying, that the investigation could not be in better hands. Having myself been familiar with the Author's writings, and in frequent correspondence with himself, for twenty years, I have here ventured to add my humble testimony to that of the great Swedish philosopher.

Let me now turn, for a short time, to the objections which have been brought against these researches, as far as they have come to my knowledge.

The first is, that his observations (in Part I.) are made on diseased subjects, and are therefore of no value. Without dwelling on the fact, that, even in Part I. the Author has produced numerous observations on healthy persons; while, in Part II., the observations are chiefly confined to the healthy; I would observe, in answer to this objection:

That it possesses no cogency in itself, inasmuch as a natural fact is not the less a fact, nor the less worthy of study, because it occurs as a morbid symptom. Everything here depends on *how* the facts occurring in diseased subjects, are ascertained and examined. And it is not likely that they will soon be better observed than by the Author.

But further, many of the best-ascertained facts in medicine rest on the statements of nervous patients. If some hysterical subjects exhibit great uncertainty in their statements, and even occasionally a tendency to deceit, this is itself a symptom worthy of investigation; and if such things render our researches more difficult, this should only stimulate us to increased care and accuracy in observing. It is clearly impossible to maintain, that the phenomena presented by nervous patients *cannot* be studied and ascertained; that they *ought to be* so studied, may be held to be equally clear.

But once more, in such of the Author's observations as were made with patients, more or less nervous, he had, even in Part I., the concurring testimony of five or six of these, affected with different diseases, and separately examined, to the existence of the phenomena which he described. This alone, in such hands as his, is a sufficient guarantee for the existence of these phenomena as objective realities.

Another objection is, that in these researches, the Author observes, not with his own senses, but with those of others; and that the results, therefore, must be unworthy of confidence.

In answer to this, I would remark, in the first place, that innumerable facts, for example those of medicine, in many cases, are thus observed. No physician ever saw or felt a pain, such as a headache, a toothache, or a twinge of neuralgia, described by his patient, nor a sensation of cold, or of heat, or of confusion of head, or of nausea, or of faintness; and yet no one doubts the existence of these sensations, which are, as the careful reader will see, almost the very same sensations as are caused in the sensitive by powerful magnets, by crystals, and by other odylic agencies, including the human hand. In like manner, numerous cases are on record, in which the patient, not the physician, saw visions, spectres, flashes of light, &c. The existence of these as *subjective* phenomena, has never been doubted; neither are we entitled to doubt of the subjective existence of similar sensations, when described by sensitive persons, although we cannot see them. And when, as in the present case, we also find that these subjective phenomena are uniformly produced, in many different persons, by the same external causes, we cannot deny to those causes, or to their effects, an *objective* existence also. We know nothing of the light of the sun, save by its effect on our senses; a blind man knows nothing of it save by its effects on the senses of others, as reported to him by them: and those who do not happen to be sensitive, can only learn any thing of odylic light by the reports of the sensitive, with this great advantage over the blind man, that they can themselves vary the conditions of experiments, and, knowing what light and colours are, can fully understand the accounts of those who see them. Everything, here again, depends on the sagacity and skill of the observer. The list of sensitive persons in Part II. contains the names of several scientific men and physicians, some of them well known and distinguished. It may therefore be anticipated with confidence, that, before long, a sensitive philosopher will be found, able and willing to devote himself to this most attractive investigation. His sensitiveness will no doubt facilitate *his* researches; but it will not add any new weight to the Author's results, which, although observed in these very difficult circumstances, have been obtained in a manner which, as far as they extend, leaves nothing to be desired on this head.

But further, this method of research, by the medium of others, is often, from the necessity of the case, resorted to in other branches of physical science. It is very common for a geologist, or a meteorologist, for example, to build up interesting and valuable speculations on what he considers the trustworthy accounts given by others, of what he himself has not seen. The Aurora Borealis is a case in point. Most of the later speculations on this point are founded on the descriptions given by those who have seen this meteor in its native polar regions; but these speculations proceed from persons who perhaps never saw, or have only rarely seen the imperfect Aurora of lower latitudes. The essential point is, to make sure of the competency and accuracy of the observer. A French commission has lately visited these regions, for the purpose of studying that and other arctic phenomena. But no one will ever dream of objecting to the use which will certainly be made of these observations, by many who may never have seen an Aurora at all, and certainly have not seen those which were observed by the commission. Now, the sensitives of Baron von Reichenbach, a majority of whom are intelligent and educated persons, have observed the whole series of auroral phenomena, invisible to him, but most beautifully visible to them, as produced in his magnetic spheres. And he has had the double advantage, denied to those who will make use of the observations of the French commission; *first, of producing, varying, and intensifying the phenomena at pleasure; secondly, of cross-examining the observers, with the phenomena actually under their eyes.* It may be possible to imagine more favourable circumstances than these, although such are not likely to be realised; but at least there is no impossibility inherent in the method.

The next objection is, that the observations are founded on falsehood, deceit, and imposture in the sensitive subjects; which of course implies, that the experimenter has been their dupe, if not their accomplice. Indeed, this has not always been left to implication. Now, it is hardly necessary to point out, that it would not be easy to deceive a practised observer, like the Author, even in one case; and we cannot possibly be more secure of the truth and honour of any observer than of his. But to believe, that from six to twelve impostors, as in Part I., or sixty, as in Part II., could all succeed in deceiving the person who had it in his power to compare their statements, and that their separate false statements should perfectly agree together, demands an amount of credulity which could not possibly hesitate at the marvels of Munchausen. The class of sceptics here alluded to are as profoundly credulous of evil, as they are irrationally incredulous of phenomena which appear inconsistent with their preconceived notions of nature.

But this is not all; for I maintain, that under such circumstances as those of the present case, to assert the existence of fraud on the part of persons of blameless character, without evidence of the fact, and simply because their statements, in an obscure and difficult matter of physical science, appear at first sight absurd or incredible, is to sin against common justice and good feeling, in a manner for which no justification can be found. And to do this, as it will be seen by the Author's Preface certain authorities have done, without any investigation of the subject, or even of the work criticised, is a mode of proceeding in questions of natural science, against which every one who values truth, and has the progress of human knowledge sincerely at heart, is bound, in the name of science, most energetically to protest.

Dr. Dubois Reymond has thought fit to express, *without quoting one word to justify it*, an opinion of the work and of the Author, the only effect of which, if it had any effect, would be to disparage them, unexamined and unheard, in the public opinion. But it will, sooner or later, have a very different effect, namely, that of proving to the world the incapacity, (from want of due acquaintance with the subject,) of the critic for the task he undertook, or rather, with most insulting expressions, declined to undertake. I shall not imitate the style of this critic. I know that the experiments of Dr. DUBOIS REYMOND on the developement of electric currents during muscular action, have justly given him a high reputation in physiology. But he will allow me to suggest, that if he could not enter into the details of the Author's researches, he could have no right to enter on a course of denunciatory criticism and virulent abuse. He is a young man of merit in physiological science, who has rashly pronounced on a work which he has not studied. But as his sentence, however unjust and rash, may yet do harm, I have thought it my duty—although otherwise I should not have felt justified in doing so—to record, as I have done above, in opposition to the hasty judgment of Dr. DUBOIS REYMOND, (formed without a careful study, or indeed any study, of the Author's work,) that which I have formed, after long and minute acquaintance with them, both of the work itself, and of the very high qualifications of its Author. I would add, that Dr. DUBOIS REYMOND'S own experiments might have taught him a more just view of the subject. If electric currents exist, as cannot be doubted, in the body, do we not know that every substance through which such a current passes, becomes, for the time, a magnet? Why, therefore, if the critic read any part of the work, should he recoil from the conclusion, that in the body may be found magnetism, or a force which in magnets is always associated with magnetism? The experiments of the critic contradict his criticism, and add new support to the previously published views of the Author.

There is one more point, on which I would offer a few remarks. This is, the bearing of the Author's researches on the subject of Somnambulism, and on that of Mesmerism.

It will be obvious to the attentive reader, that the highest or extreme degree of sensitiveness to the new influence, is found in spontaneous somnambulists, but not confined to the somnambulistic state. Somnambulism takes its place, along with nervousness, spasms, catalepsy, &c. as an indication of a state of the system favourable to the developement of sensitiveness, *but not at all essential to it.* Spontaneous somnambulism is of very frequent occurrence; so frequent indeed, that it is only wonderful that there should ever have existed doubts as to the possibility of inducing it by artificial means.

This leads us to what is called the mesmeric state, which is merely somnambulism, artificially produced. Its existence in this way is now generally, I may say universally admitted; while yet, hitherto, we could not in any way account for the fact, that one human being is able to produce this state in another, and that, too, without contact. Now, the Author's discoveries show us, that a force, or influence, analogous to heat, electricity, and magnetism, but distinct from all, exists in the human body; and that a large number of persons are more or less sensitive to this influence.

MESMER and others produced powerful effects by using magnets and other means; and this is now cleared up, by the discovery that the new force residing in the human body, does also reside, not only in magnets, but in all other bodies. The conclusion naturally is, that these forces are identical. There is a fluid (or imponderable power, or influence) which is not ordinary magnetism, and which acts strongly on the system. Thus the crude observations of MESMER and his followers, which were overlaid and discredited by their gratuitous and often absurd theories, are brought into a coherent physical shape; and the reader will find, not only that they are reduced to purely physical causes, but that the contradictions, failures, inconsistencies, and confusion of the earlier mesmeric observations, are also referred to their true physical origin. Even the mesmeric *baquet* ceases to be a mystery to us.

In short, the Author has shewn that these most obscure natural phenomena, like all others, admit of being studied as a part of physical science, and that they will well repay the investigator. But the reader will observe that the Author, preferring to begin at the beginning, has not, in these researches, studied the new phenomena on persons in the mesmeric state.

The phenomena of mesmerism, or artificial somnambulism, are natural facts. Like all facts, they ought to be studied, and if, hitherto, the subject has presented results from which many recoil, it should be borne in mind, that men of science have, in general, refused to investigate it, and that the only way to obtain true and valuable results, is to apply to these phenomena the method I have endeavoured above to illustrate, as the Author has done, to a closely allied class of phenomena. The close connection between the two classes of facts may be easily illustrated.

Every one knows, that one of the statements most frequently made by persons in the mesmeric or sleep-waking state is this, that they see flames issuing from the points of the operator's fingers; nay, some are reported to have seen this when in the ordinary waking state. These statements have been ridiculed and decried as absurd, and denounced as wilfully false; but these denunciations have been fulminated without a shadow of proof. And now these researches prove, that a large proportion of mankind possess the power of seeing these flames. The mesmeric fact, therefore, is confirmed and established by the Author's observations on persons not in the mesmeric state. The same remark applies to the strong influence exerted by magnets on the system, to the production of what is called magnetised (properly odylised) water, and to various other phenomena. But these things are only alluded to here, as it is only in Part III., not yet published, that the Author treats *fully* of odyle in its relations to the human frame.

The objections so commonly urged against mesmeric observations, are the very same as those which we have discussed above in reference to the researches of the Author; namely, the impossibility or incredibility of the facts, and the supposed existence of fraud and imposture. I need not repeat what I have before said; but I may here add, that nothing can possibly be more contrary to all scientific rules, than to reject a fact simply because it appears to us incredible or impossible, or because we cannot account for it. We can account for nothing, in any department of science, in the sense of explaining *why* it occurs; we can only observe and establish facts, classify them under some common law, and trace their mutual connections. We can as little explain gravitation as we can magnetic attraction; our nightly sleep we can explain as little as we can the mesmeric sleep; but we can study both, and determine the laws which regulate them; and it is our duty to do this, to the best of our abilities, in the latter case, as well as in the former. Nor is it possible for us to declare, *à priori*, what things are possible, and what are impossible, in any case short of mathematical impossibility. Our opinions on these points can only bear reference to what we know of nature, and even NEWTON felt how insignificant was the extent of that knowledge, when compared to the vast mass of truth which is still unknown.

As to the cry of fraud and imposture, it is, in a scientific investigation, utterly inadmissible, for the reasons formerly stated, and many others. Let us for a moment try to imagine the possibility of all the hundreds and thousands who have been, and daily are, mesmerised, being all impostors, while all the operators, if not equally deceitful, are themselves deceived. Let us bear in mind, that this accusation is commonly and recklessly brought, by those who have never investigated the subject, not only against those who have done so, but against persons of great ability, and of unblemished character and honour, and generally without a shadow of evidence; let us do this, and we must come, I think, to the conclusion, that such a line of argument is as irrational, unjust, and dishonest in itself, as it is void of all cogency in opposition to observed and established facts.

I am far from denying the possibility of the attempt to deceive in mesmerism, especially when made the subject of public exhibitions, or practised with a view to excite wonder. But this cannot apply to experiments made, as thousands have been, and as all ought to be, in the privacy of the philosopher's study, without any such motive coming into play. And were deceit attempted, there is no doubt that one who knew the art of experimenting would as easily be able to detect it in mesmeric experiments, as the Author could do in his researches. But while I admit the possibility of such a thing, and while I would look with suspicion on the exhibitions of those who only make a trade of mesmerism, yet in all the cases, private or public, which I have had an opportunity of seeing, I am bound to say that, to the best of my knowledge and belief, no deceit was even thought of; and, indeed, I believe that it was not required. Nature is more wonderful than any forgery of man's invention. I have repeatedly myself produced mesmeric phenomena in persons as incapable of deceit as any that can be found, even among sceptics. And, in short, without the most stringent proof, the charge of fraud is not for a moment to be tolerated. But even were fraud established in one or more cases, this could not affect other cases, in which no fraud was probable, or even possible.

That there is something worthy of research in these phenomena, all who have either personally studied them, or carefully read the existing evidence, admit; the best security against fraud is the utmost possible knowledge of the subject, which can only be obtained by men of scientific training and turn of mind devoting their energies to it. If this be duly done, the cry of imposture will soon die away, and mesmerism will take its natural and proper place as a branch of natural science.

If there be any phenomena which, in a certain sense, we might be entitled to call incredible, they are those of clairvoyance. I do not profess myself to have seen these; and, from their comparatively rare occurrence, a rational incredulity is here excusable. But even in this part of the subject the existing evidence, a great part of which rests on testimony which is of the highest character, cannot be disposed of by any amount of incredulity. We must carefully distinguish between the facts themselves, and the various, often absurd explanations and theories which unscientific observers seem to think it necessary to devise, and insist on giving; while others, before admitting the facts, require them to be explained, as if the truth of a fact depended on our being able to account for it. All this is unscientific. There *are* facts, of the most interesting kind, connected with this part of the subject; and we need not doubt that, if studied in the right spirit, these facts will one day admit of a natural physical explanation, as far at least as any natural fact can be explained. The example, indeed, above given, of the flames at the point of the fingers, proves that there is a physical cause for some, as we shall sooner or later find there is for all, of the mesmeric phenomena. But further, even clairvoyance has not unfrequently, as we know from the best authority, occurred, just as somnambulism and cataleptic rigidity have done, spontaneously, without the use of any mesmerising process whatever. It is, therefore, independently of mesmerism, a fact to be investigated according to the usual rules of science.

I have ventured to make this digression, because I am firmly convinced that the example set by the Author, in thus studying the more obscure and difficult natural phenomena, will be followed, and with the most desirable results, in the allied subject of mesmerism; and that, until scientific men take the trouble to investigate mesmerism in this manner, they are not entitled to pronounce an off-hand judgment, still less an injurious judgment, on the labours of those who, perhaps with fewer qualifications, but certainly with equal love of truth, have undertaken the investigation which they have themselves neglected.

To return to the Author's experiments. He has, by the minutest instructions, put it in the power of every one to repeat them. It is hardly necessary to point out, that, in repeating them, if we wish to obtain or to control his results, we must strictly attend to the conditions; indeed, this is so obviously necessary, that I should not have alluded to it, had not the proceedings of certain physicians in Vienna, in what they intended for an experimental refutation of the Author's conclusions, furnished evidence that even educated men, who might be supposed to be acquainted with science, may in the grossest manner violate this essential and simple rule. For some almost incredible details on this subject, I refer to the Author's notes at p. 354 and p. 370.

I have now to add, that I have myself repeated some of the Author's experiments, and have found, as was to be expected, that his account of the phenomena was perfectly accurate. For a few details, the reader is referred to the Appendix. (A.)

There is no difficulty in finding sensitive subjects; but it is not so easy to make good and trustworthy observations when we have found them.

I have reason to know that many persons have made similar observations, in confirmation of those of the Author. One gentleman, well-known as a lover of science, has favoured me with some of his, which will also be found in the Appendix. (B.)

Those who have thus had an opportunity, or who may hereafter have an opportunity, of thus confirming, illustrating, or correcting the observations of the Author, will confer on me a very great obligation by privately communicating to me the results they have obtained or may obtain; which, or a selection of which, with their permission, I should wish to add to the future parts of this work. But it will be still more desirable, that such observers should publish their results separately.

Finally, with regard to the translation; I have endeavoured, in every case, to adhere rigidly to the meaning of the Author, even where, from the peculiarity of the German idiom, or of the Author's characteristic style, which is not always easily rendered, it has been necessary, (and this has often been the case,) totally to change the form of the sentence. It has been my endeavour to produce a readable English work, as far as the nature of the subject, and the very frequent occurrence of technical terms and of necessary repetitions of nearly identical details, would permit.

Part III., which will treat in detail of the odylic phenomena in their connection with, and their effects on, the human frame, will be published as soon as the original appears. The subject of somnambulism necessarily comes under discussion, and most interesting and valuable results may be expected from the Author's method of investigating that subject.

WILLIAM GREGORY.

114 PRINCES STREET, EDINBURGH,
16*th April* 1850.

AUTHOR'S PREFACE

TO THE

SECOND EDITION OF PART I.

The Seven Treatises which form this work were intended to appear successively in the monthly numbers of Liebig and Wöhler's Annalen der Chemie, and were sent to the Editor in time for their publication to have commenced in July 1844. But accidental circumstances prevented this, and thus it happened that they were collected in two separate supplementary numbers of the Annalen, and published for the first time in March and May 1845. This will explain the somewhat unusual form in which they appear.

In this Second Edition, some things have been corrected; but, on the whole, the work has undergone no essential alteration. My researches, continued without interruption since it appeared, have confirmed and more strongly established the observations recorded in it. I have thought it right not to abandon the half historical, half systematic course which I had pursued in detailing my observations, and presenting the views deduced from them; for this is the real course of all natural science, in which the correction of former errors always keeps pace with the extension of our knowledge.

It was easy to foresee, that a matter of so strange and peculiar a kind, as the subject embraced in these researches, would meet with objections; and I was prepared beforehand for the necessity of defending my experiments, and the conclusions deduced from them, against both well-founded and groundless opposition and contradiction. The new field of investigation here opened up, advances its limits so near to the bulwarks of existing and established doctrines, and comes in contact at so many points with all that is known of the Imponderables, that it was not to be expected that it should be allowed to take its place in the crowded domain of science, without much resistance. Yet I was only prepared for rational judgment and criticism; for, here and there, failures in repeating my experiments, arising from errors in the necessary arrangements; for opposition to the conclusions deduced from these experiments; or for the announcement of different conclusions or deductions from the recorded facts. I was not, I confess, prepared for an attack, (which every friend to science, possessed of good feeling and good manners, must agree with me in considering ill-bred in the extreme,) made on my work, and on myself personally, by a Dr. DUBOIS-REYMOND, in KARSTEN'S *Progress of Physiology in the year* 1845. This philosopher does not think it necessary in the smallest degree to enter into my experiments and deductions; but, from his lofty eminence, entitles my labours, courtly enough, "an absurd romance, to enter into the details of which would be fruitless, and to him impossible." I have no doubt of the truth of both statements. It would be fruitless, because he did not understand the work; and a foolish judgment on that which we do not understand, must be fruitless; and impossible, because he did not read the work connectedly, and it is impossible to enter into the details of a matter, into which we have not taken pains to acquire some insight. Now, that he did not read my work, but only turned over the leaves in the manner of superficial and unconscientious reviewers, he proves by abusing it as "the new testament of mesmerism," and consequently showing that he has not seen that my work is the very first in this department, which, in most points, goes exactly in the teeth of the theories of MESMER, and places the phenomena on a totally different foundation. My critic goes on, in a feeble attempt at wit, to describe himself as having been "greeted by the magnetic *baquet*, and the wretched magical trash of Baron VON REICHENBACH;" &c., and therefore he has not read that it is precisely my work, which banishes for ever the "*baquet* and the wretched magical trash" of MESMER, by tearing down the veil that hid his mysteries, reducing them to their naked physical existence, and substituting sober scientific investigation in the room of all the old phantasmagoria. But ever since science has existed, ignorance has assumed the right of judging and condemning that which it could not understand. The polite and well-bred Berlin physiologist is then pleased to cast in my face a few common places, such as—my work is "one of the most deplorable aberrations that has for a long time affected a human brain;" my statements are "fables, which should be thrown into the fire;" and many similar learned vulgarities.

He who assumes the right publicly to sit in judgment, and to pronounce sentence on a scientific work, is, before all things, bound in duty to inform himself thoroughly of its contents; and he is further bound to support his sentence, as all public judges do, by the reasons on which he thinks himself justified in pronouncing it. This duty is the more indispensably incumbent on him, because his judgment is one-sided, and requires the control of public opinion; moreover, the Author whose work is judged, is justly entitled to have the means of defending it. But a judgment, affecting the truth and honour of the Author of such a work, and which the judge is not ashamed to send forth without fulfilling any one of these conditions, is, in one word, nothing but impudent; and that too, in a degree perhaps unexampled in modern or in ancient literature. For surely, it never before happened in any country, that a reviewer had the audacity, or rather the folly, to treat a scientific work, without giving any reasons, without quoting a syllable of its contents, merely with gross and ill-bred abuse. I say folly, because it is folly to cast stones which we may see will certainly fall on our own heads. Either my statements contain natural truths, which have a constant existence in the physical world, and must therefore sooner or later be acknowledged, and must cover the ignorant reviewer with disgrace; or they rest on great errors, and then it must be easy for him to perform his duty of discovering and explaining these, and thus secure himself against the reproach of unconscientious levity in regard to the reputation of his fellow-man. Only narrow-mindedness and folly could, without reflection, expose themselves at once to both dangers.

Dr. Dubois-Reymond further thinks, that he could

not enter into the details of my work, "because it would at least be impossible for him not to be guilt y of using *unparliamentary language* in doing so."—The insolences with which he has loaded me are not sufficiently unparliamentary, not rude enough for him; he has a store of expressions of yet coarser calibre. He has given us proofs of the delicacy and good feeling which restrain him from using these; but now I shall save him the trouble of all further hypocrisy, and tell the real truth for him. *He has not the courage to venture on an examination of the details of my treatises.* The matter is not one which lies on the surface; the recorded facts cannot be destroyed off-hand by mere flippancy; and the conclusions logically deduced from them, are not to be washed away with critical ink. But a thorough examination of such a work is troublesome and laborious; this is annoying, possibly fruitless, possibly even giving rise to serious misgivings. And as, unless he take all this trouble, he cannot attack the details of the work, without the risk of pronouncing an immature and rash decision, he dares not enter into these details, for fear of afterwards feeling the scourge of the Author. It is far easier and cheaper, to pass, with unworthy superficiality, only over the outside of a subject, daubing it here and there with mire, to lower it in the public estimation, and then to run away under cowardly and hypocritical pretences. I beg Dr. DUBOIS-REYMOND, in regard to his entering into the details of my work, not to place the smallest restraint on himself on my account; I challenge him to the arena, with all his "unparliamentary expressions" from the banks of the Spree; and I give him my word, that he shall find me there, and that he shall receive exactly such an answer as he deserves.

It is in the nature of every experimental inquiry, that it cannot be free from defects. It is *because* we feel these defects in our knowledge that we make experiments, in order to complete the former, by newly discovered and newly acquired knowledge. But while thus employed, we observe ten, nay, a hundred new gaps or blanks; and the reader, on his part, probably observes a dozen, which have escaped the notice of the experimenter; the reviewer perhaps detects many more. It is most desirable that these should all be publicly announced and brought under discussion, that the Author, or others, may investigate the matter further on all sides, or fortify what he has already won against all doubts. This is advantageous to all, even to the discoverer of every new scientific fact. My labours will as little be found to be without defects, as those of men standing much higher than I do have ever been thus perfect, and least of all in the natural sciences. No one can have felt this more strongly than myself. Every notice of any deficiency, expressed with proper feeling, I shall thankfully receive, and experimentally improve my work accordingly. But lofty insolence, stained, moreover, with the most glaring ignorance of the work reviewed, must be driven back and confined to its due limits. It is not only my interest, it is the interest of all those who labour and who write on scientific subjects, that such weeds should be rooted out and swept away, and not allowed to flourish in the garden of science.

That, in point of fact, men standing far higher than I do, are not exempt from the greatest errors in their labours in natural science, is not merely an assertion made to excuse the defects of my own work, but one which I shall at once prove. M. JOHANNES MUELLER, our great physiologist, and the pride of Germany, whose admirable writings are an oracle to his contemporaries, in his Manual of Human Physiology, (fourth edition, I. 26.) not in a notice of my work, but when speaking of the "so called animal magnetism, of magnetic passes, of laying on of the hands, and of the passage, from one person to another, of the so called magnetic fluid," uses the following words:

"These histories are, however, a *deplorable labyrinth* of lies, deceit, and superstition;* and they have only proved, how ill qualified are most physicians to make an empirical investigation, and how little they know of that principle of testing a subject, which in all the other natural sciences has become the universal method."

But what now, if it should appear, that it is, on the contrary, M. MUELLER, who himself lives and moves in that "*deplorable labyrinth?*" What if it be precisely in my work, that there are to be found adduced the very testings which he recommends, pursued exactly *according to the method universally followed in all other natural sciences?*—And what, lastly, if, by these very testings, hundreds of facts, proving the actual existence of such a fluid, or imponderable, or dynamide, or influence (whatever name be given to it), which, by means of *passes*, of *laying on of hands*, and of *transference* or *communication*, produces astounding

* Prof. MUELLER is here too hasty in his conclusion. What follows is not only quite true, but of great importance. But it does not, and cannot affect more than the mode of research usually followed in these matters, *not the facts themselves* which are to be investigated.—W. G.

physical and physiological effects,—what, I say, if hundreds of such facts have been thus as plainly exhibited and as fully demonstrated as any other physical or physiological truth can possibly be, and by that very inductive method of research?—Surely, in this case, we may say, and we must say, that even the great MUELLER has erred seriously in a matter, on which he has permitted himself to deliver a condemnatory and injurious judgment, without having previously investigated it; and that, in the next edition of his Manual, he will cancel that hasty passage. What I have mentioned must be regarded as a striking example of the extent to which men of the highest abilities, and the most distinguished in science, may be guilty of the greatest blunders, when under the influence of prejudice and of preconceived opinions; and may commit errors of an extent so incredible, that the very reproach which they bitterly and unsparingly cast on others, finds the most direct application to themselves, and returns on their own head.

Dr. DUBOIS-REYMOND is, as he himself informs us in the 58th volume of POGGENDORFF'S Annalen, under the influence, in scientific matters, of Prof. J. MUELLER, is probably also his pupil, and, in his just veneration for his highly distinguished teacher, feels bound *jurare in verba magistri*; for we may see, that his attack, in which he does me the honour to attribute to me "the most deplorable aberration that has, for a long time, affected a human brain," coincides almost literally with the "deplorable labyrinth of lies," &c. of M. MUELLER; (both philosophers appear to possess an abundant stock of compassion for deplorably erring authors;) and where M. MUELLER speaks of lies and imposture, Dr. DUBOIS-REYMOND thinks to give me pain by dark hints, tending to bring me into suspicion; such as "the hidden, but real and proper foundation, and the concealed object" of my work. But these gentlemen cannot, in this case, *see the forest for the trees.** They hear of an unconnected mass of the most strange and singular phenomena in nervous patients; but their system has no rubric for such things, and the Doctors, astonished and perplexed, immediately suspect deceit, to the amusement and often the indignation of the less prejudiced byestanders. Thus, on one occasion, a grisette in Berlin cooled her anger by applying her fingers to the ears of one of the learned schoolmen; and when he saw, in the glass, that these organs were assuming an alarming degree of longitude under the operation, he shouted, "Treason!" "Imposture!" and then every thing which, in a thousand shapes, urgently knocks for admission at the door of calm and reflective testing and study, is remorselessly kicked down stairs, without a hearing, as "lies and imposture and superstition."—This is a very convenient way of saving ourselves the trouble of a thorough investigation; but it is as hasty and one-sided as it is unscientific and unconscientious.

Every natural science, and every branch of a natural science, must at first pass through a period of darkness and error. Physics were preceded by magic, chemistry by the art of making gold, medicine by the elixir of life and the philosopher's stone, astronomy by

* "Sie sehen den Wald vor lauter Baümen nicht," a very expressive German proverb, applied to such as allow their attention to be distracted from the essence of a matter by trifling difficulties in the details. It is here truly applicable, where the bugbear of fancied imposture blinds men to facts.—W. G.

astrology, &c. &c. Philosophy, theology, and jurisprudence, have alike had a period, during which, if we may so speak, they sowed their wild oats. Our first notions have always been indistinct, confused, and hence favourable to the marvellous, to the mysterious, and to superstition, and liable to abuse. But it did not follow from this, that the enigmatical shell did not conceal a solid nucleus of truth. It is entirely according to the nature of the thing, and therefore exactly what might have been expected, that the subject of sensitiveness (as I have used the term), and the peculiar force lying at the root of its phenomena, should have had to pass through such a period of infancy in our notions and conceptions of it; and this the more, on the one hand, the less we were able to isolate or confine it; and, on the other hand, the deeper it was seen to penetrate into the, to us, mysterious and concealed sphere of nervous manifestation. But that this tottering period of infancy has, in this case, already continued for nearly three quarters of a century, certainly too long for our enlightened times, is due, for the greater part, to the almost criminal and selfish opposition of the cultivators of the exact sciences, who met the most striking and interesting *facts*, not only with deaf ears, but even with a sort of senseless hostility. Baron BERZELIUS, whose loss science must long deplore, and who, as is well known, took a warm interest in my researches, assured me that he had, during forty years, always entertained the wish that this subject, which, from the multitude of the recorded and continually recurring facts, could not possibly be destitute of all real foundation in truth, might be taken up by some one, who should make it the object of a special and thorough investigation and testing, according to the modern inductive method of research in natural science; and he added, that he rejoiced that such a person had at last been found in me, ready to devote himself to the rational investigation of the subject.

Another reason why this has been so long retarded, and why the gropings in the dark seemed likely never to end, is this: that those who attended to the subject, one after another, insisted on beginning to build the pyramid at its apex. They wished to do at first what they ought to have tried at last, namely, to cure diseases. Before taking the smallest pains to acquire the very slightest knowledge of the inner nature of the deep-lying force, the effects of which they had observed, they were bent on—making a trade of it! Then, in their operations, they stumbled on somnambulism, on clairvoyance, and even on insanity; that is, they everywhere met with the manifestations of the force in their maximum of intensity, and complicated with states of disease both intense and hitherto inexplicable. While they stood amazed at the phenomena on the large scale, and of high intensity, the explanation of which they felt to be beyond their power, they neglected to enquire into its small and feeble beginnings, on which alone the foundation for a truly scientific structure could be securely laid. We have not drawn our knowledge and our doctrines of electricity and of sound from the lightning and the thunder; nor have we obtained our knowledge of the expansive force of steam from the eruptions of volcanoes. But as our ancestors fabled concerning these phenomena, because they did not understand them, exactly so do modern philosophers, of the same category as Dr. Dubois-Reymond, give the reins to their fancy in talking about what is called Animal Magnetism, because they know nothing about it. I do not here speak of medical men ; but the cultivators of physics and of physiology have managed this matter no better. The majority of the former (medical men) reject all study of the phenomena, because they *cannot* comprehend the connection of cause and effect in them. The majority of the latter do so, because they *will not* comprehend this. But this is not the path of research in natural science ; and the latter class, in truth, sin more grievously against the progress of enlightenment than the former. It is, indeed, any thing but honourable to our contemporaries, that they obstinately adhere to the primeval condition of blind ignorance in these matters, and will not perceive how fearfully they, by so doing, expose themselves on this side.

But I have not found the difficulty of getting at the truth, in such investigations, by any means so great and insuperable as those who dread the labour of research into the subject constantly assert it to be. What is so often said of lies and imposture, is found, on closer examination, really to exist in quite a different quarter. It is not found essentially in the sensitives, but, on the contrary, it lies in the subjective personality of the experimenter, either because he is blinded by prejudice and preconceived notions, or, as frequently happens, because he is not properly qualified for the task he has undertaken. He who takes the subject in hand must know *how to investigate*, and how to put questions to nature, if he wishes to obtain clear and instructive answers. But this is not a qualification found in every man, as far as we know. I must here say, to the honour of the mixed population of Vienna, that, among about a hundred persons, whom I have examined hitherto in the course of these researches, to a greater or less extent, and of whom more than sixty are named in this and the Second Part, hardly one was found who gave me two or three somewhat highly coloured answers; and this, when it did occur, arose certainly more from ignorance than from any dishonest intention. Such answers were, moreover, instantly detected and rebuked by me. Considering the internal natural connection in which all these phenomena stand to each other, and the threads of which connection I now firmly hold, it is absolutely impossible for any one to deceive me, even for a few minutes, by false answers, which I should instantly discover to be false, were they attempted. But, in fact, these people never dream of lying or deceiving; they speak out straightforward what they see and feel, when I try any experiment on them. Most of them exhibit an honest and pleasing zeal, in trying to explain to me as clearly as possible what it is that they recognise or perceive. This is encouraging to me, and I find in it some compensation for the injurious treatment which I have met with at the hands of men of science, who had most reason to be grateful to me for my labours. All these innumerable answers to my questions agree in all points so perfectly, that every rational doubt must yield to the evidence of truth; and in this beautiful agreement lies the guarantee for their essential trustworthiness.

But when the experimenters do not know how to

put their questions; when awkwardness and clumsiness cannot use the tools; when ignorance cannot arrange the necessary conditions of experiment; when want of tact cannot comprehend the answers; and when want of acuteness or intelligence is unable to discover the mutual relations of the phenomena;—then begin confusion and perplexity; the results, being misunderstood or misinterpreted, contradict each other; and the ill-qualified observer, rather than admit to himself or to others his own deficiencies, will a thousand times sooner adopt the dishonest expedient of accusing the observed person of deceit. But the deceiver, in regard to nature and science, is no other than himself, who, in his incapacity, has the impudence and the folly to try to brand truth with the mark of falsehood.

SCHLOSS REISENBERG, NEAR VIENNA,
February 1848.

October 1849.

This work was printed and ready by the spring of last year, when the outbreak of the German revolution prevented its publication. Matters having now assumed a different aspect, the publication now takes place. It was necessary to mention this, to render intelligible some determinations of dates and times, which, being printed, cannot now be altered.

ERRATA.

P. 29. l. 10. *for* a *read* the.

P. 158. l. 3. from bottom, *after* middle *insert a semicolon.*

P. 182. bottom line, *after* this *insert* way.

P. 190. l. 13. *for* positively *read* positivity.

P. 453. l. 17. *for* crystallise *read* crystallised.

PHYSICO-PHYSIOLOGICAL RESEARCHES

ON THE

IMPONDERABLES,

IN THEIR RELATION TO

THE VITAL FORCE.

PART I.

SECOND EDITION, CORRECTED AND IMPROVED.

Einar neuen Wahrheit ist nichts schädlicher als ein alter Irrthum.

Nothing is more hurtful to a new truth than an old error.

GOETHE.

INTRODUCTION.

If we make downward passes with strong magnets, having a supporting power of about 10 ℔s., along the persons of from 15 to 20 individuals, but without touching them, we shall always find one or perhaps more among the number, who feel affected thereby in a peculiar manner. The proportion of human beings which may be found to be thus sensitive, is, in general, greater than we imagine. It sometimes happens, that, in the above number, three or four sensitive persons are found; indeed, I am acquainted with an Institution, where, on trial, out of 22 females there assembled, not less than 18 perceived the sensation caused by the passage of the magnet, more or less distinctly. The nature of this impression on such excitable persons, who may, however, often be justified in regarding themselves as perfectly healthy, is not easily described. It is rather unpleasant than agreeable, and is associated with a gentle feeling, sometimes of cold, at other times of warmth, which resembles a cool or tepid *aura* or current of air, which they believe gently blows upon them. Occasionally they experience a dragging or pricking sensation, as well as formication; some complain very soon of headache. Not only females, but also men in the prime of life, are to be met with, who distinctly perceive this influence. It is sometimes very vividly felt by children.

In order to produce this effect, it is not a matter of essential importance, but rather indifferent, whether we take a horse-shoe magnet or a straight bar magnet, or whether we use one or the other pole, provided it be only active, and of nearly the strength above mentioned. The passes should be made from the head towards the feet, and not too quickly. The magnet is carried as close to the surface as is possible without actual contact with the dress; and, to guard against the effect of imagination on both sides, the passes may be made from the occiput over the neck and back. The magnetised person is thus not aware whether, at any given moment, a pass be in progress or not, and his statements are free from prepossession.

Powerful men, and healthy strong women, usually experience nothing, when thus tried. Yet I have met with such individuals, who, although in vigorous and blooming health, felt themselves decidedly affected by the magnet. The excitability here alluded to is more frequently observed in persons of sedentary habits, while they may be, notwithstanding, regarded as perfectly healthy; more particularly in men who are continuously engaged in writing, and in girls constantly employed in needle-work; and likewise in persons who are depressed by secret grief, by the pecuniary cares of life, by the loss of relatives, or by disappointed ambition.

After such persons, who may be regarded as not in full health, come those who are sickly or in part diseased, who are very often magnetically sensitive, especially those who are commonly said to have weak nerves, who are easily terrified, or have been agitated and shaken by fright; and then many actual patients, in numberless kinds of disease, but more particularly in those cases in which local or general spasms accompany the disease, during abnormal development of the system at the period of puberty; those who are very strongly and disagreeably affected by certain odours; most especially persons suffering from catalepsy, St. Vitus's dance, and various forms of paralysis; many hysterical patients, and finally, without exception, the insane, and spontaneous somnambulists. Thus, the sensitive subjects form a chain, from the healthy man or woman to the sleep-walker, of which the extreme links are represented by a powerful man, and a feeble somnambulist. We may easily convince ourselves of this fact in every great hospital.

The magnet, therefore, appears to be an agent which *influences generally the vital actions;* and while this property has been observed, and individual physicians have endeavoured in various ways to apply it, with a view to the possibility of deducing from it a method of cure, without, however, having obtained very distinct or satisfactory results; on the other hand, natural philosophers have not yet included it in the domain of physics; and natural science in general has everywhere, up to this time, avoided it, on account of the uncertainty of the observations hitherto made. Magnetism, notwithstanding, offers, when viewed from this side, on closer investigation, a high and infinitely varied interest. If a part of the phenomena here becomes mixed with those of vital action, this takes place exactly, and especially, where the limits of the Organic and the Inorganic become confused. While natural philosophers hesitated whether to consign these phenomena to the province of physiology or of physics, they were neglected on both sides. They were thus left to practical medicine, and did not always fall into the best hands. In the following pages, I hope to succeed in unloosing some parts of the Gordian Knot; to unite a number of the phenomena under common points of view; and to bring them under fixed physical laws.

FIRST TREATISE.

LUMINOUS APPEARANCES AT THE POLES AND SIDES OF STRONG MAGNETS.

1. THOSE sensitive persons who are really or apparently in good health, observe, in the magnet, nothing particular, beyond the above-mentioned sensations; and bear close proximity to it without injurious effects. This is not, however, the case with sensitive persons in a state of disease. The action of the magnet on such persons is, according to the nature of their disease, sometimes agreeable, at other times disagreeable, at other times again painfully unpleasant, and that, occasionally, to such a degree, that fainting, and cataleptic and convulsive attacks occur, of such violence, that they may at last become dangerous. In the latter class of cases, to which also somnambulists belong, there is usually superadded *an extraordinary acuteness of the senses;* the patients acquire a singular delicacy and power in taste and smell; many kinds of food become as intolerable to them, as the usually agreeable perfumes of flowers become offensive. They hear and understand what is said three or four rooms off; and are often so sensible to light, that, on the one hand, they cannot bear the light of the sun or of a fire; on the other, they are able, *in the dark*, not only to recognise the outline of objects, but also to distinguish accurately their colours; and this in such a degree of darkness, that persons not sensitive

are utterly unable to perceive any thing. These things are pretty generally known, and require no proof here. Moreover, it is not so difficult, as it appears at first sight to many who in all such matters imagine something incredible or supernatural, to understand the possibility of such phenomena. Not only do most animals excel civilised man in the acuteness of certain senses, but even savages, consequently human beings also, frequently equal dogs and other animals in the delicacy of smell and hearing. And in regard to vision, horses, cats, and owls are familiar examples of the power of the visual apparatus to see tolerably well even in a dark night.

2. By the kindness of a physician in Vienna, I was, in March 1844, introduced to one of his patients, the daughter of M. Nowotny, Revenue Officer, Landstrasse No. 471. She was a young woman of 25, who had suffered for 8 years from increasing headaches, and had then become affected with cataleptic fits, accompanied by spasms, both tonic and clonic. In her had supervened intense acuteness of the senses, so that she could not bear either sun-light or candle-light. She saw during the darkness of night her room as if in twilight, and clearly distinguished the colours of all objects in it, such as clothes. On this patient the magnet acted with extraordinary force in various ways, and in every point of view she belonged to the highest class of sensitives, so that she was in no way inferior, in acuteness of the senses, to the true somnambulists, although she herself was not a somnambulist.

Seeing all this, and reflecting that the aurora borealis appears to be nothing else than an electric phenomenon, caused by the magnetism of the earth, the intimate nature of which, moreover, we cannot yet explain, since no direct emanations of light from the magnet are known in physics, it occurred to me to try whether such an acute vision as that of Madlle. Nowotny might not possibly, in absolute darkness, be able to perceive some luminous appearance in connection with the magnet. The possibility of this appeared to me not to be very remote; and if it should be found to be so, it seemed to me likely to supply the key to the explanation of the northern lights.

3. The first preliminary experiment I caused to be made by her father in my absence. In order to profit by the utmost degree of darkness, and by an organ for some time accustomed to the absence of light, so as to obtain the greatest possible enlargement of the pupil, I begged him, in the middle of the night, to hold before the patient the largest magnet I had, namely, a nine-bar horse-shoe, carrying upwards of 80 ℔s., and after removing the armature. This was done, and next morning I was informed that the young woman *had actually perceived a distinct and permanent luminous appearance* as long as the magnet was open; but that it had always disappeared as often as the armature was attached.

In order to obtain on this point more sure and minute information, I made arrangements to repeat the experiment myself, with some alterations. I did this the following night, and tried it at the time when she had just awakened from a cataleptic fit, and was, consequently, in the most sensitive state. To make all sure, the windows were covered with thick hangings and the candles removed, long before the termination of the fit.

The magnet was placed on a table, about ten feet from the patient, with its poles directed towards the ceiling, and the armature was then removed. None among the assistants was able to perceive any thing whatever; but the patient saw two luminous appearances, one on the extremity of each pole. When the armature was attached, the lights disappeared, and she saw nothing more; but on removing it again, they again appeared as before. At the moment when the armature was detached, they seemed to her to shine somewhat more brightly, and then to assume a permanent condition of inferior brightness. The fiery appearances were of nearly the same size on each pole, and they did not show any tendency to approach each other. Close to the steel from which the light emanated, it appeared in the form of a luminous vapour, which was surrounded by a sort of shining rays. The rays, however, were not tranquil, but shortened and lengthened themselves continually, producing a shooting and sparkling of uncommon beauty, as the patient assured me. The whole image was more delicate and beautiful than ordinary flame; its colour was purer, nearly white, occasionally mixed with rainbow colours, and more resembled the light of the sun than that of a fire. The light was not uniformly diffused. In the middle of the edges of the magnetic poles it was denser and more brilliant than towards the corners; but at the corners the rays were collected into bundles, which reached beyond the rest of the rays. I showed her a small electric spark, which she had never seen, and of which she had no conception. It appeared to her much more blue than the magnetic light, and it left on the eye a durable peculiar impression, which very slowly disappeared.

The interest which I could not but take in the subject, made it desirable to multiply and vary the observations, and to control them, as far as possible, by repetition and careful manipulation. The patient was improving in health; her sensitiveness visibly diminished from day to day, and there was consequently no time to be lost. Two days afterwards, I again, with the assistance of her relatives, tried the experiment. It was conducted in the same way, and yielded the same results. After another day, we repeated it first with a weaker magnet, without informing her of the change. She did not now see the former light as before, but observed only two fiery threads, as she expressed it. These were obviously the edges of the magnet, which alone, from the inferior vividness of the light from this magnet, were visible to her. When we now removed the armature from the strong magnet first used, she immediately recognised the former light in its shape and colour as before. Some days later, during which her health had rapidly improved, the experiment was repeated, but now even the light from the strong magnet did not appear to her as formerly. She saw the two lights less distinctly, smaller, and somewhat uncertain; they often seemed to sink, again to rise, sometimes almost to vanish, and then to return after a short interval. Next night she saw, even on the strong magnet, only the threads of fire; and on the night following, the phenomenon had become invisible to her sight, with the exception of two bright lightning-like flashes, which appeared and disappeared as often as I detached the armature.

4. So far Madlle. MARIA NOWOTNY: Her rapid improvement in health had now so reduced her sensitiveness, that further experiments became impracticable, and would have yielded no new results, had they been practicable. As she was an intelligent girl, cultivated for her station in life, and of excellent character, I had the fullest reason to regard her statements as genuine and exact. Notwithstanding, and in order to give to them consistency and scientific value, I was, above all things, desirous to obtain confirmation from other quarters. I had become acquainted, in the course of my investigations, with Dr. LIPPICH, a learned physician, Ordinary Clinical Professor in the University of Vienna, and was now, by his kindness, placed in communication with one of his clinical patients. This was Madlle. ANGELICA STURMANN, aged 19, daughter of an hotel-keeper in Prague, suffering from pulmonary tubercles, and long affected with the lower stages of somnambulism, with fits of tonic spasms and catalepsy. I found, on trial, the effect of magnets on her so powerful, that, in sensitiveness, she far surpassed Madlle. NOWOTNY. When, in the darkened ward, I detached the armature from the large magnet above mentioned, carrying 80 lbs., while Dr. LIPPICH stood near her, and she was, moreover, perfectly conscious, she suddenly ceased to reply to his questions. She had instantly, on my removing the armature, been thrown, by the influence of the magnet, into spasms and insensibility. This offered but little prospect of a good result from my experiments; but they were not without some fruits. After an interval, the girl came to herself, and told us, that at the moment when I disarmed the magnet, she saw a *fire flash from it*, as high as a small hand, white, mixed with red and blue. She was on the point of more closely examining it, when instantly the effect of the magnet deprived her of consciousness. With extreme eagerness, I should have wished to repeat the experiment, and the patient was willing to allow this; but the physician thought it likely to be injurious to his patient, and I was, of course, compelled to relinquish the idea of further trial. But I had gained my chief object, namely, a confirmation of the luminous appearance seen by Madlle. NOWOTNY. A patient, ill of

a different disease, without the remotest connection with her, had also seen the phenomenon.

5. In another ward, Dr. LIPPICH showed me a lad of 18, a shoemaker's apprentice, who had, by fright and ill treatment, been thrown into convulsions, which continually recurred at intervals. When I approached him in the dark with the magnet, he immediately spoke of *fire and flames, which he saw before him*, and which were renewed as often as I removed the armature. But this lad was too uneducated to permit of accurate experiments being made on him; and in the meantime, more interesting opportunities for following out the details of the subject presented themselves.

6. The first of these was Madlle. MARIA MAIX, aged 25, daughter of an upper servant in the imperial palaces, living in the Kohlmarkt, No. 260, to which place I was taken by the kindness of her physician. He was treating her for paralysis of the inferior extremities, occasionally accompanied by convulsions. She was in perfect possession of her senses, had never either walked or spoken in her sleep, and was an intelligent and clear-headed young woman. When the great magnet was disarmed before her by night, an experiment often repeated, she invariably saw at once *the luminous appearance over its poles*, about a hand-breadth in height. But when attacked by spasms, the appearance increased to her eyes in an extraordinary degree. The light was no longer confined to the poles, where it appeared much larger, but she now saw streams of light *from the whole surface of the steel*, weaker than at the poles, but exhibiting the whole magnet in a bright light, which produced in her, as also in Madlle. NOWOTNY, a durable dazzling sensation in the eyes, which very slowly went off. We shall soon see how all these things are related together. In the meantime, I had now obtained the fourth confirmation of the existence of the magnetic light. But it was only now that I was to meet with by far the most remarkable and the clearest of those who observed for me these remarkable phenomena.

7. This was Madlle. BARBARA REICHEL, aged 29, of stout build, the daughter of a servant in the imperial palace of Lax-

enburg. When 7 years old, she had fallen out of a window two stories high, and had ever since been subject to nervous attacks, sometimes passing into insanity, at other times into sleep-walking and speaking in her sleep. Her illness intermitted, often for long periods. She had, at this time, just recovered from a severe spasmodic attack, retaining, however, the whole exalted sensibility of her sight. She was then in full strength, looking well, perfectly clear in her mind, and was in the habit of going alone through the crowded streets to visit her relations. I invited her to my house, and she was so obliging as to come as frequently as I wished, thus enabling me to make use of her extraordinary sensitiveness to the magnet, in trying experiments with physical apparatus, such as could not conveniently be carried to other houses.

This patient combined in herself the rare advantages, first, that she saw the magnetic light as strong and bright as any patient confined to bed; secondly, that she was, in the intervals of the attacks, strong and healthy, could move about, and was very intelligent; lastly, that with the highest degree of sensitiveness to the light, she yet bore perfectly the approach of the magnet,—in fact as well as healthy persons; which was so far from being the case with most sensitive subjects, that, as we have seen in Madlle. STURMANN, and as happened also, in a less degree, in Madlle. NOWOTNY, the open magnet soon caused spasms, and even produced insensibility. With such subjects little can be done; but Madlle. REICHEL was able quietly and deliberately to carry out every investigation to the end. Persons of this temperament are invaluable for scientific researches. And thus I have been able, by means of Madlle. REICHEL, to obtain most accurate results, and of the highest value, in reference to the theory of electro-magnetism. But in this place I shall only adduce such of her observations as concern the magnetic light.

This light she saw, not merely in darkness, but even in such a degree of light, or rather obscurity, as enabled me, with difficulty, to perceive objects; in which, therefore, I could work, and alter or multiply the experiments. In such a twilight, the magnetic lights appeared to her shorter and smaller, and she saw fewer of them; that is, those parts which were least luminous were the first to be overpowered by the faint day-light. In perfect darkness, she saw the luminous emanations brightest, their extent and size greatest, their limits best defined, and the play of colours most distinct.

8. If a magnet was presented to her in the dark, she saw it *luminous, not only when open, but even when closed by the armature.* This, at first sight, appears strange, but the sequel will show that her statement on this point entirely corresponds to the intimate nature of the phenomenon. The luminous appearance was, of course, different in the two cases. In the closed magnet she perceived no parts or points at which the light was particularly concentrated, as was the case on both poles of the magnet when open. But the closed magnet gave out, from all its edges, from the lines of junction of its bars, and from its corners, short, flame-like lights, constantly in a state of undulating motion. With the nine-bar horse-shoe magnet, supporting 80 ℔s., these were not longer than the thickness of a finger, or thereabouts.

9. When opened, the magnet exhibited the beautiful appearance represented in fig. 1. plate I. The drawing was made by Madlle. REICHEL herself, as well as she was able to do it; but she regretted that she could not attain to an exact image of the phenomenon, as presented by nature. One arm of the horse-shoe was about 10 inches long, and the flaming light rose to nearly the same distance above it, and was broader than the steel. At every depression between any two adjacent bars, there were smaller flames in the edges and corners, sending out sparks at their points. These, she said, were blue; the great flame was white below, yellow higher up, passing into red, and ending above in green and blue. It was not tranquil, but flickered, swelled, and started upwards, shooting out rays all the time. In this case, as in that of Madlle. NOWOTNY, the patient saw no mutual attraction or tendency to union between the flames, not even between those on the opposite poles. She also, like Madlle. N., saw no appreciable difference between the poles, in reference to the light. Fig. 2. plate I. gives a side view, in which a distinct bundle, of a

delicate flaming aspect, springs up from each edge of every bar. This, for the sake of distinctness, was not represented in fig. 1. Along the back and the inner side of the steel, feebler lights every where stream forth. On the inside they all turn upwards; but on the outside they rise, on the upper part, for a short distance, then go out, on the middle, horizontally for a very short space, and lastly, on the lower part, turn downwards in the opposite direction. These flames have been already partially described by Madlle. MAIX. Below, at the curve of the horse-shoe, that is, on the magnetically indifferent parts, they are shortest. Fig. 10. plate I. exhibits them as seen in a single bar of the horse-shoe magnet. The condition of the flames which appear along the longitudinal edges of each of the nine bars, when connected, is remarkable. Where the edges are accurately fitted together, and almost coincide in one line, the flames from each are yet distinctly separated, although we might expect them to form but one at the base. Immediately above this they diverge, and consequently converge towards the neighbouring flames from the other side of the same bar, in such a manner that a cross section would exhibit the appearance represented in fig. 3. plate I.—Weaker magnets, the light from which was drawn by Madlle. REICHEL, gave the same image, but with shorter emanations.

10. I showed her a straight bar magnet. It was 18 inches long, and about $1\frac{1}{3}$ inch broad, like ordinary bar iron. She represented its light, as in fig. 4. plate I. At the pole, which points to the north, that is, at the negative pole of the magnet, she saw a larger, at the opposite positive pole a smaller flame, the latter about half the size of the former. They had the same swelling, starting, and shooting motion, as in the horse-shoe, and were red below, green in the middle, blue above. At each of the four solid angles of each pole a brighter light streamed forth, each of these flames making an angle of 45° with the plane of the extremity, and had a somewhat rotatory motion, which did not occur in the chief flame in the middle. There was, therefore, a five-fold division at each end of the magnet. Madlle. NOWOTNY observed something analogous in the horse-shoe, which exhibited a more intense and more

lengthened emanation of light at each solid angle of the extremity. The four edges of the straight bar, like the separate bars of the horse-shoe, were covered with weaker light, exhibiting a play of colours, red, green, and blue, but flowing tranquilly out. This appearance was of equal intensity in its whole length, and neither the edges nor the indifferent parts could be distinguished, as in the horse-shoe.

11. The form and direction of the flame did not appear to be much affected by placing the bar-magnet in the plane of the magnetic meridian, or in the magnetic parallels, with the poles directed forwards or backwards, or in the direction of the magnetic inclination; and the terrestrial magnetism seemed not powerful enough to exert any considerable opposing influence.

12. I now took an electro-dynamic apparatus, in order, first, to produce an electro-magnet under her eyes; secondly, to observe the effects which the electro-magnet and a common steel-magnet might produce on each other, in reference to the light. The apparatus consisted of a horse-shoe magnet of wide curve, between the poles of which an electro-magnet, covered with its coil, could be made to rotate. The horse-shoe, the poles of which were pointed upwards, had limbs of which the cross section was a square of about 0.69 of an inch on each side. In the dark it exhibited essentially the same appearances as the nine-bar horse-shoe. At the four solid angles of each end, flames rising at an angle; and in the middle, proceeding from the central point of the base, a longer flame rising vertically. This last, however, was not, in this case, the thickest mass of flame, but had taken the form of a thin, straight, vertical needle of fire,—a difference possibly depending on the relative power of the magnet or its size, or on other collateral circumstances. Perhaps a small depression which had been made in the middle of each end, to admit of the rotation of fine points fitted to them, may have contributed to it. The light was in this form stationary, and the same at both poles, with the exception of a slight difference of size. As soon as I allowed the current of a Grove's pair to pass through the coil, wound round the soft iron which was to form the electro-magnet, the iron gave

out flames at both ends, and instantly showed the luminous appearances of a bar-magnet. Nay, when the current was no longer allowed to pass, and the iron had thus ceased to be a magnet, it yet continued to give out light at the poles, and, in regard to the light, behaved in a manner analogous to that observed in Ritter's galvanic pile; that is, it acted after removal of the exciting cause. (I shall return to the nature and explanation of this phenomenon in one of the following treatises.) *An electro-magnet, therefore, exhibits, in its emanations of flaming light, visible to the eyes of sensitive persons, a perfect analogy with the ordinary steel-magnet.*

13. But the action of the two flames, that of the steel-magnet and that of the electro-magnet, on each other, was most remarkable. The former yielded to the latter uniformly, and that, as decidedly as the flame of a candle yields to the current of air from a blow-pipe. To avoid as much as possible long descriptions, which are tedious to read, and yet difficult to follow, I refer at once to the fig. 5. 6. 7. and 9. plate I. Fig. 5. exhibits the steel-magnet alone, with its flames; fig. 6. *a* and *b*, the electro-magnet placed below the poles of the other, with ground plan; fig. 7. the electro-magnet above and near to the poles; fig. 8. rather higher; fig. 9. still higher above them. The three last show the striking manner in which the one flame is made to recede or yield by the other. Whether the cause of this lie in the difference of power, or elsewhere, remains for future investigation.

Thus we have, in Madlle. Reichel, the fifth, and the clearest witness to the fact of the luminous appearance at the poles of magnets.

I must mention, finally, Madlle. Maria Atzmannsdorfer, a young woman of 26, living in the suburb Alte Wieden, at the Golden Lamb. She is the daughter of a pensioned army surgeon. She suffers from headaches and spasms, with sleep-walking; but looks well, and walks like a healthy person through the streets. I brought her in the dusk of the evening to my house, into a room which I could render perfectly dark by means of inside shutters. She proved highly sensitive, and saw bright flames on the ends of the magnets. She described the flame, from the nine-bar horse-shoe, as being larger than it appeared to Madlle. REICHEL, indeed more than twice as high. The light, the play of colours, and the motion of the flame, she described exactly in the same way as Madlle. R. She saw, too, like her, the whole magnet luminous, and over its entire surface clothed in a delicate light. She is the sixth witness.

14. Let us now place together the different statements briefly. The nine-bar horse-shoe magnet showed, in the dark,—

a. To Madlle. NOWOTNY, in an advanced stage of convalescence, a kind of luminous vapour, surrounded by and mingled with rays of light $\frac{1}{2}$ to $\frac{3}{4}$ of an inch long, moving, starting, or shooting, white, now and then with a play of rainbow colours.

b. To Madlle. MAIX, when free from spasms, a white flame, a handbreadth in height.

c. To Madlle. STURMANN, a white flame, as long as a small hand, with play of colours.

d. To the shoemaker's apprentice, a flame a handbreadth in height.

e. To Madlle. MAIX, when affected by spasms, a light generally diffused over the whole magnet, which dazzled her eyes, and was largest and brightest at the poles.

f. To Madlle. REICHEL, a flame 10 inches long, shooting rays, flickering, and iridescent; lateral flames from each bar of the magnet; and a general feebler emanation along all the edges of the bars over the whole magnet.

g. To Madlle. ATZMANNSDORFER, the same appearances more intense; and the whole magnet bathed in light.

15. It thus appears, that *persons who are highly sensitive, perceive, according to the degree of their sensitiveness, and to the more or less complete darkness, a smaller or larger luminous appearance of the nature of a moving flame, at the poles of strong magnets*, when examined in the dark. They differ indeed according to their individual powers of perception, as to the size of the flame, but they agree perfectly in the general statement, that such a luminous appearance, of considerable size, invisible to the eyes of such healthy persons as were present, really exists at the poles of magnets. Now since all the above witnesses,

with the exception of Madlles. MAIX and REICHEL, who were acquainted, had no communication with or knowledge of one another, lived at great distances from each other, and yet, in my numerous experiments, never contradicted each other, and still less themselves, nor ever stated any thing contrary to the ascertained laws of Electricity and Magnetism; since, finally, I am conscious of the care and conscientiousness with which I made these experiments, I can have no hesitation in expressing my conviction that the perception, by highly sensitive persons, of luminous appearances at the poles of magnets is indisputable, and is to be regarded as a fact ascertained and won for science, as far, at least, as this can be accomplished by one observer. I am certain that we shall not have long to wait for confirmation of the fact from other quarters. Sensitive persons are indeed not so common in small towns, but yet may be found everywhere if we only look for them. In large cities, they are so far from being rare, that I look on it as no difficult matter, to find, in a city like Vienna, should it be necessary, hundreds of them at one time. My statements will therefore be easily controlled in Berlin, Hamburg, or Paris.

16. Let us now turn to the consideration of some of the properties of this magnetic light. That we cannot see it with the healthy eye, is in itself far from surprising. When we reflect how vast a difference exists between sun-light and candle-light, the former of which WOLLASTON estimated as 5560 times, LESLIE even at 12,000 times more intense than the latter; when we see how feeble is the light from the flame of Alcohol, Wood-spirit, Carbonic oxide and pure Hydrogen gases, as well as other combustibles, the flame of which is not only invisible to us in direct sunshine, but even partially escapes our notice in the diffused light of day; such enormous differences in the luminousness of different flames or lights, render the step to absolute invisibility for our healthy organs no longer a wide one; and consequently the existence of such a light, invisible to ordinary sense, is not only possible, but easily conceivable. We cannot therefore be surprised to find that such lights exist, and that luminous emanations, too feeble to be, in general, visible to us, shine over the poles of magnets.

17. In order, if possible, to convince myself that it was really light, and not a phenomenon of a different nature, which was perceived by sensitive persons, I was desirous to make an experiment with the Daguerrotype, and to see whether it were possible to produce an impression on the prepared silver plate. For the performance of this experiment, I applied to my obliging friend, M. CARL SCHUH, a private gentleman in Vienna, devoted to physical science, and known by his improvements on the Gas-microscope, and his dexterity in Daguerrotype. He shut up an iodised silver plate, in front of which the magnet was placed, in a dark box; a similar plate was shut up in a dark drawer without any magnet, at the same time. After some hours, he found the first plate, when exposed to mercurial vapour, affected by light; the other not. But the difference was not very great. To bring it out more fully, he placed the magnet, opposite a prepared plate, in a box, wrapped up in thick bedding, and with the most anxious precautions to exclude light during the experiment, of which I was myself witness. It was allowed to remain 64 hours; and then in the dark exposed to the vapour of mercury. The plate now showed on its whole surface the effect of light. From this it appeared, that, provided other causes be not capable of affecting the prepared plate, when exposed to them for a long time, *it must be a real light, although feeble, and slowly acting on the plate, which flows from the magnet.*

18. *I made, with a similar intention, another experiment with a lens.* It had an opening of about 8 inches, and a focal distance of about 12¼ inches for a candle placed 59 inches from it. I now placed in a perfectly dark room, the magnet, the flame of which appeared to Madlle. REICHEL 10¼ inches high, behind the lens at the distance of about 25 inches, and directed the axis of the lens towards a wall, to which I called the attention of Madlle. REICHEL. The able mechanician, M. EKLING, was present. We found it necessary gradually to remove the lens to 54 inches from the wall, during which the observer, Madlle. R., saw the image constantly diminishing, till it reached the size of a lentil, and had consequently shrunk from 10¼ inches to the dimension of about 0.12 to 0.15 of an

inch. But, in spite of this great degree of concentration, none of the assistants could perceive even a trace of light. We obtained, however, a sure means of repeatedly testing the accuracy of our observer. She had placed her finger on the spot where she saw the focal image; I felt for her hand, and placed my finger on the spot indicated. I now desired M. EKLING, who held the lens, to shift its direction, but without saying how. The focal image of course changed its position in the direction in which the lens was moved; Madlle. REICHEL instantly pointed out another spot; which I felt for, and on which I placed my finger, desiring M. EKLING now to say, in what direction he had moved the lens. When he said, to the right, to the left, upwards or downwards, my finger was always found to have been placed in the direction indicated. The truth and accuracy of Madlle. R.'s statements, as to the place of the focal image, were thus established beyond a doubt. She described the image as red, and saw the whole lens illuminated with red light from the magnet.

19. *The magnetic light gave out no heat;* at least none appreciable by our most delicate instruments. When directed on the Thermsocope of Nobili, I could not detect, even after a long time, any motion of the astatic needle in the differential Galvanometer.

20. It was desirable to acquire some more intimate knowledge of the true nature or substantiality of the flame, light, or whatever it may be called, observed over the poles of magnets. As it did not proceed from its source in a purely radiant form, but in a flickering shape, which produced all sorts of curved undulations and ever-changing lines, it could hardly consist of a simple emanation of light alone. When, indeed, I turned the poles of the magnet downwards, or in any lateral direction, the light flowed out in the same shape as when the poles were directed upwards, but in a downward or lateral direction, according as the magnet was placed. This testified to its being imponderable, as was more than probable, but proved nothing positive as to its essence. On the other hand, I thought the answer important, which was given to my enquiry, how it behaved when blown upon? The observer

declared that it flickered to the side, like other flames. When a solid body was brought too near it, it bent, near its point, round the object. When the lens in the preceding experiments had been approached too near the magnet, she had seen the flame bend round it, exactly as happens with the flame of a candle when a piece of glass is held in it, in order to be blackened. When the hand was laid on the poles, the flame flickered up round the hand and between the fingers, &c. &c. It follows, that the *magnetic flame either is itself material*, or *has a material substratum*, and further, that the *magnetic light is something distinct*, and *the flame a compound appearance*, in which some material body is combined with the probably immaterial essence of Light. In fact, the observer saw the flame bend round the lens, while the light was transmitted and collected in the focus. Both Madlle. NOWOTNY and Madlle. STURMANN have also assured me, that the flame shed around it a light which illuminated neighbouring objects; and Madlle. REICHEL marked accurately the distance to which this illumination extended on the table on which the magnet lay. I measured it, and found it to extend over a circle of about 38 inches diameter. Whether that, which flows from the magnet in the shape of a flame, be a material emanation from it, or whether it merely represents a change of condition produced by the magnet in the surrounding air, or in the sense of the prevailing modern theory, in the luminiferous Æther, accompanied in its progress by development of light;—these are questions, to answer which, many other things, and among them, the gradual loss of power in steel magnets, must be well weighed, and which must be left for future investigation. Here, for the present, it is only necessary to establish, that the *magnetic flame*, which yields to mechanical obstacles, is not identical with the *magnetic light*, which possesses a higher, radiant nature.

21. I now return to the consideration mentioned in § 2. of the Introduction. The first practical application of these observations, would be an attempt to apply them to the explanation of the aurora borealis. We possess the beautiful observations of DAVY, who applied to the aurora the action of

the magnet on the electric light in rarefied air, and endeavoured to shew that the aurora was probably caused by such a discharge of electricity at the upper limit of the atmosphere. Since, however, we have learned, by the recent polar expeditions, how low in the atmosphere the aurora often descends, the ingenious idea of DAVY has lost much of its probability. It has lost the foundation on which it rested, namely, the rarefied space, and along with this, the diffusion of free electricity, which he derives from our storms, and applies to the explanation of the aurora. The certainty which we possess, that the aurora is only formed under the influence of the terrestrial magnetic poles; the hitherto total want of the knowledge of any direct emanation of light from the magnet; the fact, now ascertained, that luminous emanations, although invisible to the ordinary sight, and exhibiting to sensitive persons a play of colours, especially white, yellow, and red, actually do proceed from the magnet;—all these considerations naturally lead us to suspect that the aurora may either be magnetism itself flowing from the polar regions, or a direct effect of such magnetism. We know that the appearance of the aurora produces a disturbing effect on the magnetic needles of entire countries, as the magnetic flame, or the magnet from which it proceeds, also does, at proper distances. But it is probably only the emanations from the magnet, and not the magnet itself, which thus act at a distance; and if this be the case, the effects of deviation produced by the aurora, will coincide with those of the magnet. If, finally, we compare the appearances of the magnetic light, as seen by sensitive persons, in their details, with those of the aurora, the probability of this coincidence in nature is obviously increased. The aurora is well known as a white arch of light, according to some as a vaporous or misty mass of light at the polar horizon, from which a flickering, variable, shooting of rays, collected in bundles, towards the equator, takes place. Its lines have indeed one prevailing direction, but are not always straight or parallel to each other, appearing often somewhat curved, and occasionally scintillating. Their colour passes from the white of the arch to bluish, emerald green, and especially to red, and they illuminate whole tracts with the last mentioned colour. The same shooting of luminous rays, the same flickering, delicate, vapour-like flame running in curved lines, the same varied play of colours, the same red illumination of objects;—all these things are described in the same way by the observers of the magnetic light. The observations are not indeed among themselves perfectly alike, but they agree in all the chief points. The principal discrepancies concern chiefly the size of the flame, which is less essential; and they are easily explained by the different degrees of sensibility to the magnetic light found in different sensitive persons. We see, for example, in the case of Madlle. MAIX, that according as she was in her usual tranquil state, or affected with spasms, she saw two different luminous appearances. In the former state, she saw only a light of a handbreadth over the poles; in the latter, not only did this light appear much larger and more intense, but the whole magnet was surrounded with luminous emanations. In like manner, we find in the case of Madlle. NOWOTNY, that the magnetic light, in its apparent size, kept place with the progress of her recovery, and that it appeared from time to time smaller, as her disease diminished, until at last it eluded her observation. At one stage of her illness she observed immediately over the steel a kind of luminous vapour, which the far more excitable Madlle. REICHEL never saw. From this luminous cloud Madlle. NOWOTNY saw bundles of luminous rays proceed, as Madlle. REICHEL saw similar bundles proceeding from the solid angles of the ends of weaker magnets. This luminous vapour or cloud next the steel resembles in the highest degree the luminous arch of the aurora; and if Madlle. REICHEL did not see it, the cause probably is, that in her observations, the much more flickering light and flames arising from the sides of the bars, covered the luminous cloud, so that she could not, of course, perceive it. It may be expected that, as her recovery advances, a period will arrive when these lateral flames will disappear to her eyes, and then the luminous vapour will become exposed and visible to her, as to Madlle. NOWOTNY.

It is this tranquil, light, vapour-like appearance which renders the resemblance to the aurora so close, that we feel involuntarily led to admit the complete similarity of the aurora and the magnetic light. Let me not, however, be misunderstood: I do not mean to say that I regard the identity of the two phenomena as being demonstrated, for there lies between light which is visible to all healthy eyes, and that which is invisible, a chasm, not yet filled up, and which cannot be even filled up by the assumption of a difference of intensity in the two phenomena. But this much I hold for certain, and I think myself entitled to express the conviction, that a most remarkable analogy exists between them, so great, indeed, as undeniably to render the identity of the magnetic light and the aurora in a high degree probable.

22. Retrospect.

a. A strong magnet exerts on many healthy and diseased persons a peculiar influence; it is an agent, which affects the vital force.*

b. Those who are highly sensitive to this influence or irritation, frequently exhibit exalted acuteness of the senses, and are then able to perceive light and flame-like appearances in the magnet. The strength and distinctness of this perception increases with the sensitiveness of the observer and the darkness of the place.

c. The pole — M gives the larger, the pole + M the smaller flame, in the latitude of Vienna. The flame divides, according to the form and structure of the magnet, at each pole into several flames of iridescent colours. Its form and colour vary according as the magnet is open or closed, a common steel magnet, or an electro-magnet, free, or under the influence of another magnet.

d. The positive and negative flames show no tendency to approach each other, or to unite.

* See Appendix, A.

e. The flame may be mechanically bent hither and thither, like ordinary flame.

f. It gives out light, which is red, acts on the plate of the daguerreotype, and may be concentrated in the focus of a lens, but has no appreciable warmth.

g. The magnetic flame and its light exhibit so great a resemblance to the aurora borealis, that I must consider their identity highly probable.

SECOND TREATISE.

CRYSTALS.

23. The physician who attended Madlle. Nowotny had repeated, in her case, even before I saw her, certain observations made by the older physicians in cases of catalepsy; namely: first, the fact which Dr. Petetin of Lyons, in 1788, as well as others, had noticed and published, that, when a strong magnet was placed in contact with the hand, the hand adhered to the magnet, as a piece of iron does; and, secondly, that water which had been subjected to the action of a magnet, passed several times along the vessel containing it, was at once and decidedly distinguished by the patient from ordinary water. This last fact was first observed and made known by Mesmer; it has been often ridiculed, and as often reasserted. We shall see, in the course of the present investigation, how much of it, when it is subjected to the touch-stone of physical research, is found to be true or false.

The adhesion of a living limb to a magnet, is a fact altogether unknown in the sciences of physics and physiology, and few persons have convinced themselves of it by actual ocular demonstration; it is therefore necessary here to explain and illustrate it in some degree. When Madlle. Nowotny, in the cataleptic state, lay without consciousness or motion, but without spasms, and when a horse-shoe magnet capable of supporting about 20 ℔s. was approached to her hand, the hand adhered so firmly to it, that when the magnet was raised, or moved sidewards, backwards, or in any direction whatever, her hand stuck to it, as if attached in the way in which a piece of iron would have been. She remained utterly unconscious all the time; but the attraction was so strong, that when the magnet was moved towards her feet, farther than her arm could reach, she did not let it go, but, although insensible, raised herself in bed, and followed the magnet with her hand as far as she could possibly do; so that the effect was the same as if some one had seized her hand, and by means of this drawn or bent her body forwards towards the feet. When the magnet was at last moved so far that her body could not be moved or bent further, so as to enable her to follow it, she was compelled to let it go, but remained then in the manner always observed in cataleptics, fixed and immoveable in the position into which she had been brought. This I saw daily, from 6 to 8 p.m., at which time the fits of catalepsy occurred; and the fact was usually witnessed by from 8 to 10 persons, physicians, natural philosophers, chemists, and others interested in the sciences. It can be of no essential advantage to refer to these persons.

When I visited the patient at other times of the day, I found her fully conscious and cheerful; but the phenomena were the same. Her hand adhered to, and followed the magnet, exactly as in the insensible cataleptic state. The account which, at my desire, she gave of the phenomenon, added but little to my knowledge of it, or to the power of explaining its strangeness, considered physically. She described her sensations as of an irresistible attraction, which she felt herself involuntarily compelled to follow; and which, even against her will, she was forced to obey. It was an agreeable sensation, combined with a soft cooling breeze or *aura*, which flowed downwards from the magnet on the hand, which felt as if attracted to it, and drawn forwards by a thousand fine threads. Moreover, she had never, she said, perceived anything similar to it in her life before; the whole was a peculiar, indescribable feeling, which, when the magnet was not too powerful, had in it something infinitely refreshing and agreeable.*

* As this phenomenon appeared almost too marvellous for belief, and was too strongly in contradiction to the known laws of magnetism to allow me easily

I had afterwards the opportunity of observing exactly the same phenomenon in Madlle. REICHEL. Her complaint was different from that of Madlle. NOWOTNY, but was, like hers, complicated with cataleptic fits; and her hand followed the magnet, exactly as above described, both in the cataleptic and in the ordinary state. The same thing was reported to me, for I was prevented from personally making the observation, as occurring at a certain stage of the illness of Madlle. STURMANN, by Professor LIPPICH; and I have every reason to

to admit it, I confess that at first I had doubts of the perfect trustworthiness of the patient's statements, and suspected the possibility of imposture, although this idea stood in direct opposition to the evident honesty of the family, and the truthfulness of the patient. I took, therefore, various precautions, blindfolded the patient's eyes in the cataleptic state, and operated with the magnet in many varied ways, &c. The results were always the same. It is proper here to mention some of the tests to which I subjected the patient. Among others, I arranged with a friend, on my giving a certain sign, to remove and replace alternately the armature of a powerful magnet, supporting 80 lbs., on the opposite side of the wall, behind the place where the patient lay in bed. He was to point the poles of the open magnet towards her, while I stood at the bedside and observed her. The armature had hardly been removed, when the patient became restless, and complained, that surely an open magnet must be lying near her, begging her friends to look for it, and relieve her from her distress. She was always painfully affected by large magnets, whereas the effect of small ones was pleasant and cooling. The armature was now replaced without her knowledge, and she became again calm. When the experiment was secretly repeated, she became quite puzzled, and could not conceive what was the origin of these varying feelings of discomfort which attacked her, and left her, just as if a magnet had been presented to her. The magnet, therefore, had acted on her, through a stone wall, without her being aware of its vicinity, exactly as when it was opened before her, according to the known laws of magnetism, which irresistibly penetrates all bodies. At last we solved the enigma to her, and the experiment was repeated with her knowledge. It produced the same results, namely, at every time the armature was removed, the same restlessness and the varying flush which, during the secret experiments, I had seen come and go.—M. BAUMGARTNER, formerly Professor of Physics, a well-known cultivator of science, made, for his own satisfaction, another testing experiment with her, which was strikingly well devised. When the phenomena produced by the magnet and its peculiar action on the patient had been repeatedly shown to him, he produced from his pocket a horse-shoe magnet of his own, of which he said to the byestanders, in the hearing of the patient, that this was the best and most powerful of all the magnets in his collection of apparatus, and that he was desirous to see how strong an effect it would have on the patient. To our astonishment, Madlle. NOWOTNY declared that she could not confirm this; on the contrary, she felt it not only weaker than the weakest of all the magnets repose absolute confidence in the accuracy of his statement. These different cases, added to the observations reported by PETETIN, NICK, and others at an earlier period, leave no doubt of the truth of the fact, that, *in certain diseases, especially where catalepsy is present, there exists a decided attraction between the human hand and a powerful magnet.*

I made further, with Madlle. NOWOTNY, an experiment of the same kind on her feet. There also a similar attraction existed, but very much weaker than in the hands. No other part of the body could be found, which possessed the same sensibility to the influenee of the magnet.

24. The first question which naturally presented itself was this: whether the attraction, exerted by the magnet on the patient, were mutual; and also, conversely, whether it were exerted by her on bodies, such as iron, capable of magnetic induction, and which are for the time converted into magnets by the approach of a magnetic body; in other words, whether, in consequence of her diseased condition, magnetism, and with it, magnetic attraction, existed in her person. To try this, I took filings of iron, and brought her finger over them. Not the smallest particle adhered to the finger, even when it had just been in contact with the magnet, and was, consequently, more strongly differentiated than it might naturally be. A magnetic needle finely suspended, to the poles of which I caused her to approach her finger alternately, and in different positions, did not exhibit the slightest tendency to deviation or oscillation. Another experiment bearing on this point, was

that were then in the house, but almost quite inert. It had to her neither taste nor smell, produced no burning sensation, and had no attraction for her hand. M. BAUMGARTNER smiled at our surprise, and told us that he had demagnetised by stroking it the wrong way with another magnet, the horse-shoe which he had brought, and which was certainly, before he did this, his best and most powerful magnet. He had thus deprived it of all but a trace of its power, so that it was little better than a mere piece of iron. In fact it no longer attracted its armature. M. BAUMGARTNER wished to test the truth and consistency of the patient, and gave us a new guarantee for them. After such tests, and I could add many more, the reader will not, I trust, require from me any further security for the truth and genuineness of the facts, the accuracy of which will be sufficiently vouched for by their consistency among themselves, as will be seen in the sequel.—R.

made, at the request and in the presence of M. BAUMGARTNER. When we raised the magnet, and with it the hand and the whole arm, it appeared to many who tried the experiment, that the magnet was heavier than by itself, by the whole weight of the arm. This I could not perceive, but many believed that they felt it distinctly. A horse-shoe magnet was now suspended to the arm of a balance, and accurately counterpoised. After laying the hand of the patient, with its back downwards, on a flat support, I confined the points of the fingers with my own, and brought her hand near the suspended magnet. The hand of the patient strove to reach the magnet, and I had to exert some force to keep it in its place; yet the tongue of the balance never moved in the slightest degree, not even when the magnet almost touched the hand, and the fingers made such convulsive efforts to close, that I had difficulty in keeping them extended.

While I was engaged with these investigations, there appeared in the journals the statement of THILORIER, who had, as he said, converted steel needles into magnets, by contact with nervous patients during their fits. Whether the fit can in this respect do more than the disease itself, is a question I do not here enter upon; but the result of this announcement was, that I was told by their physician, soon afterwards, that both Madlle. REICHEL and Madlle. MAIX converted every steel needle, which they held for some time in their hands, into a magnet. I went to see the patients, who assured me of the accuracy of the report, and shewed me knitting needles, which attracted and lifted sewing needles. I made the experiment myself, first selecting such knitting needles as were not already magnetic, and removing all magnets from the vicinity of the patients, and after these precautions, gave them the needles. These they held in their hands, first as long, then at my request twice as long, as in the previous experiments, in which their physician had converted needles, as he supposed, into magnets. But the needles did not become magnetic, and all efforts to produce this effect proved fruitless. Doubtless the state of the needles had not, in the previous experiments, been examined before giving them to the patients; for out of a dozen knitting needles, half will always be found already more or less magnetic. At last I was assured that Madlle. STURMANN was so powerfully magnetic, that her finger caused the needle to deviate 25° or 30°. I was invited by Professor LIPPICH to an experiment, and actually saw the suspended needle deviate considerably. As it was not sufficiently protected against currents of air, I myself tried, next day, the experiment, with the precaution of placing the needle in a vessel, covered by a glass plate, through which the result might be seen. I made in the side of a vessel, a perforation just sufficient to admit the finger, which filled it. In this way, the patient's finger could be brought close to the needle, without much disturbing the air, and the respiration of those standing near could no longer affect the needle. When the finger was introduced, it seemed as if some degree of adhesion took place. I examined the point of the finger, and as it seemed to me rather moist, I rubbed it with a little flour. All adhesion had now disappeared, and the needle remained motionless. In the former experiment, the finely suspended needle had obviously adhered to traces of perspiration on the finger, and when the viscidity of these was destroyed by the flour, all attraction instantly ceased. To make still more sure, I introduced the patient's finger into the spiral of a differential galvanometer. But neither when it was introduced, nor when it was withdrawn, was any induced current perceptible, and the astatic needle remained motionless.

25. From all this it follows, that there is nothing ponderable in the attraction exercised by the magnet on the hands and feet of cataleptics. This attraction has no supporting power, cannot even attract iron filings, and is equally incapable of acting on the needle, and of inducing an electrical current. The arm, which in the cataleptic state followed the magnet, therefore, supported itself, and its passive attraction had obviously quite a different significance from that of the attraction of iron towards a magnet, or more correctly, towards magnetically and oppositely polarised matter, in the sense hitherto attached to that expression.

It is sufficiently known that, in physics, we are not acquainted with any kind of attraction which is not mutual. But it is equally certain, that a person in a state of cataleptic insensibility, a state which cannot be feigned, not only has no free volition, but has no volition of any kind. Now, since the mechanical magnetic attraction exercised on persons in that state by the magnet, is a fact which is not only here established by sufficient experience, but which may easily be confirmed and controlled in every large city, where such patients may always be found, it acquires, in spite of its apparent anomalousness, an established scientific existence, and imperiously demands further investigation. Not in order to explain it, but in order to render its character in some degree intelligible, I may here refer to all the attractions and repulsions which vegetative life, both in animals and plants, continually effects in a thousand forms, without our being, at present, enabled either to perceive or to infer the existence of mutual attraction. A root penetrates with force deep down into a hard soil; it breaks and splits powerful mechanical obstacles. We perceive no indications of mutual attraction or repulsion which may impel it so forcibly; and yet the result occurs. Analogous forces carry the hand of a cataleptic patient towards the magnet; and this occurs, whether, in the first instance, we can comprehend and account for it, or not.

26. When I took, instead of the medium magnet, supporting 20 ℔s., the large one, which carried 80 ℔s., and laid it on the extended hand of Madlle. Nowotny, she seized, whether in the insensible or in the ordinary state, the end offered to her, and grasped it so firmly, that it could not be torn from her without the strongest effort. She herself could not let it go. The whole hand was spasmodically clenched, and the spasm bound the fingers round the pole of the magnet, and contracted the whole hand so violently, that all power of voluntarily moving the hand ceased.

27. I have mentioned above (§ 2), magnetised water, which the patient instantly distinguished from ordinary water, even when she knew nothing of what had been done. Nothing could be more repulsive to me than the reappearance of a thing apparently so absurd, the mere mention of which fills all che-

mists and natural philosophers with disgust. But, in spite of this prejudice, I could not deny, or annihilate to my own mind by denying, that which I saw before my eyes, as often as I tried it; namely, that Madlle. Nowotny always accurately, decidedly, and without one failure, distinguished a magnetised glass of water from one not magnetised. We cannot contend against the force of facts by any argumentation; I was, therefore, obliged to admit the fact, although I was unable to comprehend it. But when the same fact presented itself, at a subsequent period, in the cases of Madlles. Sturmann, Maix, Reichel, Atzmannsdorfer, and others, nay, when I saw it, in some of these, in a still higher degree, I relinquished my doubts and my resistance to a phenomenon, the actual existence of which admitted no longer of any rational doubt.

28. The strangeness of the phenomena was yet to reach the acme of incomprehensibility; when it appeared, namely, that not only the magnet, but a mere magnetised glass of water, possessed the power of carrying with it the hand of Madlle. Nowotny. This was indeed the case, in a far less degree; yet her hand was unmistakeably solicited, as well in the cataleptic state as at all other times, by a glass of water previously acted on by the magnet, so that an effort to follow it in every direction was manifest.

29. Observing this, and feeling convinced that so strange a phenomenon could not exist, as an isolated one, in nature, I was desirous to try, whether that which presented itself in the case of water might not also be effected by other bodies; and if this should happen, I hoped to see it in such a variety of instances as should admit of laws being deduced from them. With this view, all sorts of minerals, preparations, drugs, and objects of all kinds, were subjected to the action of the magnet, and tried on the patient in the same way as the glass of water. I found, in fact, that all of them immediately acted on her, just as the magnetised water had done; they excited the hand more or less powerfully to follow them, but with various differences in their mode of action. Some caused general spasms through the whole body; others only in the arm; others only in the hand; and finally, others hardly excited spasms at all,

although all had been equally magnetised. There must, therefore, exist a difference in the different kinds of matter employed, which it was necessary to take into account.

30. To investigate this difference, I now tried the experiment of bringing the same substances into contact with the hand of the patient, *in their natural state, and without having previously magnetised them.* To my surprise, they now acted on her, with a force which was often but little inferior to that which they had exhibited when magnetised. The action, however, was not always accompanied by the solicitation of the hand to follow the object. On the other hand, that other action, in virtue of which the patient had (§ 26.) convulsively grasped the magnet in her hand, occurred frequently in various degrees of force. The method I pursued was this: I placed the different substances in the hand of the patient, while she was in the insensible cataleptic state, and observed the effects. I then repeated the experiments when she was in her usual conscious state, and free from the cataleptic affection. In this comparative trial it appeared that, in both cases, the effect was qualitatively the same, but was beyond comparison quantitatively more powerful during the catalepsy than at other times. The effect essentially consisted in this: that the different substances, when laid in the hand of the insensible patient, either

a. Caused tonic spasm of the fingers, just as the magnet did, and forced them to clench themselves involuntarily, so that the substance was grasped in the hand. The substances which thus acted again subdivided themselves into such as at the same time solicited the hand to follow them, and such as did not produce that effect: or

b. Appeared inert, excited no spasm, and left the hand motionless.

The first effect occurred with various degrees of energy in different substances. It either occurred instantly on contact, or was more or less slowly and gradually produced. In the latter case the fingers began to bend, and gradually contracted more and more, till the hand was clenched in a state of tonic spasm. This was precisely what was observed when very weak magnets were laid in the hand of the patient while she was insensible.

31. When I compared the substances employed, they arranged themselves, in relation to this effect, not according to the nature and composition of the bodies, not according to the electro-chemical series; not even according to their power of conducting electricity. Nay, the same substance, in different specimens, acted differently, in one case producing the effect, in the other not. This was observed in carbonate of lime, sugar, quartz, and many others. I next perceived that, *among the amorphous substances, there was not one* which acted so that the patient grasped it in her fingers; and on the other hand, *that all the bodies which produced that effect were crystallised.* Among the crystalline bodies tried, however, there were a good many which had no action. When I now, omitting the amorphous substances, arranged all the crystallised bodies in two opposite groups, in one of which I placed those which were inert, in the other those which acted on the hand like a magnet, I found that the former, the inert group, included *all substances confusedly crystallised*, such as loaf-sugar, Carrara marble, and dolomite; and also all substances *existing in masses, composed of many groups of small crystals, running in various or opposite directions*, such as prehnite, wavellite, lumps of sugar-of-lead, masses of crystallised silver from Kongsberg, &c.; and that the other, the active group, contained *all single, detached crystals*, and *those where the crystals, if in groups, had their principal axes parallel or nearly so*, such as celestine, many kinds of gypsum, and fibrous red and brown hematite.

For the sake of greater clearness, I give here the list of the substances tried :—

I.—Inert Bodies.

a. Amorphous.

Ivory, wood, &c.
Anthracite.
Cannel coal.
Mineral pitch.
Amber.
Glass of all kinds.
Osmium. Rhodium.
Palladium.
Mercury.
Silver and gold, in coins.

Copper and brass.
Bar iron.
Zinc, cadmium, lead.
Compact limestone.
Compact red copper ore.
Potassium. Sodium.
Dry hydrate of potash.
Chrome iron ore.
Selenium.
Liver of sulphur.
Cast sulphur.
Compact talc.
Gurhofian.
Magnesite.
Pumice-stone.
Obsidian.
Menilite.
Common opal.
Petrified wood.
Egyptian jasper.
Compact quartz, with fatty lustre.

b. Crystalline bodies.

Granular limestone.
Dolomite.
Orpiment.
Wavellite.
Kakoxene.
Irregular crystallised masses of native silver from Kongsberg.
Loaf sugar.
Native sulphuret of antimony.
Prehnite.
Natrolite.

II.—Active Bodies; all crystallised; fine detached crystals, many of them large and splendid specimens from the Imperial Private Cabinet of Natural History in Vienna.

a. Such as, with hardly sensible spasm, induced the fingers to close on the object:

Native rough diamond, very small.
Metallic antimony.
Mesotype.
Witherite.
Tinstone.
Mica.
Corundum.
Prussiate of potash (yellow).
Sugar-candy.
Leucite.
Garnet.
Augite.
Hornblende.
Staurolite.
Blue vitriol.
Foliated graphite.
Wolfram.
Metallic bismuth.
Rutile.
Lievrite.
Asparagus stone.
Sphene.
Iron pyrites.

Analcime.
Adularia.
Felspar.
Boracite.
Celestine.
Topaz.
Apatite.
White lead ore.
A crystal of gold, ½ inch thick.
Alum.

b. Such as caused the hand to close on them, but did not solicit it to follow:

Pistacite.
Cobalt glance.
Zinc blende.
Specular iron ore.
Magnetic iron ore.
Rock salt.
Rock crystal.

c. Such as acted so strongly, that they caused the hand to be firmly clenched, and when brought near the hand, attracted it.

Meteorite from Macao.
Crystallised sandstone from Fontainebleau.
Calcareous spar.
Arragonite.
Tourmaline, both when cold and warm.
Beryl.
Gypsum.
Fluor spar.
Heavy spar.

All these comparisons admitted of being reduced to this; that granular crystalline limestone, compact quartz, and loaf-sugar, were inert; whereas a crystal of calcareous spar, a rock crystal, or a fine crystal of sugar-candy, that is, detached crystals in general, when laid on the hand of the patient, instantly excited involuntary contraction, attracted the hand, caused it to become clenched, and to grasp the crystal, and, in part, with the strongest tonic spasm. Here, therefore, we perceive, *in single crystals, a peculiar power, a fundamental force,* which had hitherto remained unobserved. This force, as far as it has yet been recognised, *belongs to matter, not as such, but in virtue of its form and state of aggregation.* POUILLET, in MÜLLER'S translation of his Manual de Physique, p. 167, expressly says: "that it has not hitherto been observed in ponderable matter, that its form, or the arrangement of its molecules, is a cause of new forces acting at sensible distances.

But we see, that the supposed case is here the actual one; matter must be crystallised, else it does not act in the way here described.

32. When I now tried an active body, a crystal, by itself, I observed that the power of exciting spasm in the living organism was not possessed in an equal degree by all parts of its surface. There were points, and the patient easily recognised these, which had this power in a very feeble degree, or not at all; others, in which it was found of increased intensity. It appeared that each crystal exhibited especially two such points, which were the proper seat of the power. And these two points are, in each crystal, diametrically opposite to each other; they were, in fact, poles of the principal axis of the crystal. The action of both was similar, but that of one was stronger than that of the other; and there was also this difference, that from one a cool, from the other a warm soft *aura*, or apparent current of air, seemed to flow.

33. While I was endeavouring to pursue the manifestations of this force in different directions, I made the experiment of drawing downwards, along the inner surface of the patient's hand, the stronger pole of a crystal of moderate size, namely, a rock crystal of nearly 2 inches thick, and 7¾ inches long; making passes exactly as with the magnet, when it is intended to cause ordinary sleep. I found the action *exactly equal to that of a small compass needle*, which I used for comparison. It was 5½ inches long, less than ¼ inch broad, and about 4-100ths of an inch thick. It weighed about 184 grains, and supported about twice its own weight. The patient felt, when I drew the point of the crystal, without contact, slowly down from the wrist over the palm of the hand and along the fingers, an agreeable, gentle, cool *aura*, which she said I led along her hand. The needle produced the like effect. If I inverted the experiment, drawing the needle upwards from the point of the middle finger to the wrist, it produced an unpleasant lukewarm sensation, obviously disagreeable to the patient. The crystal produced exactly the same sensation, when I used it in the same way. On another occasion I tried a crystal of three times the size. With the downward stroke it produced

an effect equal to that of a bar-magnet capable of supporting 2 ℔s.; but the upward or back stroke acted so powerfully, that it produced a spasmodic state extending up the arm and nearly to the axilla, and lasting for several minutes, and could not with prudence be repeated, on account of its violent action.

34. Furnished with these facts, I went to the Clinical Hospital of the University, with the intention of trying whether the observations which I had made on Madlle. NOWOTNY could be repeated and confirmed on other similar patients, so as to enable me, in some degree, to generalise them. By the kind assistance of Professor LIPPICH, I was again enabled to have recourse to Madlle. STURMANN. I drew along the hand the point of a rock crystal, about 6 inches long, and nearly 2 inches thick. The patient instantly perceived the warm and the cool sensation very vividly. When I used, in the same way, the other pole of the crystal, the sensations were the same, but weaker and inverted. The two patients, therefore, coincided in their statements. The effect on Madlle. STURMANN was so strong, that it drew into sympathy with the hand the whole arm up to the shoulder, over which the warm and the cooling sensations gradually spread. When I, on a subsequent occasion, tried a crystal of thrice the size, the first stroke of the crystal down the hand acted so powerfully, that she suddenly flushed and then became pale, so that here also I could not venture to make a second trial of it.

I now tried making the passes from the head over the face. The results were exactly similar, and the sensations particularly distinct along the temples. Corresponding trials with the needle gave the same results; only this time the needle appeared sensibly weaker than the crystal. I subsequently tried the same experiments with Madlle. MAIX. In the case of this very sensitive patient, who, however, was always perfectly conscious, the crystals acted not only along the line of the stroke, but also over a broad tract up and down the hand, and this effect extended up the arm. Two months later, I tried Madlle. REICHEL. This strong and healthy-looking young woman was so sensitive to the poles of crystals, that she perceived their approach by the sensations they excited, even at a certain distance. Like the previous patients, she found the *aura* from the — M pole (that pointing towards the north in the needle,) cool when it was drawn downwards, warm when drawn upwards; and, on the contrary, that from the + M pole (pointing south in the needle,) warm when drawn downwards, cool when drawn upwards. At last I became acquainted with Madlle. ATZMANNSDORFER, (§ 13.) and found in her a sensitive,* who perceived the stroke of crystals most strongly. Small crystals of fluor spar, gypsum, and iron pyrites, an inch or more in length, she felt, when drawn downwards over the hand, very cold. With thin acicular crystals I could, as it were, draw lines of this sensation on the hand, but the upward stroke of these was warm, and affected her so disagreeably, that it attacked her whole person in an unpleasant manner, and began to excite spasms, as often as I repeated the experiment.

35. While I was one day explaining all this to a friend, and, in order to show him clearly how I operated on the patients, had passed the same rock crystal along his hand, he looked at me with amazement, and declared *that he felt, himself, that which I was describing as the sensation experienced by the patients*, namely, a very distinct cool *aura*, as often as I made the downward motion over his hand. He was a strong healthy man, in the prime of life, and permitted me to refer to him. It was M. CARL SCHUH, a private gentleman residing in Vienna, and distinguished by his great scientific acquirements. From that time I was in the habit of trying the same experiment on all my family and friends, and on many persons, strangers to me, among whom were physicians, natural philosophers, chemists. I am permitted to name here our celebrated naturalist, Dr. ENDLICHER, chief of all the public botanical institutions in Vienna. I found that not only the suffering patients, but also very many persons, ex-

* I borrow the term "sensitive" for the magneto-physiological reaction, from vegetable physiology, in which it is applied to plants possessing a peculiar irritability, such as some Mimosæ, Berberis, Dionæa, Hedysarum, &c.; in distinction from "sensible" in animal physiology, which, as is known, is used in a somewhat wider sense.

perience these sensations ; and that a large crystal of quartz, gypsum, heavy spar, fluor spar, and other substances, when made to pass near enough over the palm of the hand, excites and renders distinctly sensible, in many persons, certain feelings, usually those of coolness and warmth. This took place even when the crystal had been previously warmed to about the temperature of the hand, which was necessary in order to meet the objection, that, for example, the sensation of coolness might depend on radiation from the hand towards the cooler crystal.* Many could tell, with averted face, that is, by the sensation alone, whether I had used the positive or the negative end of the crystal, that is, the stronger or the weaker pole.

From all this it follows, that the following laws prevail in nature.

a. *There resides in matter a peculiar force, hitherto overlooked, which, when the crystalline form has been assumed, is found acting in the line of the axes.*

b. *At the poles of the axes it is most powerful ; but its effects are different at the two poles, and are opposite.*

c. *The influence of this force on sensitive persons, coincides exactly with that exerted on the same persons by a magnet and by its poles.*

36. In order to enable every one to repeat these experiments, I would state expressly, that a large detached crystal with a natural termination is necessary ; and that it must be larger, the less sensitive the person is. Heavy spar, fluor spar, and gypsum, are best adapted for the purpose. It is of no moment whether the hand be coarse or fine in its texture, for I have occasionally found the coarsest hand of a mechanic more sensitive than the most delicate hand of one whose occupation is that of writing. The crystal should be drawn over the inner surface of the hand, from the wrist over the palm and down to the point of the middle finger, as near as possible, but without contact, and at such a rate of motion that one pass occupies about five seconds. The crystal is held verti-

* This objection appears also to be sufficiently disposed of by the fact, that one end of the crystal caused the cool, the other the warm aura ; and also that the upward and downward strokes cause opposite sensations.—W. G.

cally over the hand. Among my family and friends I have found more than one-half to be sensitive. I never told them my object, but asked for the hand, drew the crystal several times over it, and then asked whether they felt any thing, and what? The usual answer was, a cool or warm *aura*. That the sensation is very slight and delicate, it is hardly necessary to say. Had it been so strong as to require no particular attention to detect it, it would not have been now, for the first time, observed and pointed out, but would long ago have been generally known. Some persons who do not perceive it on one day, do so on the morrow, or the day after, or after a week. I saw a remarkable instance in the case of my own younger daughter, OTTONE. She felt the action of a long crystal of gypsum always very distinctly, while her brothers and sisters could hardly perceive any thing. She suffered, during some time, from headache, which gradually increased, and at last she was confined to bed. Two days after this she was attacked, for the first time in her life, with violent spasms, was ill for a fortnight, and then recovered. But she now felt the action of the crystal no longer, or only in the faintest degree. It would appear that the development of the disease had destroyed the sensitiveness. Thus the sensitiveness varies in the same person according to the time and state of health. The effect is often not perceived on the first pass, but becomes distinct on the second, third, or fourth. It occasionally happens that a person perceives the first stroke distinctly; it is not perceived on the second and third; and again comes out with the fourth or fifth. The stroke must not be made too fast; for the full effect requires a certain time for its production. It also sometimes happens that the sensation precedes the crystal, and is perceived at the point of the finger while the crystal is yet over the palm. In other cases it only becomes distinct when the crystal is leaving the point of the finger. In some places, persons have been blindfolded, and when tried, have given uncertain answers. This cannot surprise us after what has been said. Very sensitive persons, even blindfold, will always give consistent answers; those who are less sensible will be uncertain, and possibly inconsistent, the more so, as blindfolding in itself produces an unnatural and disquieting state, in which the attention is divided and scattered, and the tranquillity necessary for such delicate observations is generally absent. If, also, many persons be present, if they speak much and ask all manner of questions, move backwards and forwards, it is natural that we should obtain more or less incoherent answers concerning a sensation which does not sweep over the hand like a hard brush. Many persons feel the upward stroke more distinctly than the usual downward one. But there are also persons who are absolutely devoid of this sensitiveness; and these are, perhaps, the most healthy of all. One pole, that which points towards the north, generally acts more powerfully than the opposite pole.* The warmer tepid sensation is usually less distinct than the other. It generally contributes to the strength and distinctness of the sensation, if the hand be held in the direction of the meridian,† the fingers pointed towards the south. Nothing whatever must be allowed to come into contact with the patient during the experiment, because the delicacy of feeling is thus disturbed, and the attention distracted, whereas for these experiments we require the whole power of attention of the patient. It is worthy of notice, that the power of observing these sensations is, like other senses, improved by practice. He who never tastes wine, coffee, or tea, is usually unable to dis-

* On account of the well-known difficulty of expressing the poles of the magnet with reference to those of the earth, in so far, namely, as the latter possess an opposite magnetism to that of the poles of the suspended needle which respectively point to them; and in order to do away with all periphrases and misapprehensions, I shall, in this work, call that pole of the needle which points to the north, the northward pole (den gen Nordpol), and the opposite one the southward pole (den gen Südpol). These names may be here and there found fault with, but they will be easily understood, and are short. Even in the newest German Manuals, or Systems of Physics, that of Pouillet, translated by Müller, and that of M. Baumgartner, the names of the poles are directly opposed; what the former call the north pole of the needle, the latter calls the south pole. It is, besides, well known that French and Germans on one side, and Germans on the other, use opposite terms. Hence the difficulty and confusion which exist in this part of science, and which may excuse this suggestion of a method of getting rid of them.—R.

† Query? The magnetic meridian?

tinguish the varieties of these beverages; while a practised taster can instantly, and with great delicacy, discover the slightest differences. M. SCHUH, after some time, arranged for me the crystals which he tried, in series, which perfectly coincided with those derived from the observations of the diseased sensitives. M. STUDER, of Zurich, a healthy young man, residing in Vienna, was very soon able to do the same, and his arrangement coincided almost exactly with that of M. SCHUH.

---

37. It had now been ascertained that a polar force resides in crystals, which they possess in common with magnets. The next point for investigation was, whether this was identical with, and perhaps only quantitatively different from, what we commonly call magnetism, or whether the two forces were qualitatively distinct. In the former case, the new force must admit of being brought under the same laws. It must attract iron filings, as is the case not only with the magnetic metals, nickel, (cobalt, manganese, and chromium,)* but also with oxidised bodies, even impure ones, such as magnetic iron one. It must further, if identical in kind with magnetism, not only act, as magnetism does when residing in steel, on the living organism, when it is found in quartz, gypsum, &c., but also convert into magnets, when applied to them in the same manner, such bodies as are capable of becoming magnetic, steel, iron, nickel, (cobalt, &c.?) When I, to test this, *dipped the point of my crystals in fine iron filings, I could observe no attraction.* With the view of providing the force residing in crystals in the greatest possible degree of energy, I purchased the largest crystal I could procure—a rock crystal from St. Gothard, eight inches thick. It was a colossal six-sided prism with terminal acumination, which I could with difficulty use, and the effect of which on the hand the most sensitive among my healthy subjects, M. SCHUH, described as being as strong as if I had gently blown on his hand cold air through a straw. I placed a fine steel sewing needle opposite this powerful crystal, and treated it exactly as I would have done with a bar magnet to

* It must be remembered, that this was written more than a year before the appearance of Faraday's Researches on Magnetism.—R.

render it magnetic. The stroke was repeated a dozen times, and I then tried it with iron filings, but nothing was attracted. I multiplied the strokes to more than a hundred, but still not a particle of iron was attracted. As a counter-trial, I drew once along the needle the compass needle which Madlle. STURMANN had felt to be weaker than the small rock crystal first used, and the sewing needle immediately attracted a long beard of filings. *Hence, a sewing needle, stroked with the point of a crystal, received from it no power capable of acting on iron.* But it was notwithstanding possible, that such a power might be excited in the needle by a process analogous to magnetic induction, and that it might be rendered capable of attracting iron filings, as a bar of soft iron is, when placed in contact with a magnet. In order to test this, I fastened a fine needle to the points of several large crystals, and dipped it, while thus in contact with the crystals, into iron filings; *but the needle never attracted even a trace of iron.*

38. The approach of any crystal, even of the large rock crystal, to a finely suspended magnetic needle of the most delicate kind, had no disturbing influence; the needle remained motionless. I then suspended a large crystal, free from iron, as for example, a crystal of gypsum, by means of a very fine twisted fibre of natural silk, so that the crystal hung with its longer axis horizontal. It was placed under a bell jar, and a magnet was brought close to it; but the magnet exerted no action, *and the direction of the crystal was not in any way affected.*

39. I wished to see whether a magnetic wire would act on a suspended crystal as on a magnet. I prepared a small voltaic battery of several elements, each having a surface of 8.60 square inches in surface, and joined the poles by means of a thick copper wire. I then suspended, as before, a crystal of gypsum, free from iron, and about 4 inches long, so that the poles could freely move in a horizontal plane. When brought near the copper wire, which was also horizontal, the crystal and the wire appeared perfectly indifferent towards each other. *Not the slightest visible mutual action occurred.*

40. It was necessary to enquire whether crystals, when ap-

proached to a wire, induced in it, as magnets do, a momentary current. I formed a coil of 25 windings of thick copper wire, covered with silk, and connected it with the terminal pair of an exceedingly sensitive SCHWEIGGER's multiplicator, or differential galvanometer. When I now rapidly introduced a crystal of gypsum or a rock crystal, several inches long, into the interior of the coil, *the astatic needle exhibited not the slightest deviation.* The same was the case when it was withdrawn. The weakest magnetic needle which I tried in place of the crystal, instantly caused a deviation of 25°.

41. The relation which might exist between the force residing in crystals and terrestrial magnetism, was now a point of much interest. We possess some researches of HAUY, BIOT, and especially an investigation by COULOMB, in which it is rendered probable, although not demonstrated, that perhaps all bodies either possess magnetism, or are capable of becoming magnetic. We may pass over the researches of HAUY, which only extend to iron ores and substances containing iron. M. BIOT, in his treatise on this subject, (GILBERT'S Annalen, LXIV. p. 395; 1820), is uncertain whether it be magnetism, or, as he expressed it, "another analogous force" which acts on the substances. But in this memoir also, only such substances are mentioned as were obviously more or less contaminated with iron. One experiment was made with two needles of silver, one of which was of silver as pure as it could be made by chemical means; the other of silver which had been melted with iron, and which, as no trace of iron could be detected in it by chemical re-agents, might also be regarded as chemically pure. The latter, notwithstanding, acted on the magnet 416 times as powerfully as the former. It was concluded from this that it really contained a minute trace of iron, although chemistry could not detect it, and that, in general, inconceivably small proportions of iron were yet sufficient to give to another substance the capacity for magnetism; nay, that even the needle made from so-called chemically pure silver, may still have retained a trace of that metal. The most detailed investigation is that of COULOMB, read in 1802 to the French Institute, (of which an abstract is found in

GILBERT's Annalen, XII. p. 194). COULOMB leaves undecided the question, whether the substances he employed were free from iron; and GILBERT with some reason remarks, that they could hardly have been so. The experiments consisted essentially in placing between magnetic poles needles, and causing these to vibrate, both before they were placed between the poles, and when in that position. The number of oscillations in like times was always found smaller in the previous experiment, than when the needle was between the poles, and the action of the magnet was thus established. But we must inquire, What was this action? and every one will perceive that it admits of a three-fold explanation. The first ascribes it to the presence of iron in the needles; the second, to magnetism inherent in the substances; the third, "to another analogous force," as BIOT guessed. The decision of these questions, as far as they refer to my investigations, seemed to me to be most likely to be attained by a new and direct experiment. I had crystals, especially of gypsum, which acted as powerfully on all my patients as a bar magnet, capable of supporting from 4 to 6 ℔s. Such a bar, suspended by a linen thread, instantly placed itself in the plane of the magnetic meridian. If, now, the peculiar force residing in the crystal, which acted on the patients with equal power, were identical with the magnetic force residing in the bar, the crystal, when suspended, ought also to place itself in the plane of the meridian. To test this, I suspended various crystals, free from iron, and especially one of gypsum, 1.5 inches long, by a fibre of silk from the coccoon, about a foot in length, and covered the whole with glass cylinders. After several hours had elapsed, the crystals had become quite motionless, but never in one direction more than in another, neither pointing to the north, nor in any uniform direction whatever. If I made the support revolve to the extent of a quadrant, the crystal, after it had come to rest, was always found in a position inclined to an angle of 90° to its former one. I could therefore cause it to assume any direction I pleased. There did not exist, therefore, the very slightest resistance to the torsive power of the finest silken fibre; *and the peculiar force residing in crystals does not give to the bodies in which it exists even the slightest tendency to assume any particular direction, and stands in no relation to terrestrial magnetism, capable of causing them to assume such a direction.*

On the one hand, therefore, the relation of magnets and of crystals to the animal nerve was entirely alike ; while, on the other, the relation to iron, to the electric current, to magnetic poles, and to the magnetism of the earth, was, in magnets and in crystals, totally different.

42. Thus much, then, is ascertained : *That polar force residing in crystals,* and which is recognised by its peculiar irritative action on the healthy and diseased animal nerve, *is not identical with the magnetic force,* as at present known. It does not attract iron ; has no effect on the magnetic needle ; is, even when powerful, unable to excite the slightest trace of magnetic attractive power for iron in the finest steel needle ; is without action on the polar wire ; produces no induced current when placed in the coil of a galvanometer ; and does not obey the attraction of terrestrial magnetism.

43. But conversely, it has been shewn, *that with the magnetic force, as we are acquainted with it in the loadstone and the magnetic needle, that force is associated, with which, in crystals, we have become acquainted.* For, since the magnet acts on the animal nerve in the same manner as crystals do, it possesses, along with those properties which are not found in crystals, that also which resides in them.

44. From which it necessarily follows : *That the force of the magnet is not,* as has been hitherto taken for granted, *one single force, but consists of two, since, to that long known, a new, hitherto unknown, and decidedly distinct one, must be added, the force, namely, which resides in crystals.*

The phenomena presented by the magnet, therefore, divide naturally into two portions, which in their manifestations complicate each other ; and it will be necessary to subject to revision a part, at least, of the vast store of observations which science has collected in reference to the magnet.

---

45. I now proceeded to make some investigations into the nature of these newly discovered properties of crystals. I first

wished to ascertain, whether the new force could be conducted, whether bodies could be charged with it, and whether it could be concentrated; whether, and in how far, it had any analogy with Magnetism and Electricity, which may be conducted, transferred from one body to another, and concentrated. Since, for the present, I possess no re-agent for it, except the nerves of healthy and diseased sensitives, and the action on the healthy subjects hitherto known to me* was so feeble, that I could not with sufficient certainty distinguish different degrees of it in such persons, I was compelled to make use of the exalted sensitiveness of the patients formerly mentioned. I had reason to be entirely satisfied with the results of this mode of investigation. For as the persons, with whose aid I conducted it, although in the most widely different circumstances, and suffering from various diseases, not only agreed in the account of their sensations, but also made statements, which, considered theoretically, were harmonious and consistent, I had every reason to consider these statements exact and valuable. I know well all the objections to which I expose myself, in declaring this. But these have no weight with the sober investigator of nature, who only walks with firm and sure step within the limits of experience. All that we can collect of observations on the external world we must finally learn through our senses, because by means of them alone we can observe. We have only five senses and no more; but we know, with certainty, that in nature, as well in us as around us, changes are at every moment going on, which we cannot perceive, only because we have no sense fitted for the perception. In every, even in the shortest interval of time, numberless electrical motions take place around us; but we do not perceive them in the slightest degree. If now, a being were to descend from the clouds, gifted with a sixth, or electrical sense, which should enable him to observe and to point out the most delicate electrical changes, with the same acuteness, as we, with our

* The reader will observe, that this was written and published in March 1845, at a very early stage of the author's researches. We shall see that, at a later period, he was more fortunate in finding very sensitive yet healthy persons in great numbers.—W. G.

organs of vision, can do so in regard to the phenomena of light;—should we not eagerly listen to his instructions, and ask him a thousand questions, in order thereby to clear up and extend our knowledge? A person born blind, who never has had a conception of light and colour, causes himself to be led by one possessed of sight; and when he invariably finds, by touch, a stone where his guide has previously declared that one was to be seen, he believes that his guide has organs, by means of which he sees. Now, a person affected, like most of my patients, with nervous disease, is to us, in fact, such a guide, possessing a sensibility for electric and magnetic movements, to whom, therefore, up to a certain point, as it were a new sense has been unlocked, which, as it appears, is wanting generally in the healthy. But I do not here mean, by persons affected with nervous disorders, somnambulists, lunatics, or sleepwalkers, &c.; but only, generally, the majority of those who suffer frequently from spasmodic attacks. Somnambulists are persons, in whom these disturbances of the normal nervous state have reached the highest degree, and in whom sensitiveness has attained its maximum. They present to us the reactions in their strongest form, and exhibit the finest discrimination of differences; but they are by no means indispensable to the researches to which I have devoted myself. Madlle. NOWOTNY, with whom I made many experiments, was, during the six weeks of my observations on her, far removed from somnambulism, and suffered only from cataleptic spasms. Madlle. MAIX never showed a trace of somnambulism. M. SCHMIDT, surgeon in Vienna, on whom the magnet, the crystals, and the magnetism of the earth, produced most remarkable effects, was a young man, perfectly healthy in other respects, who merely suffered from spasms in one arm, for a short time, in consequence of a chill. And the sensitiveness diminishes as we come to those in full health, who only perceive the stroke made with large crystals as a cool aura on the most sensitive parts of the hand or person. And among them also we find this difference, that some perceive the aura very distinctly, others much more feebly, and some not at all.—If, then, all these things stand in regular connection, we cannot annihilate, by disputing or denying them, the cause and the effects; and in my opinion it would not be well done, to reject phenomena, which may become so valuable a key to the investigation of natural truths, and that, too, in those very branches of physics and chemistry, in which nature seems disposed the most obstinately to keep her mysteries concealed from us. The strange or new sense, the peculiar irritability, of nervous patients, is *directed chiefly* to magnetic phenomena; they are, for such phenomena, a truly invaluable re-agent, the like of which we do not otherwise possess. Their sensations are not vague, as has hitherto been generally believed, although many physicians and enthusiasts have given them a bad name; but everything follows regular laws, which are easily discovered, if we will only thoughtfully look for them, pursue them with the arms supplied by physical and chemical knowledge, and apply to them the touchstone of experimental criticism.

I could not help making this digression, which was indispensable to the appreciation of the point of view from which these researches are undertaken. I now return to the question, whether the peculiar force residing in crystals can be conducted, transferred, and concentrated. When I stroked, as with a magnet, the most multifarious objects, with the pole of a crystal, such as a piece of wood, a glass of water, leather gloves, scraps of paper, or any other, every patient could distinguish these objects, if presented without delay, from similar ones not so treated. The sensation was sometimes cool, at other times warm; it was perceived in the hand when the objects were laid upon it, and, in different cases, differed in degree, reaching in some to the point of becoming unpleasant. Paper was found to admit of this transference in the weakest degree. Mlle. STURMANN perceived no effect from a book, once stroked with the large rock crystal; hardly anything, when several strokes were given; but at last, when I fastened it to the crystal for some time, and quickly laid it in her hand, she perceived a feeble warmth. A bit of porcelain, stroked with the crystal, caused a cool sensation. A rod of German silver, belonging to an electro-magnetic apparatus, when treated in the same way, produced a lively warmth.

Soft iron, a piece of blue tarnished steel from a saw plate, and a hard steel file, when acted on in the same way, all produced, although previously indifferent, a feeling of warmth in the hand. When I placed my hand in hers, allowed her to become accustomed to it, and then passed it along the crystal several times, she instantly perceived, on again taking hold of my hand, a marked difference. It appeared much warmer, and this apparent warmth continued, gradually diminishing, for four minutes, during which time I caused her repeatedly to try it. A similar series of experiments was made with Mlle. MAIX, and subsequently with Mlle. REICHEL. The substances tried were copper, zinc plate, linen, silk, and water. They yielded perfectly similar results. Mlle. ATZMANNSDORFER could instantly tell whether I had placed the rod of German silver in contact with a crystal of gypsum or with an amphorous substance; and even whether I had acted on it with the warming or cooling end of a crystal. The warmth was stronger or weaker, or passed into coolness. *The new force*, therefore, *could be transferred to other bodies, so as to charge them with it.* It could be transferred to iron and steel, and yet, *these substances, so charged, did not attract iron filings*, as I have already stated.

I endeavoured to ascertain, whether this transfer could be effected, like that of magnetism, by stroking from pole to pole, or like that of electricity, at a single point. I found that it was indifferent, whether I stroked the substances lengthwise, or whether they were left for a short time in contact with the crystal. No difference could be observed in the sensations of warmth or coolness produced. A large rock crystal, placed with its point in contact with a glass of water, produced magnetised water, quite as well as a horse-shoe magnet.

46. The next question was, whether this force was subject to a coercitive power residing in matter, whether it were permanent or transitory when thus communicated, and if transitory, after what interval. I charged various objects, the rod of German silver, the steel file, soft iron, porcelain, and the book. The last soon lost its charge. The porcelain, tried on Mlle. STURMANN, retained it for two minutes; the rod five minutes;

the soft iron five minutes; the steel ten minutes. I proceeded in the following manner. I did not take the objects in my naked hand, but carried them on a sheet of paper. When the patient had tried them, I laid them aside till the sensation disappeared in the hand, which usually occurred in about a minute. I then caused her again to try, without myself touching the object, and continued this, till it no longer caused warmth or coolness. It appears from this, that the charge, in the objects tried, is not, even by night, permanent, but of short duration, not exceeding ten minutes; *that it soon disappears*, and cannot, like magnetism, be permanently communicated to steel. But since iron filings instantly fall off from a bar of soft iron, which had attracted them by magnetism induced in it when in contact with a magnet, as soon as the magnet is withdrawn, whereas here the effects produced by transference last for a time, although a short one, we must answer the question proposed in this paragraph by saying, *that all bodies possess a coercitive or retaining power for the new force*, although that power be a weak one. This is not precisely the case with magnetism, as far as we know it in its polar action, for there the retaining power, according to the most recent researches, is found in very few substances, perhaps, strictly speaking, in iron (steel) and nickel alone.*

Is the new force capable of isolation? Can it be confined? The first experiments on these points were made with Mlle. STURMANN, who was sensitive, indeed, but not always clear in her distinction between tepid and cool sensations. When I laid a book on her hand, and placed on it the point of the large rock crystal, she perceived no effect in the hand. As she always felt the effect of the crystal, when alone, when it was held pointed towards the inner surface of the hand even at a distance of 17 to 19 inches, while now the crystal was only separated from her hand by the thickness of the book, about $\frac{3}{4}$ of an inch, it appeared by this experiment that a thick mass of paper acted, at least for a short time, as an obstacle to the action of the force on her nerves. A board of deal had a similar effect, but less perfectly; after a short interval she perceived

* Written a year before the recent investigations of Faraday.—R.

the sensation through it. Eight folds of printing paper allowed the action soon to pass; four folds of woollen stuff hardly opposed an obstacle to its passage. A porcelain saucer was laid on her hand, and touched from above with the crystal. She perceived coolness when I turned the saucer. On the other hand, a sheet of iron, placed on her hand, caused a warm sensation, as soon as I approached the crystal. When I touched the sheet with it, the effect flashed, as it were, like a shock through the elbow and up to the shoulder. Various metallic wires, when one end was held in her hand, and the other placed in contact with the points of crystals, instantly caused warm or cool sensations. I placed one end of a rod in her hand, and connected the other with the point of a small crystal; the sensation instantly darted through the hand to the elbow. When I tried in the same way the large rock crystal, the sensation reached the shoulder, and caused spasmodic feelings.

When I subsequently repeated the same experiments with Mlle. MAIX, the action was found to pass through all bodies without exception; only with metals the passage was more rapid, as it were, more shock-like, than with vegetable matters, cloth, &c., which required a short time to allow the full effect to be perceived through them. I tried wool, silk, glass, and zinc, with Mlle. REICHEL. The conduction through woollen threads more than three feet in length required a very short time; with silk, glass, and zinc, it was instantaneous, and appeared to take place with a velocity too great to be measured. My experiments with the very sensitive Mlle. ATZMANNSDORFER yielded the same results. Brass wire, the German silver rod, glass tubes, rods of lead, platinum foil, bar iron, gold (gilt?) threads, copper plates, held in her hand, were instantly traversed by the force derived from the crystal with which they were placed in contact. The result is, *that the new force acts through all bodies, but in various degrees.* Paper, wool, and wood, retard its passage, at least for a short time; porcelain does so much less; silk and glass are perfect conductors. Metals not only allow it to pass, when in contact with the crystals from which it emanates, but in some degree, before contact, that is, when the crystals are merely approached very near to the metal. With actual contact, an instantaneous effect is observed. As far as the above experiments can justify a conclusion, they point to this, that the conducting power of bodies depends less on the nature of the substances than on their continuity. All woven stuffs were inferior in conducting power to entire, continuous masses; cotton and wool, to wire and silk thread, for example. The perfection of the conduction is differently perceived by patients differing in sensitiveness; so much so indeed, that while all bodies act as conductors for the more sensitive, sensible differences are observed, by those who are less sensitive, in the conducting power of different substances.

48. I was desirous to ascertain the amount of capacity for the charge of the new force. I stroked with the crystals, and placed in contact with them, at different times, the rod of German silver and the steel file. One stroke produced a feebler effect than several; but when I had charged for about a minute, the charge reached a point, beyond which, under the given circumstances, it could not be increased. That is, I could not increase the strength or duration of the effects produced by the charged object on the patient's hand, such as the warm or cool sensation; which, however long the charging was continued, never lasted longer than about five minutes, except in water and steel, in which it lasted about ten minutes. The charge, therefore, was not instantaneously completed, but increased during the contact for a short time, and then reached its maximum. *The capacity for the charge was therefore satisfied within a few minutes.*

49. With regard to the amount of the force, and its relation to the size of the crystal, the experiments indicated, that small crystals, from the size of a lentil to about two inches in length, of gold, rock crystal, gypsum, diamond, or hornblende, produced a feeble sensation, and were only vividly felt on the backward or upward stroke. Above that size, the power increased with the size of the crystal; but of course the exponent of the rate of increase could not, for the present, be determined.

50. The difference between the poles in their action on the animal nerve was exhibited in an opposition of cold and warm

sensations. Crystals almost invariably caused, when passed over the hand, a cool sensation from one pole, a warm from the other. Mlle. NOWOTNY and M. SCHMIDT, Surgeon, felt, as persons in health did, a pleasant cool aura from the downward pass; but a disagreeable tepid warmth from the upward stroke. With Madlle. STURMANN I tried crystals of tourmaline, arragonite, rock crystal, gypsum, and cleavage forms of Iceland spar and tellurium. In all, one pole, the stronger, gave on the downward stroke a cool, on the upward a more feeble warm impression. This difference was very vividly perceived by Mlle. REICHEL, who could even distinguish the poles of all crystals at a certain distance, and very accurately, by the cool and warm sensations they excited. It was observed, however, most distinctly, by Mlle. ATZMANNSDORFER, as I have already stated. But even healthy subjects, for example, Professor ENDLICHER, as already mentioned, M. STUDER, the carpenter JOHANN KLAIBER, residing here, and many others, clearly and decidedly distinguished the poles of all crystals, even of very small ones. The opposition of the poles was therefore, in the first place, exhibited in their producing a tepid or cool sensation on the sensitive nerve. I shall mention, further on, some other points of opposition between the poles.

51. The high degree of distinctness of the impression made on diseased sensitives is very remarkable. They not only perceive the effect produced by the mass of any substance presented to them, but feel distinctly that there are points in which the force is accumulated or concentrated. Mlle. NOWOTNY pointed out to me in every crystal, with great precision, those points where the two active poles lay, which she soon found by means of the points of her fingers. In the case of double crystals, the axis joining the poles always passed through the junction of the two crystals. The same was done, with even greater facility, by Mlles. MAIX, STURMANN, REICHEL, and ATZMANNSDORFER. Even MM. SCHUH and STUDER perceived accurately and distinctly the polar points in large crystals; and all these observations, separately made, exhibited a perfect agreement among themselves.

52. The electricity of crystals, as excited in tourmaline, boracite, and other minerals, by heat, had no perceptible influence on the effect of the new force on the animal nerve. I heated these bodies to different temperatures, but *this produced no distinct change in the results.*

53. Is there no relation whatever between the magnetism of the earth and the direction of the force peculiar to crystals? If we consider the manner in which crystals are often grouped, we might be led to suppose that the two forces are, at least to a certain degree, independent of each other. Any one who has ever seen a drusy cavity of quartz opened in a mine, and has examined it from all directions while *in situ*, cannot have failed to observe that the hollow is lined on all sides with crystals, the axes of which lie in all possible directions. But even in mineralogical cabinets we can make the same observation, as in the nodules of chalcedony, the interior hollow of which is covered with crystals of quartz and amethyst. I have never been able to detect in them the slightest preference for any one direction. Other crystallisations, again, which form radiating lines, grouped round a common centre, as in natrolite, zeolite, mesotype, arragonite, pharmacolite, form radiating spherical bundles, the rays of which run in all directions, and do not indicate the influence of any directing force. Artificial crystallisations, as they occur in our laboratories and manufactories, are in the same way thoroughly confused, as in prussiate of potash, alum, sugar-of-lead, sugar-candy, which, in large crystallising pans, deposit their crystals without preference for any one direction. This appears to agree with the indifference exhibited by a suspended crystal towards the magnetic needle and the polar wire. In this point of view, therefore, the new force, as far as concerns the direction of the crystals in which it resides, is independent of terrestrial magnetism.

54. If, then, the new force residing in crystals is on every side free from that attraction towards inorganic matter, which distinguishes the magnet in so high a degree; on the other hand, it is so much the more surprising, and appears to claim the highest scientific interest, *that this force shares with that*

of the magnet the singular property of attracting living organic bodies. For the same effect, which I have already described as produced by the magnet on the cataleptic Mlle. Nowotny, I found to occur when the points of large crystals were used, as has been (§ 31) already detailed. The crystals caused her hand to contract, produced in some cases convulsive clenching of the fist, and solicited the hand to follow them; that is, they attracted it, not so strongly as a large magnet, but exactly like a weak one. I am convinced that a crystal of sufficient size would have caused her hand to grasp and adhere to it, just as it adhered to a powerful magnet. This *elective affinity of the new force,* in virtue of which it attracts living and not dead matter, is the most astonishing phenomenon which it presents, and indicates the important relation in which it stands to that which we call life or vitality; on which mysterious subject, unless I am greatly deceived, the study of the force residing in crystals holds out a near prospect of the most valuable results.

55. In the first treatise, I have reported my investigations on the subject of the light which flows from the poles of powerful magnets. It was obviously natural to imagine the possibility of the same phenomenon at the poles of crystals; indeed I was compelled to regard its occurrence, *à priori,* as even probable. I made, therefore, an experiment with the exalted powers of vision of Mlle. Sturmann. A room was darkened as completely as possible. The patient entered it, and remained there till her eyes were accustomed to the darkness. I then showed her the large rock crystal. She instantly saw a flame-like brightness over it, of the size of half a hand; blue, passing into white in its upper part, very different from the magnetic light, which she described as much yellower and redder. The experiment was twice repeated in the following night. To make sure, I placed the large rock crystal, eight inches thick, in a part of the room unknown to her, before she entered it. As soon as the darkness was rendered complete by shutting the door, she each time and instantly discovered the place where the crystal stood, and gave, in all the three experiments, the same account of its light. She described

it as somewhat of the form of a tulip, and, like one of its petals, or like the flame of a candle, beginning below with an arch directed outwards, but soon turning upwards and extending to the length of her finger. The colour she again described as blue, passing above into perfect white, while a few scattered stripes, or threads of red light, ascended into the white. The flame was in motion, undulating and scintillating, and cast around it, on the surface of which the crystal rested, an illumination extending over a circle of more than $6\frac{1}{2}$ feet in diameter, exactly as in the case of a magnet; in which a flame-like appearance, and light radiating from it, can be distinguished with certainty. I now tried Mlle. REICHEL with various crystals in the dark. She considered the flame-like appearances singularly beautiful, surpassing even those of the magnet in beauty of colour and regularity of form. She perceived the light, not only over the poles, but within the mass of the crystals. The flame over the poles she described almost exactly as Mlle. STURMANN had done; but the light in the crystals was, according to her, essentially different. It appeared in star-like forms, which, when the crystal was turned round, took other shapes. It was no doubt the crystalline structure, and the planes of cleavage in various directions, which caused the production of internal light and internal reflections; phenomena which cannot occur in the same way in a steel magnet. She made drawings of the lights in large and small crystals, which represent most surprising appearances. I intend to collect at a future period, for the sake of comparison, all the magnetic luminous phenomena with which I have become acquainted, and to give, with this description, copies of Mlle. REICHEL'S drawings of the light from crystals. Mlle. MAIX, also, whose quiet and accurate method of observing I particularly valued, admired, during many sleepless nights during which I allowed the large rock crystal to stand on her stove, the beautiful spectacle of a whitish star of light of the size of half a hand, at its point. Mlle. ATZMANNSDORFER pointed out, in a number of different crystals, in the dark, the luminous poles, and arranged them for me in a series, according to the intensity of the light.

Since, then, all the very numerous crystals which I tried, exhibited the same re-actions of the peculiar force manifested at their poles, as those which occur in their maximum degree in a large rock crystal, we are led to the conclusion, *that, in general, all crystals, like the magnet, send forth delicate flame-like light from their poles,* not usually visible to the healthy eye, but visible to the exalted sense of nervous patients, all whose senses are in an exalted state of acuteness. It is hardly necessary to mention, that this phenomenon is probably connected with that of the development of light during crystallisation, frequently observed by chemists, and long since admitted into the systems of chemistry. The nature of these emanations is not yet understood; they were taken for electrical, because they somewhat resembled electric light, but without any direct proofs of identity between them. M. HEINRICH ROSE, however, has recently shewn, that this light is not associated with development of heat or of electricity, neither the air thermometer nor the electroscope being in the slightest degree affected when immersed in crystallising solutions of substances which give out, on crystallising, the greatest amount of light, such as the double sulphate of potash and soda. (POGG. Ann. LII. pp. 443, 585.) Now that we are acquainted with permanent luminous emanations from the poles of crystals, which as yet exhibit no coincidence with electric light, but, on the contrary, are decidedly distinct from it, it becomes even more than probable, that the light evolved by bodies while crystallising belongs, not to electricity, but to the development of the proper light of crystals; and that, during the sudden formation of solid crystals from the molecules suspended or dissolved in the liquid, circumstances may occur, under which the light of crystals becomes so concentrated as to be visible to ordinary eyes. I do not here enter into the question of what is the nature of this light, which, like the sun, shines without intermission, while the body from which it flows does not in the smallest degree diminish; whether it be a vibration communicated to fluid media surrounding the body, &c. The atoms, and still more the molecules of matter, are believed to be polar; and we regard them as the elements of which crystals are built up. Their arrangement to form a large crystal, which, as a whole, is also polar, is the summation of all these minute or molecular polarities; and, we may ask, are the poles of the crystal an expression of this, just as the terminal plates of an open voltaic battery represent each the summation of one of the components of the electricity of all the single elements? Is a crystal a pile, in reference to the new force, as a voltaic pile is, in reference to the electric force? These are obvious questions, which must be reserved for future researches. In the mean time, the observations made here on five different patients, and on many occasions, which agree so well among themselves, will soon be confirmed elsewhere, and by other observers. But I would warn others not to make the experiments with somnambulists, in the sleep-waking state, but either with patients not subject to somnambulism, or, when such cannot be procured of sufficient sensitiveness, with somnambulists only in the ordinary fully conscious state; and to employ the sleep-waking state either not at all, or, at most, as a means of controlling observations made in the ordinary state. In my researches, I have never experimented with patients in the state of somnambulism, but have left them to the physician, and have contented myself with the part of spectator. To prevent mistakes and failure, I repeat, that where it is desired to repeat my experiments, the darkness must be complete, and so dense, that, even after remaining in it for a long time, we cannot perceive any trace of even a glimmer of light. Lastly, the crystal must be very large, if possible not less than eight inches thick, and of proportionate length. In very sensitive patients, smaller crystals may indeed answer the purpose; as we have seen that Mlles. Reichel and Atzmannsdorfer saw light from almost all crystals, especially sulphates and fluorides, which invariably surpass rock crystal, when of equal size.

56. The preceding researches lead to the conclusion, that the peculiar force of crystals here described opens up a new leaf in the history of the Dynamides or Imponderables; that it obeys the general laws of those influences, but yet possesses within those laws a special code of its own; to study which,

and to reduce its propositions to fixed quantities, ought henceforth to form one of the problems of physics. It will be especially desirable to discover a universal inorganic re-agent for this force, a means of recognising and measuring it, which may liberate us from the frequently more than painful dependence on diseased persons, hospitals, and uncultivated people of every kind.

57. Retrospect.

a. Every crystal, natural or artificial, exerts a specific action on the animal nerve, which is feeble in the case of healthy subjects;* powerful in those affected with disease, especially of the nervous system; and most powerful in cases of catalepsy.

b. This force is seated chiefly in the axes of the crystals, and is most powerful at the opposite ends of these: It is therefore polar.

c. Light, visible in the dark to sensitive eyes, is sent out from the poles.

d. In certain diseases, this force solicits the hand to a kind of adhesion, resembling that of iron to a magnet.

e. It does not attract iron; causes no tendency in any bodies to assume a direction related to the magnetic polarity of the earth; has no action on the magnetic needle; induces no galvanic current in a wire; and is, therefore, not magnetism.

f. It may be transferred to, and collected in, other bodies by mere contact.

g. Bodies possess some degree of coercitive or retaining power for it; but only for a short time, during which the charge gradually disappears.

h. It is conducted by different substances in different degrees, and better in proportion to their continuity of structure.

i. The capacity for being charged with it is in direct proportion to the strength of the new force.

k. It is exhibited at the opposite poles of different quality;

* This conclusion, as to healthy subjects, is much modified in the sequel. —W. G.

that is, like the magnet, the pole corresponding to the — M pole usually excites a cool sensation; that corresponding to the + M pole a warm sensation. Quantitatively, the pole pointing to the north (— M) is the stronger; the opposite pole (+ M) the weaker of the two.

l. Warming the crystals caused no perceptible change in the results.

m. This new force is one of those residing in, and exhibited by the magnet; it is, therefore, a part of the influences of the magnet, which may be separated and isolated.

THIRD TREATISE.

AN ATTEMPT TO ESTABLISH FIXED PHYSICAL LAWS IN REGARD TO THE VARIABLE PHENOMENA, HITHERTO CLASSED UNDER THE NAME OF ANIMAL MAGNETISM.

58. I SHALL now try to apply the laws ascertained in the two preceding treatises to another class of observations, related to those already described, and to give to these observations greater extension and a firmer foundation. From time immemorial certain enigmatical phenomena have been known, which are produced by the magnet in its action on certain patients, and especially on somnambulists. In the last century, and probably earlier, it was observed *that similar effects could be produced without the magnet, and with the hands alone.* It was not possible, in the state of knowledge hitherto attained, to discover any sure connection between that power of the magnet and the analogous one of the human hand or foot. As little was it possible to recognise in the phenomena any regularity, or to detect any law by which they were governed. The consequence was, that all natural philosophers laid aside and neglected the subject, and did not admit it to a place among the physical sciences. Individual physicians and a few amateurs in part kept up the tradition, and in part increased the mass of incoherent observations. They called it, for want of a better name, Animal Magnetism, an expression so much the more unfit, as the phenomena included coincided

less perfectly with that which is properly called Magnetism. In the mean time, a number of works have appeared on the subject, chiefly of a medical character. Few of these are good; many are entirely one-sided; and many are downright unreadable.

I have, in the first instance, avoided going deeply into this literature. I wished to keep my view and my judgment free and unprejudiced, and for the present, to build up my work on the foundation of my own experience. It appeared to me better to select my own path in the direction which natural philosophy usually pursues, and which is not always that of medicine. The physician is chiefly concerned with the discovery of a means of cure; the natural philosopher with that of natural truths. The former has a tendency to the concrete, the latter to the abstract; and this original divergence is the reason why they have hitherto been so little able to combine in their researches.

59. After I had proved the existence in crystals of a force, which, in spite of all its differences, yet exhibits an unmistakeable analogy with magnetism; while, on the other hand, the so-called animal magnetism, with essential differences, yet also allowed us to catch glimpses, in certain resemblances, of a surprising parallelism with magnetism; this apparent analogy led me to investigate whether any thing, and how much, common to both, could be ascertained, and whether laws might not be discoverable to which animal magnetism is subject, as the force in crystals appears to be. Since we may imagine crystallisation to be the link uniting dead to living matter, I thought I might entertain some hope of giving, from experimental research, to animal magnetism a point of union with other departments of natural science, and perhaps of supplying it with that firm scientific foundation, to attain which it had hitherto striven in vain.

60. In order to clear the way, it appeared to me above all things necessary to ascertain, as far as possible, the part played in these phenomena by the magnetism of the earth. If the magnet, or a crystal, exerted so marked an action on sensitive persons, the power of terrestrial magnetism, which gives direc-

tion to the magnet, could not certainly fail to influence the animal nerve. And I clearly perceived, that it was impossible to obtain a pure unmixed scientific result from any experiment whatever, as long as this powerful factor, which must in every quarter influence the phenomena, was not attended to, measured, and taken into account.

With this view, I therefore tested healthy and diseased persons, and among others, M. SCHUH; M. SCHMIDT, Surgeon; and Mlles. NOWOTNY, STURMANN, MAIX, REICHEL, and ATZMANNSDORFER, under various circumstances and at various times.

61. M. SCHUH, in the house he then occupied, had the singular custom, that, when he awoke early in the morning, he regularly turned himself in bed, so that his feet came to be where he had laid his head during the night; and always fell asleep again. This second sleep was invariably much more refreshing to him than the whole previous night's rest, contrary to the usual rule, that the first sleep, especially before midnight, is the most refreshing. When he failed to obtain this morning sleep, he felt wearied the whole succeeding day; and this strange habit had long become a necessity to him. I enquired the position of his bed, and found that it was so placed, that the head was towards the south, the feet towards the north. By my advice he laid himself, on going to bed, in the opposite direction, namely, with his head towards the north, and his feet towards the south. From that day forth, he never felt the necessity of turning himself in bed in the morning. His sleep was good and strengthening; and he for ever abandoned his old habit.

62. M. SCHMIDT, Surgeon in Vienna, had experienced a chill in his right arm, while travelling on a railway, and had for some time suffered in consequence, from severe rheumatism in the limb, with most painful spasms from the shoulder to the fingers. His physician employed the magnet, which quickly subdued the spasms; but they always returned. I found him lying with his head towards the south. In consequence of my remarks on this, he was so placed as to lie in the magnetic meridian, with his head towards the north. As

soon as he came into this position he expressed instantly feelings of satisfaction, and declared that he felt, generally, refreshed in a singular degree. The previously existing chilliness and rigours were instantly exchanged for an agreeable uniform warm temperature; he felt the strokes of the magnet now beyond comparison more agreeably cooling and beneficial than before; and before I left him, the rigid arm and fingers had become moveable, while the pain entirely disappeared.

63. When I tested, with the magnetic needle, the direction in which Mlle. NOWOTNY lay, I found it exactly in the plane of the magnetic meridian, her head being towards the north. She had herself instinctively sought out, and insisted on occupying, this position; and it had been necessary to remove a brick stove to allow of her wish being gratified. I begged her to make the experiment of lying with her head towards the south, in order that I might observe the result. It was no easy matter to persuade her to do this; for I had to reiterate my request on three or four successive days, and to explain to her the importance which I attached to the experiment, before I could induce her to try it. At last, I found her one morning in the desired position, which she had assumed shortly before my arrival. Before long, she began to complain. She felt uncomfortable and restless, became flushed, and her pulse became more frequent and fuller; a rush of blood to the head increased the headache, and very soon the disagreeable sensations affected the stomach, producing nausea. We hastened to change the position of the bedstead on which she lay, but stopped when we had turned it round to the extent of a quadrant, her head being now towards the west. Of course she now lay in the plane of a magnetic parallel. This direction was to the patient absolutely intolerable, far more disagreeable than the former, that, namely, from south to north. This was at half-past 10 A.M. She was afraid, from her sensations, that she would soon faint or become insensible, if kept in this position, and entreated to be quickly removed from it. She was now placed in her own original position, her head towards the north, or the direction from north to south. Instantly, as soon as this was done, all

the painful sensations yielded, and in a few minutes they had so completely vanished, that she was again quite cheerful. But this intense feeling of discomfort was not the only effect produced on the patient by the changes made in her position in reference to the points of the compass; her sensations from the action of all bodies were remarkably altered, and even inverted. The usual application of passes with a magnet by her physician, which had always hitherto been very beneficial to her, were now, while she was in the changed position, most disagreeable, and in the case of a powerful magnet, intolerable to her. Substances, such as sulphur, formerly offensive, were now almost indifferent to her; others, such as lead, even became agreeable; in short, all the diseased manifestations of sensitiveness assumed an altered shape.

These observations were so important, and likely to admit of so valuable and direct an application to medical practice, that I felt bound to pursue them further and with greater attention. I therefore arranged with the physician attending Mlle. NOWOTNY, to follow out the investigation on another day. This took place, April 24. 1844. When we came in the morning to the patient, we found that she had been already for half an hour in the position from south to north. She was anxiously longing for our arrival, and urgently entreated that we would soon release her from her painful situation. All the accidents above enumerated again appeared in the same succession. Her hand did not, as in her ordinary position, follow the magnet, and was only feebly solicited by it; even the strongest magnet caused no spasm or clenching of the hand; and the reactions, or sensations, with different substances, were disturbed as in the former trial. In order to be able to experiment more conveniently, I induced the patient to dress and rise out of bed. I then placed her alternately on four chairs, arranged so across a circle, that one of them was in the line north to south (magnetic), another in that of south to north; the third lay east and west, the fourth west and east. She was placed on these in a half-reclining posture, with her feet advanced, and her head leaning backwards. The *north and south* position, that is, with the head towards the north, she found comfortable, as it always was; the *south and north* position produced successively the same results as in the two former experiments; these results being gradually and one after another developed in about half an hour. But when she was placed in the *west to east* position, the same results occurred immediately, and so rapidly, that she could hardly be retained in that position for a minute. The action of the magnet upon her became almost null; at the moment of her assuming this position, she was affected all over with an unpleasant warm sensation; and then there followed, in rapid succession, general outward and inward trembling or shuddering; restlessness; flushing of the face; acceleration of the pulse; a rush of blood to the head; headache; and finally, pain of stomach, ringing in the ears, failure of the senses, and the approach of fainting. We were compelled to bring her in haste into the north and south position, in order to restore her, otherwise she would have fallen from the chair. When this was done, the rapidity with which all these painful sensations disappeared was astonishing. After the lapse of but one minute, her countenance, which had just before expressed the most painful feelings, became cheerful. When she was somewhat restored, we tried the *east and west* position. I had a watch in my hand, and observed that not more than a minute elapsed, before all the phenomena appeared, in the same form and in the same order of succession, as in the west to east position; only somewhat more mildly.—For the sake of greater certainty, all these experiments were repeated, as we prevailed on the patient once more to place herself in the four different positions. The result was exactly the same as before.

As the illness of Mlle. Nowotny had gradually and slowly increased during the eight preceding years, I enquired whether, during the earlier stages of the disease, it had already been observed, that she felt more or less comfortable in different places. Her relations now recollected, that in other houses, in which she had lived during that period, her state had been at one time more tranquil, at another almost intolerable to her. I gave her brother a compass, and begged him to ascertain in these various dwellings, in what position her beds, sofas, or benches and working chairs had been situated. It was found that, in the Wohllebengasse, her bedstead and sofa had accidentally been very nearly in the magnetic meridian, and that she had there lain with her head towards the north; whereas, in the Marokanergasse, she had lain in a north-east and south-west direction. Now, in the former place, she had always felt tolerably well; but in the latter she had uninterruptedly felt most uncomfortable, and had constantly to struggle against the most painful sense of illness. But even now, she could never tell why, she could not endure to sit across her bed, nor across her sofa, nor to lie on the latter, but was only able to obtain comfort when lying on her bed. In the first case, she sat in a west and east position; in the second, in that of east to west; in the third, she lay from south to north; and only in the last could she attain the indispensable direction of north to south.

As between north and south, so also between west and east, a not inconsiderable difference was afterwards observable. In June, when she had so far recovered as to be able to pass the greater part of the day out of bed, I again tried her in the four positions formerly mentioned. She was able to remain some time in that of south to north; and could even bear tolerably well, for a time, that of east to west; but still she could not hold out longer than a minute in the position of west to east, without feeling the former effects, down to the pain of stomach. It required several minutes of the north to south position to do away the bad effect of one minute spent in that of west to east. *This last position, therefore, was by far the most disagreeable and injurious of all.* I would add, with reference to the place of the sun, and to terrestrial thermo-magnetism, that this last experiment was made about 5 P. M.

64. Provided with these experiences, I now visited Mlle. STURMANN in the Clinical Hospital of the University of Vienna. She suffered from pulmonary tubercles; and her fits were called eklampsy. By her own account they had come on, in the sixteenth year of her age, three years previously, after some nights of violent and long-continued dancing. I found

her on a bed, lying from west to east. I then tried on her a powerful magnet capable of supporting 80 ℔s.; made passes with it; laid it on her head, and in contact with her feet. It produced only some feeble reactions. I now begged her physician, Dr. Lippich, to allow it to be placed in the magnetic meridian from north to south, which he very obligingly did. Everything became instantly changed. The patient immediately gave signs of satisfaction; the previous restlessness left her; a painful smarting of the eyes, from which she had constantly suffered, disappeared. Instead of the intolerable heat which had before tormented her, she felt refreshing coolness, and a general sense of relief pervaded her frame, while we observed her. There followed a night of such quiet refreshing sleep as she had not for a long time enjoyed. From that time forward, her bed was kept in the same position, and this she earnestly entreated. On another occasion I prevailed on her to turn round in bed, so as to place herself with her head towards the south. This time, just as rapidly as before everything had changed for the better, everything changed for the worse. General restlessness and heat appeared; flushing of the face and oppression of the head followed; *and the peculiar painful smarting of the eyes immediately returned.* But all these feelings were, as it were, stripped off, as soon as I again placed her in the north to south position. In this normal position, I again tried the magnet. But what a difference! She, who had before hardly perceived its action, could now not bear it, when I removed the armature, even at some distance. I had placed myself four paces from her head; and when I spoke to her, she gave me no answer. When I examined her, I found her insensible and in a state of tonic spasm. After her waking from this fit, I stood at seven paces from the foot of her bed, and detached the armature.—In this case also, she had hardly uttered a word, when she became speechless, and I again found her in the same state as before. A third time, I removed from her, in the line of the magnetic meridian, to a distance of more than thirty feet from the foot of her bed. She did not at once perceive any effect when the armature was detached; but after I had remained about a minute in the same position, she suddenly stopped in the middle of a word, and became speechless. She had uttered half the word; the other half died upon her lips. She was suddenly attacked by spasms, and I found her lying rigid, with clenched hands. Her eyes were open and turned upwards; and so completely devoid of sensation, that I could lay my finger on the eyeball without causing any motion of the eyelids. How unexpected a difference of effect was here! The same magnet, which I had before placed on her head and feet, while she lay in the magnetic parallel, with hardly any perceptible effect, now, when she lay in the meridian, struck her instantly down insensible at the distance of thirty-two or thirty-three feet.

65. Mlle. MAIX, who could not walk, was kind enough, at my request, to allow her chair to be placed in the four different positions. She is neither cataleptic nor a somnambulist, but is affected with paralysis in the lower limbs. Yet, in this very different case, I obtained the same results. She could bear only the position north to south, and that of west to east was the most intolerable to her. But in this case the experiments were tried, not as with Mlle. NOWOTNY, in the morning, but about four P. M.

66. In the case of Mlle. REICHEL, her physician paid no attention to the direction of her bed; and when I remarked this, he said this patient was strong enough, so that it was a matter of no consequence. I was not of that opinion; and when I tested her, by causing her to recline in the four chairs successively, I found as marked a difference as with the other sensitives. As her bed stood in the south to north position, I advised her to change it for that of north to south. This she did, and found her nightly rest very much improved; now, for the first time, she could sleep easily and comfortably, whereas, before, she could only obtain sleep with the utmost difficulty.

67. I tried the experiment with Mlle. ATZMANNSDORFER in two different houses; the first time in the morning, while her illness was on the increase; the second time in the evening, during its height. In both cases she found the north to

south position the most agreeable; that of west to east the most intolerable.

68. All these patients now recollected, how painful it had always been to them to remain for any length of time, in church. All Roman Catholic churches are built from west to east, so that the members of the congregation find themselves, when opposite the altar, in the position from west to east; consequently in that position, which is, to sensitive persons, of all others the most intolerable. In fact, they often fainted in that position in church, and had to be carried out. At a later period, Mlle. NOWOTNY could not even bear to walk in the street, or in the garden, in the direction from west to east, if her walk lasted but for a short time.

69. These eight different cases all agree, therefore, in leading us to this conclusion, that, to sensitives of various kinds, any other position than that in which the head is towards the north, and the feet towards the south, is highly distressing; but that lying in the magnetic parallel, with the head towards the west, in our northern hemisphere, is hardly endurable by sensitive persons. In the southern hemisphere these relations are probably different. The causes of these phenomena, it is obvious, cannot be found elsewhere than in the action of that vast magnet, which is formed by the earth, its surface, and its atmosphere; in other words, in terrestrial magnetism. It acts here like any other magnet; and we are thus led, by the present investigation, to a law, which may be thus expressed:—*Terrestrial magnetism exerts on sensitive persons, whether healthy or diseased, a peculiar action, strong enough to destroy their comfort; in the case of healthy sensitives, to alter their sleep; in the case of diseased sensitives, to disturb the circulation, the functions of the nerves, and the equilibrium of the mental powers.*

70. And since the magnetic condition of the earth is subject to variations, which, among other things, have a relation to the *lunar phases*, so that, as is known, the intensity of terrestrial magnetism, so far as affected by the moon, reaches its maximum at the period of full moon; it is obvious that one of the causes, *to which the phenomena of lunacy or insanity must be referred*, is thus brought out of darkness into the light

of dawn. But I can only enter on this subject, after I have gone on to the more special development of the matter at present under consideration.

71. Since, therefore, terrestrial magnetism stands forth, in this manner, as a re-agent of surprising power on our bodily feelings; since its reaction in the eight cases above mentioned is so energetic, that it decides, in great measure, as to the sensations of health and sickness,—we are surely entitled, nay compelled, to draw conclusions from those cases concerning other cases of disease in sensitives; and we must admit, that in many, perhaps in all of these cases, it will be impossible to effect a cure by means of magnetism, unless the patient be placed in the right position as regards terrestrial magnetism; that, in every mode of cure, this must be investigated and attended to; and that all magnetic phenomena in patients suffering under diseases of the nervous system, and perhaps in many other diseases, are essentially influenced by it. The above observations furnish the key to a multitude of errors and contradictions, which appear in the whole range of animal magnetism, from the time of PARACELSUS and of MESMER down to our own days, which have staggered the soundest heads, and which have everywhere introduced contradictions among facts, and discord among opinions. For, when the same disease was treated magnetically in Vienna, in the position of north to south, in Berlin in that of east to west, and in Stuttgard in that of south to north, different results were obtained in each case, and no agreement or harmony between the observations was attainable. Nay, if *the same* physician at different times, or at the same time in different places, were to treat the same disease with the same magnetic means of cure, and if the beds of his patients happened to lie in different directions, he must necessarily observe widely different effects. He would in that case be quite confused in his own mind, and still more in his ideas of magnetism as a remedy; he would regard it as a thing full of caprice and uncertainty, and, finding it impossible to predict and to regulate the phenomena, would cast aside magnetism as an impracticable and unmanageable agent. Such is, in fact, the melancholy history of magnetism.

It has been, times innumerable, since the earliest period, seized, and as often again rejected; and now it lies there, hardly ever employed; while yet it is a truly wonderful, penetrating, nay, it may be said incomparable, means of affording relief in diseases, on which, hitherto, human art has very rarely succeeded. Even physicians themselves call the diseases of the nervous system the *scandala medicorum.* But I confidently hope and expect, that this will, in no long time, cease to be the case in the same degree. Henceforth, the powerful influence of terrestrial magnetism will be measured and taken into account; and the whole subject of magnetism, in a medical point of view, will now admit of regular investigation. Progress will be made, and we shall come to understand each other; while the world may at last hope to derive some benefit from the application of these remarkable facts; a result which has been, with justice, long looked for. If a physician here and there has noted the observation, that his patients generally felt better when lying with the head towards a certain point of the compass, yet, so far as I know, the matter has never been thoroughly scrutinised; and least of all has its true and very important signification been pointed out, or referred to its ultimate physical causes. In this place, where my object is merely to point out its relations to physics, beyond which I do not here wish to go, I have only to say, looking back to § 60, that, after I had ascertained, by the observations above recorded, the existence of a powerful influence, derived from terrestrial magnetism, acting along with that of magnets when applied to sensitive persons, *all the subsequent investigations were made with the patients in the position from north to south, which I regard as the normal direction for all the reactions of magnets, crystals, or other bodies on the living, sensitive frame, whether affected with disease of the nervous system or not, but more especially in the former state.*

72. Having been led, by the researches mentioned in the paragraphs from § 60 to the present one, to the theoretically and practically important proposition, that terrestrial magnetism everywhere and continually exerts a powerful influence on all sensitive persons; and having been so fortunate as to

bring under rule and law these new disclosures concerning the interior movements in dead and living matter, we may go back to § 59, and follow out the thread there indicated, in a different direction. This we shall now do, by studying the action of magnets and crystals on sensitive organisations.

It is known, that a piece of pure iron, free from carbon, however often subjected to stroking with the strongest magnet, yet, when separated from the magnet, does not attract iron, nor even lift a particle of iron filings. It has therefore acquired from the magnet no permanent magnetic power, and all observers agree in this, that the iron returns to its original condition as soon as the magnet is detached from it. But this is not entirely the case. We had, indeed, hitherto no means of detecting any change in the state of iron which had been in contact with a magnet; but such a means is now furnished to us by the sensitive human nerve. When, for example, I caused Mlle. NOWOTNY to take hold of a bar of pure iron, before contact with a magnet, and avoiding contact between it and my own hands, she felt it perfectly indifferent. But if I now, taking care not to touch it with my hands, placed it in contact with a magnet, separated it again, and caused her to take hold of it, the result was very different. The bar was no longer indifferent, but excited in her hand the same sensation as a weak magnet, namely, slight warmth, and contraction of the fingers; and this power it retained for a short time, gradually losing it, till, after eight or ten minutes, the bar had again become indifferent. Mlle. REICHEL perceived the action of a steel bar magnet, 19 inches long, at the distance of the length of several rooms. This bar was now connected, by means of cross bars, with a bar of soft iron of the same form and size. When the soft iron bar was detached from the magnet, and tried, by itself, on Mlle. REICHEL, I was not a little surprised to find, that, when just detached, its action was perceived by her at about the same distance as that of the bar magnet of the same size. I tried similar experiments at various times with the other sensitives. The contraction of the fingers did not occur, but all the other reactions of the magnet were perceived in the soft iron by the patients, in a feebler degree than in the magnet, but yet of considerable energy, when a powerful magnet was used. *Something, therefore, derived from the magnet, must have remained in the iron; which something is not magnetism, and with which we are not yet closely acquainted.*

73. When a glass of water was placed between the poles of a horse-shoe magnet, that is, in the course of the magnetic current in the manner described in all the works on Animal Magnetism, and thus magnetised, as it is called, not only could every sensitive patient instantly distinguish it from ordinary water, but the glass of water, when placed in the hands of cataleptic patients, immediately after being magnetised, attracted the hand like a magnet, and even solicited it to follow, as described in my treatise on the peculiar force of crystals, § 27 and 28.—*Something, therefore, must have passed from the magnet into the water, and remained in it; something that is not proper magnetism, which we have no chemical means of arresting or detaining, and the presence of which we cannot by means of any of our ordinary senses recognise.*

74. The distinguished botanist, Professor ENDLICHER, paid a visit to Mlle. NOWOTNY, and in his presence her physician made the following curious experiment. Professor ENDLICHER requested him to allow himself to be magnetised by stroking with the magnet, and then to act on the patient. To his surprise, he found, what had never previously been the case, that now he was able with his hand to attract that of the patient, to attach it to his own, and to cause it to follow in every direction, exactly as the glass of water had done. This power he retained for nearly a quarter of an hour, after which it had gradually disappeared. *The same unknown something*, which had remained in the iron bar after contact with the magnet, and which had also passed into the glass of water, *must therefore have entered into the whole person of the physician.* The same cause had produced in his fingers the same power of causing certain sensations.

75. This experiment was subsequently repeated in various forms. Sometimes the same physician laid his hand in that of Mlle. NOWOTNY, while I drew a powerful horse-shoe mag-

net down his back. The patient declared that she felt the influence from the hand of her physician, as it were swelling, almost in jerks, at every stroke of the magnet. I repeated this with Mlle. MAIX, and obtained the same results by laying my hand in hers, while the magnet was drawn down my back. I would here repeat, that this latter patient was not, and never had been, a somnambulist.

76. In § 29, I have already mentioned, as the history of my researches compelled me to do, that a multitude of objects of all kinds, after being stroked with the magnet, exerted an action on the patient, which, although feebler, was exactly of the same kind as that of the magnet itself. I there spoke of only one patient; but since then, I have had the means of trying many persons affected with nervous complaints, and among them several who consider themselves healthy, and are quite able to follow their proper occupation. They are easily discovered, for they all instantly perceive the effect of the magnet when a single pass is made down their persons with the horseshoe. Now, *all these persons*, who may be found by hundreds in a large city, if we only look for them, *felt themselves acted on exactly as when a magnet was used, only not so powerfully, by all substances which had been several times stroked with a magnet.* Any one who chooses may confirm this; for there is hardly a hamlet so small as not to yield one or more persons who will be found so far sensitive.

77. Since, then, it is fully established, by observations and experiments of many and various kinds, that all persons possessing a certain degree of sensitiveness or irritability of the nervous system, perceive the effect of a magnet without touching or seeing it, when it is merely brought near to the hand, or when passes are made with it as close as possible to the surface, and usually in the form of a distinct cool or tepid aura; since, further, the same persons perceive, in the same manner, but less powerfully, the same effects from all material substances, no matter of what nature, when they have previously been placed for a time in the line of the magnetic current, or, as it is usually called, magnetised; there follows necessarily from these two propositions a third, which people have hitherto

been unwilling to deduce from them, nay, against which they have, *à priori*, struggled with all their might; namely this: *All magnetised objects have, by contact with the magnet, undergone some temporary change, whatever the nature of that change may be.* Thus, even magnetised water, however strange this may sound, is really and actually a changed water.

78. If we now compare the effects of the force residing in crystals, as described in the preceding treatise, with those of the magnet or other bodies, as above related, we perceive, *that the influence of both on third bodies is the same, and so thoroughly identical, that no mark of distinction between them is left.* I have there shewn that the magnetic force and that of crystals, each considered in its entire nature, are essentially different, and are related to each other as a compound is to one of its component parts or elements; for example, as solar rays are to calorific rays, or as alum is to sulphuric acid. But the changes which are found for a time in other bodies, after they have been removed from the sphere of action of these forces, are in the two cases perfectly alike. Now, since these changes are perfectly produced by one part, namely, the force residing in crystals, alone, we are forced to conclude that they are also caused, in the case of the magnet, by the same force (that found in crystals) residing in the magnet, and from this part alone of the whole force or influence present in the magnet. We find, consequently, magnetic poles and crystalline poles, as far as concerns their action on the sensitive nerve, to coincide entirely, to be perfectly alike.

---

79. And now our investigation has brought us to the portal of what is called Animal Magnetism. This *noli me tangere* we shall now be able to seize.—When I made a few passes down with a magnet the person of Mlle. STURMANN from head to foot, she became insensible, and was attacked by spasms, generally rigid. When I performed the same passes *with my large rock crystal*, the result was the same. *But I could also produce the very same effect by using, instead of the magnet or the crystal, my hands alone. The peculiar force* (we shall call it *crystal-*

line) found both in magnets and in crystals, must therefore reside also in my hands.

80. In order to test this more fully, I tried the experiments which I shall presently describe. If this were the case, the force residing in my hand must produce all those effects which the crystalline force is capable of producing, as described in the preceding treatise; I could conclude as to difference or similarity, according to the degree of resemblance in the properties observed. It was, first of all, necessary to ascertain whether there existed a coincidence, and to what extent, between the action of crystals on the healthy or diseased sensitive nerve, and that of the human hand on the same re-agent. When, in the case of persons sufficiently sensitive to perceive distinctly the passes made with a large crystal along the inner surface of the hand, I drew along the left hands of the patients the points of the fingers of my right hand, turned laterally, so that one finger followed the other, and all passed over the same line, which was drawn from the wrist down to beyond the point of the middle finger, there was not one among them who did not perceive the effect, exactly as from the point of a crystal. It was generally described as a cool aura, more rarely as a tepid aura; and was not only as powerful, but usually considerably more powerful than from a crystal. I need not here speak of the diseased subjects, since all of those I have hitherto mentioned, perceived the effect with the same singular distinctness with which they felt, as a general rule, every magnetic pass; and Mlles. Maix and Nowotny were even able to distinguish the effect of each finger separately. But there were not a few healthy persons, who were quite sufficiently sensitive for this re-action. Indeed, some of these, who only felt indistinctly the action of crystals, perceived that of the fingers, used as above described, so plainly, that they could always distinctly point it out while the eyes were averted. I am permitted here to refer to my friend M. Carl Schuh, who is a strong, healthy man, and perceives the action of crystals with unusual distinctness. When, to make assurance doubly sure, and contrary to my own rule, I blindfolded him, and made slow passes with the fingers of my right hand, as

before described, over his left hand, he experienced so strong and distinct a sensation, analogous to that produced by a crystal, that he could distinguish each individual pass, and was able, for example, at all times exactly to tell when I had made exactly two-thirds of the whole pass. M. STUDER, already mentioned, also perceived this quite as plainly, as well as numerous other persons, among whom I have permission to name one of the finest, most powerful, and hardiest men I have ever seen, who has travelled through Persia and Kurdistan, and twice penetrated from Egypt into the heart of Africa; who is therefore a rare example of iron health and strength of constitution; namely, M. KOTSCHY, who accompanied M. RUSSEGGER in part of his travels. He perceives the effect most distinctly when the temperature of the air is agreeable, and less distinctly when it is cold. *The fingers, therefore, act on the sensitive nerve exactly in the same way as a crystal of middling size.*

81. I now compared the two forces, with reference to their conductibility. I caused Mlle. STURMANN to take hold of one end of a rod of German silver with her right hand, taking care previously to avoid touching it myself. I allowed her some time to become accustomed to the sensation caused by the rod, taken alone. I now placed on the other end the points of the fingers of my right hand, which were rather moist. She instantly perceived a warm sensation where the rod touched her hand, and this passed upwards as far as the elbow. I now added the fingers of my left hand: the sensation became much stronger, and reached to the shoulder. I removed my fingers: the sensation rapidly diminished, without however instantly disappearing. I next attached and removed my fingers alternately: the sensations kept pace with the changes, increasing and diminishing regularly.—On another occasion, I requested Dr. LIPPICH to do the same: his fingers produced exactly the same effects. I tried the same experiments on Mlle. MAIX. I caused her to take hold of one end of the same rod, and after a short interval, I first applied five, then ten fingers to the other end. The warm sensation was instantly perceived; and it rose and fell as I applied or

removed the fingers. With the whole ten, it was so strong as to pass through the whole arm and into the head. I begged her physician to try the same experiment. He did so with the same results; only, although he was more than ten years my junior, the effect produced by his fingers was distinctly less powerful than that caused by mine. Father LAMBERT, the confessor of the patient, was accidentally present, and I begged him also to try. She found his fingers as powerful as mine. The nurse of the patient, Mlle. BARBARA PSCHIERL, also made the trial. Her fingers caused similar sensations, but much more feebly than those of men. I repeated these experiments, substituting for the rod of German silver an iron wire, about five feet in length. When one end was held by the patient, and I applied five fingers to the other, the patient perceived a current of decided heat; and with my ten fingers, the sensation was stronger. It always quickly disappeared when I dropped the wire out of my hand. This fact was controlled by frequent repetitions. I next caused the sister of the patient, whose nervous system was also in some degree diseased, to apply her ten fingers to the end of the wire. The effect produced was strikingly feeble. The fingers of another female were added to hers: the effect was sensibly stronger; but the whole twenty fingers together did not produce nearly as much effect as my ten fingers alone, although I have long been grey and bald. I tried also a copper wire, nearly ten feet in length. It conducted the force, but less rapidly, and somewhat more feebly than the iron wire. The same experiments, with many variations, were repeated with Mlle. REICHEL, and with similar results: The action was very powerful in the case of Mlle. ATZMANNSDORFER: even M. STUDER, in perfect health, was so sensitive, that he perceived quite distinctly the action of my hands through metallic wires. It follows from all these experiments, *that the force derived from the human hand may be conducted through other bodies, exactly like the crystalline force, and that such bodies are conductors in the same way for both forces.*

82. I now wished to try whether bodies could be charged with the force from the hand. I began with Mlle. STURMANN.

I laid the German silver rod near her, and allowed it to lie for a quarter of an hour. I then begged her to take it into her hand, and thus to become accustomed to the sensation it might cause. After doing so, she laid it down; and I then took it in my hand for some seconds, and laid it down. When she now took hold of it, she felt it warm, and so strongly charged, that the well known sensation, caused under similar circumstances by crystals, rose through the hand as far as to the elbow. This was, of course, repeated, with many variations, for the sake of control. Her physician, Dr. LIPPICH, made a similar experiment. At my request, in another room, he took into his hand for a short time, one of two precisely similar porcelain saucers, not touching the other. They were now presented to the patient, who, with the greatest facility and accuracy, distinguished that which had been held in the hand from the other. After about ten minutes, the effect was dissipated, and both saucers felt alike. The experiment with the rod was soon after repeated with Mlle. MAIX in the same way as above. It yielded the same results; the rod was charged by my fingers, and the charge, which Mlle. STURMANN had felt for five minutes, was perceived by the more sensitive Mlle. MAIX to last, gradually diminishing, for twenty minutes. In both patients, the sensation was the same; one of warmth, rising into the arm, and coinciding exactly with that caused, under similar circumstances, by the rock crystal. I observed the same phenomena, some months later, in Mlles. REICHEL and ATZMANNSDORFER. The most surprising result is that obtained with a glass of water. If it be taken in one hand, and grasped below by the fingers, and if this be continued for about ten minutes, it then possesses, for sensitive patients, the smell, the taste, and all the well marked and curious properties of what is called magnetised water. Those who have never examined the matter experimentally, may exclaim irrationally against this. I was formerly myself one of this number; but all those who have tested the fact by experiment, and witnessed the effects, as I have done, can only speak of it with astonishment. The water, thus changed, which is exactly similar to that treated by magnets or crystals, has, therefore, received from the fingers an abundant charge of the peculiar force residing in them, and retains it for a considerable time. I could, after a time, produce similar effects on all possible substances, by holding them for some time in my hand. The patients, who had tried them all before I touched them, now perceived in all of them the same change, as if they had been stroked with the poles of magnets or crystals; and this, whether they knew of my having touched the objects, or had been kept in ignorance of my doing so. It follows plainly from all this, *that bodies may be charged with the force residing in the hand, exactly as with the crystalline force.*

83. That the charge *is gradually dissipated*, is obvious from what has been said, and requires no special proofs. From the two last propositions it follows, that those substances, which admit of being charged, and gradually lose their charge, *possess for the force residing in the hand the same coercitive power which they exhibit in reference to the crystalline force.* The *amount of charge* increased with the power of the hand employed, and the *capacity for the charge* seems, as far as we have studied it, to have no limits other than the varying power of the charging body.

84. It was necessary to decide, by a comparative investigation, the question, whether the force residing in the animal body possessed a dualism or polarity, such as was found in crystals. Crystals, as is well known, have more than one crystallographical axis, principal and secondary; in the tessular or regular system, they have three principal axes. When I tested them in this point of view by my sensitive patients, they all soon found out for me, as already stated, the principal axes and their poles; that is, the two opposite points, where the action of the new force was most strikingly concentrated. But in many cases, as, for example, in iron pyrites, gypsum, fluor spar, heavy spar, sphene, garnet, and manv others, they could point out also other axes, which, although the opposition between their poles was less striking, yet exhibited distinct polarity or dualism. All the patients agreed perfectly in these observations; and a crystal of gypsum which I presented to

all of them successively, and which, while it lay on a table, they felt and tried with the fingers of both hands, yielded, with all, the same results. All the patients pointed out the powerful principal axis, with its poles, the one feebler than the other; and also weaker secondary axes with their poles, in exactly the same lines and points. It often happened that the most powerful principal axis was not the longest, but a shorter one; this was often the case with gypsum; and the perfect agreement of so many observers, unknown to each other, in these statements, was the best guarantee for the reality and accuracy of their observations. These observations may be easily controlled elsewhere, because, in populous cities, sensitive nervous patients are at all times to be found. But even healthy persons, such as M. STUDER, could often, without difficulty, detect the poles of crystals by the finger. Now these poles and axes invariably coincided with the crystallographic poles and axes, and it became, therefore, more than probable, that the crystalline force has a share in the formation of crystals, if it do not alone determine that process. It may possibly be, to crystals, what the vital force is to organised tissues. But I shall not here venture to indulge in conjectures, but adhere to that which, as a matter of ascertained fact, bears on the subject; namely, *that the crystalline force exists in crystals in a polar form, and is found simultaneously in several axes in the same crystal, but in various degrees of intensity.*

85. Now I find analogous relations in animal life itself. We assume, in man, a principal axis, from above to below, and we regard the brain and the genital organs as the poles of that axis. But if I may draw a conclusion from the above recorded facts of what is called Animal Magnetism, this is not a principal, but only a secondary axis. It has been shewn, in the first place, that patients sensitive to the action of the magnet, cannot endure especially that position in which the length of the body lies in a magnetic parallel, or from west to east. In this position, the body is magnetically differentiated (under the influence of terrestrial magnetism) in its breadth; a state which it seems unable to endure. Something analogous is known with respect to chills. When a chill comes

from the side, it is at the same time more efficient and more injurious, than when it comes from before or behind. But this was rendered clearer to me by some observations which bear on the point in question. When I placed my right hand in the left hand of the very sensitive Mlle. MAIX, she perceived the same effect as when I brought in contact with her hand, in a vertical direction downwards, the north pole of a small bar magnet, or of a crystal of gypsum four inches long. But when I placed my right hand in hers, she felt a much more disagreeable sensation. If I placed my right hand in her left, and my left hand in her right, as we do when taking a friend by both hands, she declared that she felt, as it were, a circulation going on in her person, which she compared to riding at the ring in a circus. The sensation passed up her right arm, through her breast and shoulders, down the left arm, and seemed to her to pass through me, and so on continually in a circle. It was painful, and made her feel giddy and faint. When I crossed my hands so that my right hand lay in her right, and my left hand in her left, she refused to continue the contact, and declared, that it caused so painful a sensation, of a new and strange kind of struggle and contest in her arms and through her breast, a sort of undulation up and down the arms, that it was absolutely intolerable. She objected also so firmly and decidedly, after she had torn away her hands from mine, to my again taking hold of them in the same way, that I was obliged to renounce the idea of repeating and controlling the experiment, as I was invariably accustomed to do in all my other experiments.*

86. Since, then, it clearly appears from these experiments, that it is very far from being a matter of indifference, which hand is used in touching a sensitive nervous patient, it plainly follows, that the hands of a human being, in regard to the hidden force residing in them, are not in the same state; and, unless I give an entirely erroneous interpretation to the last experiment, there occurs a kind of circulation or current analogous to the galvanic current, from my left hand to the patient's right hand, through her body, and from her left hand to my right; a movement, which cannot take place, or meets

* See Appendix, A.

with great obstacles, in spite of which it tends to force its way, when I connect like hands, that is, my right hand with her right hand, and so forth. This difference between the right and left hand can depend on no other cause than the known one of polarity, as we see it in the magnet, and have learned to recognise it in crystals. Thus regarded, the principal axis passes, in man, and probably in all animals, across the body, and the longitudinal axis must be regarded as secondary. In fact, we are only formed of two symmetrical halves in the transverse line.—Every part, the organs of the brain, the organs of sense, the apparatus of mastication, &c., the arms, hands, testicles, and feet;—all these are transversely opposed in pairs, and thus we are, in general, transversely polar.

87. I examined these interesting relations at a subsequent period in Mlle. ATZMANNSDORFER, in the same way. The results were the same as those above described. She perceived, when I took hold of her hands with my opposite hands, (right with left, &c.) the current up her right arm and down the left, still more distinctly than Mlle. MAIX. When I crossed my hands, hardly a minute elapsed before she was so affected as to become sick. When I placed in her hand a rod of German silver, or a brass wire, and touched it with my right hand, she perceived, besides the usual feeling of warmth, a peculiar secondary sensation which I had also observed in Mlle. STURMANN, namely, that these bodies appeared to become light, and, as it were, downy; on the contrary, when I applied my left hand, they seemed to become heavy, and apparently much heavier than they naturally were. Without here entering into details on this peculiar sensation, I must mention it, as it furnishes another mark of the polar opposition between the right and left hand, analogous to attraction and repulsion. But in the same way as the patient recognised a difference between my two hands, she was able, no less clearly, to distinguish between her own. When I placed different bodies, such as iron pyrites, fluor spar, gypsum, reguline metals, or charcoal, in one of her hands, they excited sensations very different from those which they caused, when I begged her to transfer them to the other hand; although in

her case, one side of the body was in no respect weaker than the other.

88. Very recently, I went through a series of researches on this part of the subject with Mlle. Reichel, and was able to develope it further than with any of the earlier sensitive patients. She found not only the right hand, but the whole right side of the body, from head to foot, in every point, opposed in properties to the left side. Even when I only approached her with my right or left side, she was affected in an essentially different way. I shall enlarge on this in one of the treatises to follow at a future period; here, where my object is only to establish by observations, the fact of a polar difference, analogous to that of the magnet, between the opposite sides, transversely, of the human body, I must content myself with stating that I have, in the case of Mlle. Reichel, repeated, and thus once more confirmed, all the observations made on this subject by Mlle. Maix.

89. It appears, then, from all these investigations, that all the symmetrically double organs of the animal body, as far as examined, and especially *the hands, exhibit a difference, which is caused by a polar opposition, analogous to the magnetic; and that, consequently, there exists in these parts or organs, a dualism of the new force here under consideration, exactly as we have found such dualism to exist in crystals.*

90. I have shown, (§ 41 and 53) that terrestrial magnetism exerts no sensible action on crystals, and does not affect in the slightest degree their direction, when they are freely suspended. The same is the case in regard to the force residing in the hand. Terrestrial magnetism appears not to exert the smallest influence on perfectly healthy animals. The force which I could exert on sensitive patients with my hands, is always equally efficient, in whatever position I am placed; at least I have hitherto been unable to detect any difference. I am also unable to perceive, passively, any effect from the magnetism of the earth. I have tried the effect of lying down to sleep in different positions; but in all, I have slept equally well, in whatever direction I lay. The thoroughly healthy man, who is, perhaps, never sensitive, doubtless never expe-

riences any effect from terrestrial magnetism, although that force acts so violently on many patients, as well as on sensitive persons who are, to all appearance, quite healthy. I cannot observe, even in animals, any thing indicating the slightest dependence on terrestrial magnetism. If any animal should possess a sense enabling it to perceive the action of the earth's magnetism, we might expect to find such a sense in larvæ, which have no organs of sight. Now, as silkworms are bred on my property, I had frequent opportunities of observing the behaviour of animals of so low an organisation, at different ages and under different circumstances. But even in spinning and covering the coccoons, the animal never selects any one direction, but works indiscriminately in all possible directions. Not even in choosing their sleeping places, does a majority of the animals show a preference for any one position. *In this indifference towards the general magnetic force of the earth, the force residing in the hand coincides, therefore, perfectly with the crystalline force.*

91. With respect, now, to the remarkable direct attraction of the patient's hand by that of other persons, it has been already stated (§ 74), that the hands of one man had this effect, but only after he had been stroked with a powerful magnet. He could never produce this effect by his own inherent force. But it appeared that he was not, in this respect, powerful. At least Mlle. MAIX felt his hand much less powerful than that of Father LAMBERT, or my own. I did not, at the proper time, try the proper experiments with the cataleptic Mlle. NOWOTNY, because, at the period of her illness, I was not sufficiently aware of their importance. But I have often seen this phenomenon in Mlles. REICHEL and ATZMANNSDORFER in the more intense stages of their disease, especially in the former, and that in the presence of many other persons. In the cataleptic state which usually preceded her attacks of convulsions, her hand readily followed that of any strong young man, as well as my own. I have often caused her, when she was quite unconscious, to rise from her chair, and to follow my hand for a considerable distance. Even when I presented to her, in that state, bodies incapable

of polar induction, such as a lump of chalk, I could raise her hand with it, and if she were standing, cataleptic, on the floor, lead her forward some paces. Here it was the force from my fingers which acted; being conducted by the chalk, at the point of which it was concentrated (according to the laws developed in § 81); in which case the chalk represented the whole of my fingers, and that so completely in their force and action, as to attract and cause to follow it the hand of the patient, when I moved backwards, as my fingers would have done, without the chalk. In the same way I observed the attraction of my hand during the attacks of Mlle. ATZMANNSDORFER. I did not witness, personally, this phenomenon in Mlle. STURMANN; but, according to Dr. LIPPICH, who observed it, it took place in her in full measure. I can confide as securely in the statements of that gentleman as in ocular demonstration of the fact. From all these observations it follows, with certainty, that a power of mechanically attracting the hands of persons in a cataleptic state resides in the hands and fingers of healthy persons, just as in the poles of crystals.

92. In this comparative investigation, however, the luminous phenomena, which I have now to describe, form, both literally and metaphorically, a brilliant point. One day, when I was observing Mlle. REICHEL, who, after severe spasms, when in a sort of half-sleep, with closed eyes, was playing with the magnetic flame, an amusement in which she particularly delighted, I introduced my outstretched hand, in the dark, between her and the flame. She instantly began to play, as before, with my fingers, and to speak to the bystanders of five little flames, which jumped about in the air. She did not notice my hand, but took the motion of my fingers, at the points of which she saw little flames, for a spontaneous dancing of the flames. All those who were present now held up their hands, and asked whether fire could possibly be also flowing from their fingers. The flame was visible to the patient on the hands of every male person present, more or less brightly; but not one finger of a female had sufficient light to show a flame, and exhibited at the utmost a feeble glimmer. This was even the case with her own fingers. As long as her

illness continued, these experiments were often repeated, in order to amuse her after her fits, or for the gratification of many other persons. But when she had recovered, it was found that she had the power, which still continues unchanged, of perceiving, not only during her illness, but also in the intervals of apparently strong health, the magnetic flames, the light in crystals, and the flames on human hands, if the room were only dark enough. Nay, it appeared that she had possessed this power from her childhood. When a child, her mother had often been obliged to raise her in her arms, that she might convince herself that there was no fire proceeding from nails and hooks in the wall, as she often spoke of such appearances with exclamations of wonder. There were even two of her brothers and sisters, who, in the same way, saw everywhere luminous appearances, where other persons could see nothing. Now, while I am writing this, Mlle. Reichel daily assists me in researches, which I am making, in this direction, on electricity and magnetism; and we shall see, at a future period, from my reports, to what conclusions they have already led and will yet lead. I was thus enabled, with the aid of this sensitive patient, to study in the most sober and comprehensive manner, and for a long time, the luminous phenomena seen on the human hand; an investigation which I still daily pursue.

93. Experiments with Mlle. Atzmannsdorfer gave essentially the same results. But she saw the flames of a larger size. While Mlle. Reichel, in her peculiar degree of sensitiveness, described them as being from 0.8 to 1.2 inches in length, Mlle. Atzmannsdorfer, in the dark, saw them from 2 to 2.5 inches long, that is, almost as long as a finger. Mlle. Reichel made drawings of these beautiful appearances, which I shall give in one of the subsequent parts of this work. The fact, established by several observers, *that fiery bundles of light flow from the finger-points of healthy men, in the same way as from the poles of crystals*, is sufficient for my present object.

94. I have now compared the properties of the crystalline force, described in the preceding treatise, with the force residing in the human hand, without omitting any property. The

parallelism between them is, it will be seen, complete; and the coincidence of the two forces in their collective manifestations, so perfect, that they are obviously identical. For the sake of clearness, I here place the results briefly together :—

Passes made over sensitive persons with the hand, act on them like passes made with the poles of crystals. § 79.

The force which here acts is conductible through all bodies, like the crystalline force. § 80.

Other bodies may be charged with it, as with the crystalline force. § 81.

The charge is soon dissipated, as with the crystalline force. § 82.

Bodies have a coercitive power for both the one and the other force. § 83.

The capacity of receiving a charge, possessed by all substances is, in both cases, alike. § 83.

The force in the human body is polar, like that in crystals. § 89.

It is indifferent to terrestrial magnetism, like the crystalline force. § 90.

It exerts, like the crystalline force, a mechanical attraction on the sensitive hand. § 91.

It exhibits luminous appearances, in kind and intensity, like those of crystals. § 93.

And thus we are led back to the opening proposition of this section, § 79; namely, that, in fact, the same force resides in the human hand which is found in crystals; and *that, consequently, the crystalline force and animal magnetism are thoroughly identical; so that the same laws which regulate the former apply also to the latter.*

95. Retrospect.

a. Not only do crystals exert a peculiar influence on healthy and diseased sensitives, but the same is true of Terrestrial Magnetism. This influence is so powerful, that very sensitive patients can only remain in one position, that namely of north to south, that is, with the head towards the north, and that every other position is painful to them; in many cases, that from west to east is quite intolerable, and even dangerous.

b. All re-actions with magnets, crystals, or other bodies, on such patients, are essentially affected by a change in the position of the patient.

c. Pure iron, free from carbon, and from particles of steel, receives, as is well known, no durable magnetism when treated by stroking or contact with a magnet; but it has, notwithstanding, after separation from the magnet, acquired a peculiar force, by virtue of which it has become capable of exerting a strong and decided influence on sensitive persons.

d. The magnet communicates this unknown something, not to iron alone, but to all metals, stones, salts, water, plants, and animals, nay, living persons; in short, to all solid material objects without exception.

e. This something acts, from all bodies directly charged with it, or rendered active by what is called induction, on sensitive nervous patients, exactly as the magnet itself does, and as crystals do; and must, therefore, be identical with the proper agent in those bodies.

f. Healthy persons, especially by means of their hands and fingers, are able to act in precisely the same way on sensitives, healthy or diseased.

g. This last force, which has been called, by physicians, Animal Magnetism, has the properties of conductibility through all bodies, and of communicability to all bodies, by direct charging or by induction; of soon disappearing from them; of being retained in them for a short time by their coercitive power;—it is polar in the human body, in virtue of its dualism; has no perceptible relation to terrestrial magnetism; exerts a mechanical attraction on the hands of cataleptic patients; and is associated with luminous phenomena:—all exactly as the crystalline force is; with which it is, consequently, identical, and in all respects subject to the same laws.

h. One part of the collective force residing in the magnet; the crystalline force; and the force lying at the foundation of what is called Animal Magnetism.—These three forces, in their essence, when regarded from a common point of view, coincide, or are identical.

FOURTH TREATISE.

FURTHER SOURCES OF THAT PECULIAR FORCE WHICH RESIDES IN THE MAGNET, IN CRYSTALS, AND IN THE HUMAN HAND.

96. In the preceding researches, we have not become acquainted with all the sources of the enigmatical force, which is the subject of our investigation. Nay, the sources of it, hitherto mentioned, are not even the chief ones. Pursuing the matter, I have been led to other, and very important sources of this influence. It is known that natural philosophers have long discussed the question, whether the solar rays are, or are not, capable of communicating magnetism to a steel needle. After Morichini, who published the first observations, the subject was treated chiefly by Mrs Sommerville, M. Baumgartner, Configliaghi, and others. The recollection of their researches necessarily drew my attention to the part which the *sun* might possibly play, in reference to the subject of these enquiries. The undeniable and long known influence of the moon on patients affected with certain nervous diseases, rendered it in some degree probable that a similar influence might belong to the sun.

97. I took advantage of the first clear sky to try some experiments in this direction, with Mlle. Maix. I placed in her hand the end of a copper wire, about thirty-three feet in length, and allowed her to become accustomed to it. I then thrust a long portion of the opposite end, before the window, into the sun's rays. The patient instantly perceived the usual effects of

the crystalline force, in no great degree of energy, but quite distinct. I now attached to the wire a copper plate, having a surface of 64 square inches; and after allowing the patient to become accustomed to the sensation caused by the end of the wire, placed the copper plate in the rays of the sun. This had hardly been done, when I was greeted with an unexpected cry of delight from the sick-bed. As soon as the rays had fallen on the plate, a strong influence of the crystalline force was felt in the hand, in the form of the usual sensation of warmth, which rose rapidly through the arm to the head. But along with this well known, and not unexpected feeling, there was perceived a peculiar coolness, and this, so powerful and predominating, and with such a sensation of restoration and refreshment in all her limbs, that the patient expressed herself as being highly refreshed and rejoiced by it. Heat and cold were here felt at the same time.

98. To vary the experiment, and at the same time to obtain the result less complicated with the effect of heat, I substituted, for the plate of copper, white linen. I first allowed the patient to become accustomed to the wire in the shade, and then placed in the sun's rays the opposite end, hung with a piece of linen. The room was warm, the external air somewhat cool. The patient, notwithstanding, immediately perceived the same sensations as with the copper plate, only not so powerful; namely, first a dull feeling of increasing crystalline force giving warmth to the hand, from the wire; and then the striking grateful coolness over the whole person, this latter sensation being tolerably vivid.

99. I again varied the experiment, by attaching to the wire wet instead of dry linen, and proceeding as before. The action was in this case accompanied by a disagreeable secondary sensation, like that which, according to the patient, moist air produced in her. But the chief sensation, peculiar to the sun, namely, increasing warmth in the wire, and the peculiar refreshing coolness which was perceived, and spread from the wire over her whole system, was experienced with great vividness.

100. I now endeavoured to obtain confirmation of this ob-

servation, as a guarantee for its exactness. I tried the same experiment a few days afterwards with Mlle. NOWOTNY. She was at this time so far recovered, that she had for some weeks been no longer confined to bed; but still I wished to try the effect on her. One end of a wire was placed in her hand, and the other end placed in the sun's rays. She instantly perceived a change in the end she held, which became, to her, cool. I drew the wire back into the shade; the coolness disappeared. I placed it again in the sunshine; the coolness returned. I now attached to the end of the wire several feet of tinned iron plate, and placed this, with the usual precautions, in the sunshine. The coolness not only instantly returned, but increased, in the course of three or four minutes, to such an extent, that the patient assured me that it had become ice-cold, and began to render her fingers rigid. The plate was removed to the shade, as a means of controlling the observation; the feeling of cold instantly began to diminish, and after a few minutes, was gone; but it as quickly returned, and increased to the same point as before, when I again brought the plate into the sun's rays. I have already pointed out, that the peculiar effect produced in the hands of sensitive persons by the crystalline force, at one time appears as a sensation analogous to heat, at another as a sensation analogous to cold. The particular differences in this respect, I shall, in a subsequent part of this work, describe in detail. In the above experiments, in which the coolness proceeds from the rays of the sun, which are the source of heat to universal nature, this sensation is peculiarly characteristic of a new and specific agency. It was so strong and distinct, that a difference was observed, according as I allowed the sun's rays to fall obliquely on the plate, in which case the effect was less powerful; or perpendicularly, when it became much stronger. There was also a difference according as I made the experiment in the morning or in the evening, in July or in November, under similar circumstances in other respects.

101. With Mlle. ATZMANNSDORFER, I had no opportunity of trying detailed experiments on this subject, but in conversation she told me, that the sun, in general, produced in her

a grateful sensation, not however of heat, but of a refreshing coolness over her whole person.

102. In like manner I had, at an early period of my researches, before I had recognised this peculiarity of the solar rays, heard frequently from Mlle. STURMANN the statement, at that time to me incomprehensible, that the sun made her feel cold.

103. It was, however, with Mlle. REICHEL, that I was enabled to investigate the subject most accurately. Not only did a wire, when connected with iron plate, copper plate, zinc plate, tinfoil, leadfoil, strips of silver, gold leaf, German silver or brass plate, produce in her, when the metallic plate was exposed to the sunshine, the peculiar sensation of cold; but the same effect was observed when linen, woollen cloth, cotton or silk, attached to the wire, were placed in the direct rays of the sun. Nay, all other bodies, such as porcelain, glass, stone, wood, water, lamp oil, alcohol, sulphur,—in short any substance I might select, when attached to one end of a wire, of which the patient held the other end, and exposed to the sunshine, caused in her the same striking sensation of increasing coolness, to which all the sensitive nervous patients unanimously testified; they being as much surprised as I was myself at the unexpected and apparently contradictory nature, as compared with ordinary experience, of the facts observed. This apparent contradiction will be clearly accounted for in the sequel.

104. If it really were the peculiar force of magnets, of crystals, and of the human hand, as described and shown to be identical in the preceding treatises, which here resided in the solar rays, this could only be established by the same method which I had pursued in similar cases with crystals, &c., namely, a comparison of the effects. It was therefore necessary to state and discuss the questions; whether the solar ray can produce in bodies the same change, or the same state, which is produced by magnets, by crystals, and by the human hand? Can the mere solar ray communicate to a piece of iron the force which the latter can derive from a magnet? Can it charge all bodies with the power of acting on sensitive patients? Can it produce (so called) magnetised water? Can the solar ray, a thousand times studied and analysed, contain

yet another new and powerful force, which has up to this hour escaped the attention of those engaged in physical researches? I hardly ventured to entertain such ideas; but my desire to clear up the subject became every day more lively.

105. The first experiment which the desire of further knowledge induced me to make, was that with a glass of water. I placed it for five minutes in the sunshine, and caused the nurse, whose magnetic power was very feeble, to hand it to Mlle. MAIX, without the patient having any idea of what had been done. As soon as she put it to her lips, she declared it, without any questions having been asked, to be magnetised water. On the tongue and palate, in the throat, down the gullet, and in the stomach,—at all these points it excited the peculiar peppery burning, well known to all sensitive patients, and a tendency to spasmodic attacks. I allowed another glass of water to remain for twenty minutes in the sunshine, and it was then given to the patient, being touched only by the magnetically weakest female hands, to avoid the stronger action of my own hands. She found it now as strongly charged with the magnetic force as it was possible to charge it by means of the great nine bar horse-shoe magnet.

106. It was possible, that a considerable portion of the force might adhere rather to the glass containing the water, than to the water itself. To test this, as well as to ascertain with reference to the internal state of the water, whether it were analogous to that of iron filings contained in a glass tube, when acted on by the magnet, I caused the water, which had been sunned, to be poured into another glass, and to be again handed to the patient. I then found, as I had often done with magnetised water in the cases of Mlles. STURMANN and NOWOTNY, that the water was nearly as strongly magnetic (so to speak) as before, and that consequently the total disarrangement of its molecules, in pouring it from one glass into the other, had altered little or nothing in that internal state, in virtue of which it is said to be magnetised. Even after the lapse of an hour, when the patient drank the remaining portion of it, the so called magnetism had not entirely disappeared. The water was weaker than at first, but still perceptibly and pretty

strongly impregnated. In this, as in all its other properties, therefore, the water exposed to the sun's rays agreed exactly with that which was impregnated with the new force by means of magnets, of crystals, or of the human hand.

I repeated these experiments, for the sake of greater certainty, with Mlles. STURMANN and REICHEL. The reader will readily excuse me from enumerating the results obtained, which would only lead to tedious repetition, since they coincided with those above detailed.

107. To pursue the parallel further, I placed the rod of German silver in the hand of Mlle. MAIX, and, after she was accustomed to the sensation it produced by itself, (a precaution I invariably adopted, and which it is therefore unnecessary always to specify,) took it from her, exposed it to the sun's rays for some few seconds, and gave it again to the patient. She described it as being imbued with power, exactly as when it had been in contact with the magnet, with a crystal, or with human hands. But along with this well-known sensation she immediately recognised the grateful sunny coolness, the power of exciting which had thus been communicated to and retained by the rod. In several trials, the force continued to excite this feeling for five or six minutes, after which it was not perceptible in the rod; whereas the crystalline force, in its more usual manifestation of warmth, was found to last, in the rod, much longer; about twenty minutes, or just as long as it had continued in the same rod, when charged by my two hands. (§ 82.) The solar ray, in this case, therefore, produced an effect equal to that of my two hands, and the charge given to the rod was retained for the same time as in the case of the hands.

108. I caused Mlle. REICHEL to feel carefully one of my hands. I then went into the sunshine, and, after ten minutes, during which I took care to expose myself on all sides to the sun, I returned to her room, and gave her the same hand. She was much surprised at the rapid change and increase of power which she perceived in it; of the cause of which she was ignorant. The sunshine had obviously impregnated, with its force, my person, just as a magnet had done that of the physician, and in another experiment, my own person.

(§ 74.) Already, at an earlier period, Mlle. MAIX had told me, that she could not bear any one, coming out of strong sunshine, to approach her bed. Some time previously, a whole company of friends, who had been walking in bright sunshine, had come to visit her; but the effect of their presence was so painful and distressing to her, that she was unable to hold out, and was compelled to beg her friends to leave the room. This effect, she said, was simply that of the sun; not the pleasant coolness, but that which causes a wire to appear warm, in short that usually obtained from magnets, crystals, &c.

109. After these experiments, and when I had left Mlle. MAIX, the girls of her family amused themselves by trying some. When I next saw them, they told me, that the patient had found an iron key, which had been laid for some time in the sun's rays, as strongly magnetic as a bar magnet which was in the house. It did not attract iron, but the patient assured me that it affected her exactly as a magnet does. It had therefore been charged with the new force by the sun. The power was not permanent, but had disappeared from the key after some time, just as the crystalline force disappears from bodies charged with it.

110. This observation led the girls to try another experiment, the result of which was remarkable. They placed in the sunshine a horse-shoe magnet, which had become weak in its physiological action, instead of restoring its power by stroking it with another magnet; and had the satisfaction of being completely successful. The horse-shoe was so remarkably strengthened in its action on the patient, that, from that time forward, when a magnet lost its power, or became weak, it was always placed in the sunshine, to restore its activity. This is, in some sort, a confirmation of the observations of ZANTEDESCHI.

111. I endeavoured to complete these experiments by trying, on Mlle. REICHEL, the effect of sunshine on crystals. I found that when a rock crystal, and a crystal of gypsum, were exposed for hardly five minutes to the rays of the sun, Mlle. R. perceived that their peculiar action on her was very much augmented.

112. These facts, when combined, lead us to the law; *that that force of the solar rays, which corresponds to the crystalline force, may be communicated to other bodies.* And since these bodies become charged with it, and retain the charge for a certain time, *they possess a certain degree of coercitive power in reference to the force.*

113. Its conductibility is shown by the fact of its reaching the patient's hand through copper and iron wire. I shall only here add a few facts. When I put one end of a piece of linen, with the usual precautions, into the hand of Mlle. REICHEL, and moved the other end into the sunshine, the sensation of crystalline force soon penetrated to the hand, and there excited coolness. The same result took place with woollen, cotton, or silk cloth, when tried in the same way. When these bodies were moved into the shade, the coolness disappeared in a few minutes, but always returned when they were again exposed to the sunshine. The conduction was most rapid through silk, less so in linen; still less in woollen, and slowest of all in cotton. A bar öf wood, nearly twenty inches long, conducted the force of the solar ray pretty rapidly. With a measure 6½ feet in length, it required more than half a minute before the action had passed from one end to the other. But even a glass tube, one end of which was in the sunshine, rapidly conducted the force of the sun to the other end. *Bodies, therefore, of all kinds*, whether good conductors, bad conductors, or non-conductors of electricity, *possess, without exception, the power of conducting the force derived from the solar rays.* Those bodies, which are continuous in their structure, such as metals, glass, &c., conduct it easily and rapidly; those composed of detached parts, such as woollen and cotton cloth, do so somewhat more slowly, and with more difficulty.

114. I pass over here the experiments made in order to control the above results; and, not to dwell longer on this part of the comparison, I hasten to the luminous appearances. It was here of importance to ascertain, and I was bound to do so, whether the force derived from the sun could give, to objects charged with it, the power of yielding flames visible in the dark to sensitive persons. We know the laws of phos-

phorescence, and according to these, it was out of the question to bring into a dark room an object which had been exposed to the rays of the sun. Indeed, HEINRICH has shown, that in these circumstances a large proportion of all solid bodies gives out light, visible even to the ordinary eye. The arrangement which I adopted, in order to make experiments, during sunshine, in perfect darkness, was the following. From my study, a private stair leads to the lower story, where my collections and instruments are kept. I shut up completely the window which lighted the stair, and when both doors were closed, the stair was rendered perfectly dark. It was easy to communicate with any one shut up in this dark receiver, so to speak; and to hear and understand what was spoken aloud, either on the stair or in either of the two rooms connected with it, above and below. Mlle. REICHEL was so obliging as to allow herself to be shut up in this dark stair; and I mention these details particularly, because a great number of experiments on luminous phenomena, all to be given in their proper place, were conducted here. In describing these, I shall have to refer to the locality now mentioned. This arrangement also supplied the best test of the accuracy of the sensitive observer; who, when in her prison, at a considerable distance either above or below the room in which I was at work, could never know what alterations were there made in the experiments, but could only observe and describe the effects, when and in whatever form she perceived any to be manifested. In my room above, I had, ready prepared, several plates of about 800 square inches, of copper, iron, and zinc; of gilt metals; large pieces of lead foil, and steeped in melted sulphur, and other substances. I attached these successively to an iron wire, 33 feet long and 0.0794 of an inch thick. I then carried the wire through the key-hole of the upper door, (in which only a sufficient aperture to admit the wire had been left,) down the stair to the patient, and caused her to take the wire in her hand, and to point the end of it upwards. After she had remained so long in this position that her eyes had been brought into the state best adapted for seeing in the dark, I moved the plate at the other end into the sunshine, and did this successively with all the substances prepared for the purpose. After rather less than a minute, she saw a slender column of flame rise from the end of the wire, of about from 10 to 13½ inches long, with a breadth of only 0.8 of an inch. It became, towards the upper end, gradually narrower, and at the point little broader than a knitting needle; and it diffused around it a most grateful coolness. When she disturbed the air in speaking, the flame flickered backwards and forwards, as I have described in the flame from the magnet. According as I moved, in the room above, the plates into the shade or the sunshine, the flame fell and rose at the end of the wire; rather more than half a minute elapsing before these changes were effected.—I now removed the plates, and gave the wire to a person, who was to hold it in the left hand. This was my own daughter. When standing in the shade, she caused a small flame to appear at the end of the wire, which diffused warmth, according to the observations previously recorded. But when she stood in the sunshine, the flame soon rose to a height of nearly nine inches; and it now diffused the grateful sunny coolness. As often as she went into the shade, the flame sank to its former size, and gave out warmth. I now tried the effect of bringing metallic plates and other objects, as rapidly as possible, out of the sunshine to the patient in the dark. Without entering on the subject of flaming lights, with which she saw them covered, because these may have been, more or less, if not caused, yet influenced by phosphorescence, it is proper here to mention, that at every sharp corner, particularly at those which pointed upwards, bundles of flame, like those from magnets and from crystals, flowed out. Those from copper were green and blue; from gold and silver, bright white; from tin, dull white; from lead, a dirty blue; from zinc, reddish white; from mercury, white; and from a group of crystals of sulphate of potash, blue with white points. At last I brought a glass tube, forty inches long, and about two inches wide, from the sunshine into the darkness. When it stood vertically, she saw it, at the top, surrounded by a delicate white downy flame, ending, above, in bundles, about 3.2 inches long, which moved or darted upwards all round the tube.—I tried, alternately, polarised light, that which passed through the window panes at an angle of about 35°, and the direct solar rays, for which purpose I used a large balcony, accessible from my study; but no difference in the results could be observed. All these experiments show, *that the force flowing from the solar rays on bodies, produces the same beautiful luminous phenomena as that proceeding from the magnet, or from crystals, &c.*

115. In every point here considered, therefore, *the effect of the sun coincides with that of crystals, of magnets, and of the human hand; and this, the centre of our system, must be admitted as the fourth source of the new influence found in crystals and elsewhere.*

---

116. This observation, as is obvious in its further connection, leads us far. I avoid, however, in the first instance, entering into the infinite variety of its relations to the whole of external nature, because I am desirous, first of all, to mention and to establish all the sources of the force under consideration, one after another; and only then to enter into details on each of them separately, as far as I have hitherto been able to study their peculiarities. Yet I cannot refrain from here alluding at least to one of the many sides on which the subject must be examined, because it is precisely that which establishes more precisely the mode of action of the sun. This is *the Spectrum.* If the sun's rays exert the power here discussed, the question at once occurs, whether the influence resides in all the rays of the spectrum, or only in some, or in several more or less powerfully. I made a preliminary trial with Mlle. MAIX. I threw the spectrum from a glass prism on a wall, gave the patient one end of a copper wire which I held near the other end in my hand, and allowed her to become accustomed to the sensation. I then slowly carried the end of the wire along the spectrum from colour to colour. She could not see me, a partition being between us. Repeated experiments, as well with her as with other sensitive patients, led to the uniform result, that the violet and blue were the chief seats of the grateful sunny coolness which diffused itself

over the whole body of the patient. This part of the spectrum is that in which the intensity of the light is the least. But the more common manifestation of the crystalline force, on the other hand, namely, the sensation of warmth, nay, of heat in the wire, which was more than sixteen feet long, increased steadily from the yellow in the middle towards the orange, and became strongest deep in the red. Here we are at the point where the maximum of calorific rays is found, the true heat of which, however, was quite unable to reach the patient. These observations support those of MORICHINI and Mrs SOMMERVILLE, and tend to increase the probability of their results, which are not yet generally admitted.

117. *The two ends of the spectrum had, therefore, each its own specific, strongly marked action on the sensitive nerve.* The further and more detailed investigation of this part of the enquiry cannot fail to give us much interesting information, and will form the subject of one of the subsequent treatises.

---

118. It was only a step from the sun to the moon, the study of which was so much the more indispensable, as it is known that numberless phenomena on the earth, both among healthy and diseased persons, are more or less dependent on our satellite, from causes yet unexplored. I tried the first experiment with Mlle. MAIX. It was not to be managed without difficulty, for her windows looked north, and it was impossible from them to see the moon. In this difficulty, I resolved to carry an iron wire, nearly 0.08 of an inch thick, through two rooms, then across a court, and then again through three other rooms, to a distance of about a hundred feet. Only thus were we enabled to catch some lunar rays. I gave the patient one end of the wire, and attached the other to a large copper plate, which, with the usual precautions, was exposed to the rays of the moon. After a short time, the sensation in the hand began to change remarkably. Iron and copper wire alone had always caused warmth. The effect of the moon, associated with the other, was described by the patient as of a mixed kind, so that her statements had not their usual clearness. Without dwelling on the details of this preliminary experiment, let it suffice that

a powerful action of the moon actually took place, and was conducted to the patient through a long wire. The sensation in her right hand was much more agreeable than in her left. But that which did not occur with the sun, and appeared peculiar to the moon, was a decided attraction towards the wire in the whole arm, so that she was, as it were, solicited to carry her finger along the wire. She passed her finger slowly along it when under this influence; and had she not been confined by paralysis to bed, would have followed it, yielding to this desire, along its whole length. We here meet with something of that strange attraction which the magnet exerts on cataleptics, as we have seen; and it can hardly be doubted that this is the irresistible attraction which acts so violently on lunatics, and which can be conducted along a wire. The patient declared it to be purely magnetic; but the attraction was, in this experiment, much stronger than with the magnet. I must, however, repeat, that the hand of Mlle. MAIX never perceptibly adhered to, or followed the magnet. The peculiar difficulties of the locality, which rendered the continuance of such nocturnal experiments impracticable, prevented me from more minutely investigating these interesting phenomena in Mlle. MAIX. I was therefore compelled to seek for confirmation and further elucidation by means of other sensitives.

119. I first tried Mlle. REICHEL, who greatly assisted me in making many varied experiments. When I placed any object in her hand, and begged her, after a short time, to move the opposite end into the moon's rays, she instantly perceived the same sensations as when I had placed on the object the point of a crystal, or a magnet, or my finger, or as when it had been exposed to the sun. *This susceptibility and conducting power were found in all bodies;* but the sensation was not cool but tepid; and the sequel will show that this patient distinguished more clearly than all with whom I experimented, and also more consistently with objective reasons, between the sensations of coolness and warmth, in regard to which sensitive persons often speak doubtfully. When I caused her to take hold of the German silver rod, after a little to lay it down, then to push it into the moon's rays, then to move it into the

shade, and, after a time, to take it up again,—*she found it charged with the force flowing from the moon, that is, warm.* That a charge could be *communicated* and *retained*, was therefore proved. When I allowed copper plate, leadfoil, tinfoil, zinc plate, silver and gold surfaces, to remain some time in the moonshine, and presented them to her on the darkened stair, she saw at their corners bundles of flame, white, red, green, and blue. When I arranged a metallic plate of eight hundred square inches, so that I could at pleasure place it in the shade or in the moon's rays, and attached to it a long wire, which was carried through the key-hole to Mlle. Reichel on the dark stair, *she invariably saw, when the rays were allowed to fall on the plate, a narrow flame, hardly a finger's breadth, rising vertically to a height of about ten inches;* and this disappeared, after a short interval, as often as the plate was moved into the shade. She found this flame always warm. These experiments were repeated with the same results during three full moons.

120. From all this it follows, that the moon's light is not mere moonshine; that, although it yields us no heat, yet, along with its light, *it possesses a powerful force, which exhibits the same properties as that residing in crystals, &c. The moon is therefore a fifth source of this influence.*

---

121. Since the calorific rays of this spectrum had so remarkable an effect in increasing the peculiar action of the force now under discussion, I endeavoured to pursue further the investigation of the influence which heat might have upon it. I had at first studied the force in a kind of equilibrium, as in the magnet, in crystals, and in the human body. But now, in the sun and moon I had met with it, no longer at rest, but in a state of motion. It appeared to flow down directly from these heavenly bodies, just as we may imagine, leaving out of view the undulatory theory, the rays of light and of heat as flowing from them. I was therefore naturally led to the consideration of analogous natural influences, and first of all, of heat. To this end, I placed a copper plate on a wide vessel of earthenware, connected it with the hand of Mlle. Maix, by means of

a long copper wire, in the usual way, and laid on the copper a hollow brass ironing apparatus, with its iron heater, the latter being cold. I placed my right hand on it, and allowed the patient to become accustomed to the sensation. The heater was then removed, and replaced by another, which was at a low red heat. I first held the apparatus near the copper plate, but without contact, so that radiant heat alone could reach it. The patient immediately felt, proceeding from the wire, an increase of the usual feeling of warmth caused by crystals, &c. When I now laid the hot apparatus on the plate, and moved it slowly over the surface, to diffuse the heat over a larger space, the sensation increased rapidly and strongly as the heat extended. The patient, at the same time, complained of a singular feeling of weight in the hand. When the heating apparatus was removed, and cooling took place, the sensation diminished, and so on alternately as I applied it again, or removed it.

122. In another similar experiment, I placed in the patient's hand one end of a stout iron wire, which I held myself near the other end, and allowed her to become accustomed to it. I then applied the flame of a candle to the further end, and heated it gradually till a blue tarnish appeared. The conducted heat did not reach as far as my hand, and from my hand to that of the patient there were nearly four feet of wire; there could therefore be no communication of ordinary heat. But the sensation so often mentioned very soon appeared, increased rapidly, and soon attained such a degree, that it passed through her arm up to her head. It slowly disappeared when the flame was removed, and appeared again as before, as often as the flame was applied to the end of the wire. I repeated the experiment with copper wire, but varied it by making at the further end a compact coil of ten turns. To this coil I applied the flames of two wax candles. The result was quantitatively greater, but qualitatively the same as before; and this was observed on every repetition. I filled a wooden vessel with cold water, introduced into it the end of the wire; and when the patient had become accustomed to it, I poured off the cold water, and replaced it by hot. She instantly per-

ceived in the hand the impression of a powerful current of warm crystalline force.

123. I now tried the opposite experiment of introducing into the hot water, from which the wire went to the patient's hand, a lump of ice. Immediately the phenomena changed their character. The sense of warmth and accompanying sensations rapidly diminished; there arose a sense of dragging through the hand and arm; the unpleasant feeling belonging to the warmth was replaced by the refreshing coolness caused by the sunbeams, and this gradually diffused itself over the breast, the back, and the whole person. Ice, when placed directly in the hand of the patient, instantly caused spasms, and prevented further observations.

124. I made confirmatory experiments with Mlle. REICHEL. I heated, with the usual precautions, by means of the flame of a candle, one end of an iron wire, six and a half feet long, the other end of which she held. Ordinary heat could not, at that distance, reach her, and least of all from so feeble a source, and in the course of a few minutes. Yet the wire immediately appeared to her *to become first warm, then hot,* in such a degree, that she wondered how I was able to hold it so much nearer the flame. I felt, however, not the slightest rise of temperature where I held it. At the same time, she perceived a *cool aura flowing from the end of the wire.* All these characters are, as we have seen, found in the active crystalline force, &c. I repeated this experiment with an iron wire of the same thickness, but fifty feet long, and obtained the same results, yet with a perceptible retardation.

125. The next question was that of the luminous phenomena. I placed Mlle. REICHEL in the dark stair, conducted a thick copper wire to her hand, and heated the other end of it by means of an Argand's lamp. The flame which appeared at the end of the wire was composed of red and green light; it rose to the height of nearly four inches when the heat was strongest, and sank and rose alternately, as I removed or applied the lamp. I tried a similar experiment with a thick iron wire, nearly five feet long, the end of which I heated to redness by means of the lamp. There arose at the other end a

flame of nearly six inches, which slowly sank as the wire cooled. An iron wire of fifty feet in length, heated to redness at one end, gave, according to the patient, at the other a flame of a finger-length, on the dark stair. *The existence of luminous phenomena, visible to sensitives, caused by heat, is thus placed beyond a doubt.*

126. I was, for the present, satisfied with these proofs. They establish the existence of *effects produced by heat, both ordinary and radiant, which entirely coincide with those which demonstrate the existence of the force residing in crystals, &c.;* effects both felt and seen by different individuals. *Heat is, therefore, a sixth source of the new force.*

---

127. Friction, or the rubbing of bodies on one another, is, in its action, complicated; the results being influenced by the heat, electricity, galvanism, &c. produced in the process. But I thought it, nevertheless, right to enquire into its relation to my subject. I first laid a copper plate on a deal floor, connected it with the hand of Mlle. MAIX by a long copper wire, and then gently rubbed the plate with a board of wood. She immediately perceived rapidly increasing warmth, which rose to apparent heat when I pressed the board more firmly down, and rubbed it more rapidly on the plate. The sensation increased and diminished regularly as I rubbed more or less. When I rubbed the plate with a woollen coat, the result was the same, but the sensation stronger. With silk the effect was more powerful still.

128. I connected with the hand of Mlle. REICHEL, by a brass wire, a copper plate which lay on a floor of oak, waxed, as is usual on the Continent. I placed on the copper a piece of wood, and rubbed the former with the latter. She immediately felt the sensation of the excited force in the wire held in her hand. Zinc plate, treated in the same way, acted like the copper, but not so powerfully. In the dark the wire from both exhibited a flame, visible to the patient. I now sawed a bar of wood with a fine saw, in the dark chamber. The observer saw nothing particular in the particles of saw dust, as they flew about; but the saw, as far as it was exposed to

friction, very soon appeared luminous and red, as if red hot, while she saw a small flame sprouting from each tooth of the saw. Copper and zinc plates, rubbed together with the hand, only shewed here and there a spark. The same took place when zinc was rubbed on zinc, or copper on copper. When two pieces of gypsum were rubbed together, no light appeared. When two pieces of charcoal were rubbed, they appeared to her, at the points of contact, as if red hot, to the depth, on both sides, of a finger's breadth. Two pieces of sugar yielded the well-known light visible to all; but besides this, she saw a flame of about 0.15 of an inch long surrounding the light. Two bottles of glass, when rubbed together, appeared to me at the points of contact fiery; but she saw these points clothed in a flame as large as the first. Rough unglazed porcelain crucibles also yielded, but only at the points of contact, a bright light, visible to me; the patient saw, besides, flames as large as the outstretched hand. At this time she was in such good health as daily to go about her professional occupations through the crowded streets of Vienna.

129. I crossed two glass tubes, each about forty inches long, and rubbed them on one another. Along the line of friction I saw, in the dark, a long stripe of fire. Besides this, Mlle. REICHEL saw delicate flame-like lights, of a finger's breadth, playing round the tubes, where they were rubbed, and producing the appearance of a ribbon of fire. She found the end of the tubes, twenty inches from the point of friction, apparently very hot, as long as the rubbing was continued. This sensation immediately disappeared when I ceased rubbing. At the end or edge of the tubes she saw flames of a finger's length flowing out, which diffused far around a tepid aura. Two bars of iron yielded similar results, namely, fire on the line of friction, not, however, visible to me; a sensation of warmth during the friction, ceasing when the friction stopped; and flaming emanations from the ends of the bars, diffusing a tepid aura.

130. In all these experiments the objects were never isolated, but either lay on the floor, or were held in my hands, or those of an assistant, so that the electricity excited by

friction had always a ready means of escape. The heat caused in the objects by friction could not possibly, when the friction was arrested, disappear so rapidly as the flames sank to the eye of the patient. The electricity of contact, which could be excited in these cases, in which, generally, similar substances were rubbed together, must have been so trifling that it might be neglected. Indeed, in the instance in which zinc and copper were rubbed together, and where contact electricity must have been excited, there was hardly a trace of light produced. Galvanism, therefore, can as little have been concerned in these phenomena as friction electricity. For the same reasons, I consider the possible thermo-electricity as being here not powerful enough to account for the phenomena in the amount and degree observed; but I rather hold that, along with the partial influence which the above-mentioned forces may have exerted, the greater part of the peculiar luminous phenomena here perceived by sensitive persons, is to be ascribed to the friction in itself. And thus I consider, although with less confidence than in the preceding cases, *that friction must be regarded as a seventh source of the influence residing in crystals, &c.*

---

131. That light occupies an important place, when we consider the origin of this peculiar force, we have already seen in the researches on the solar and lunar rays. But whether the new force be inherent in light, or only associated with it, or whether it proceed from the radiant forces which accompany light;—these are questions which are no doubt of essential importance, but which would be premature in this place, where we are chiefly occupied with the sources of the new force in general. The study of the special characters of these sources must be left for future investigations. I wish, therefore, here only to ascertain, whether light, in general, be one of these sources. I have hitherto only examined natural, not artificial light. When, in bright day-light, I brought a lighted wax candle near to Mlle. Maix, she perceived a peculiar coolness caused by it. Several candles increased this coolness, so that it pervaded her whole person. I removed, step by step, the

candles to the end of two adjoining rooms, together twenty-four feet long. The coolness was at this distance much diminished, but still in some degree perceptible. She described it as obviously analogous to that produced by a wire, the end of which was in sunshine. This observation, which to herself was quite unexpected, led her to remember, that, in attending certain ceremonies of the Catholic church at certain periods, and which consist in powerful illumination at night by means of hundreds of wax candles, as in the illumination of the representation of the sepulchres of saints, &c., she had never been able to hold out. The lights had invariably so chilled her, to the marrow of her bones, as she expressed it, that she was compelled to leave the church. Now, Mlle. Maix has suffered during nearly the whole of her life, in a less degree, from the disease which has now become so severe, and is to be regarded as a born sensitive, who was subject to the sensations peculiar to that state at every period of her life, even when she seemed healthy, and was able to walk about. This peculiar effect, on her, of light from distances at which the radiant heat could only have been very feeble indeed, producing, besides, an action on the nerves exactly opposite to that of heat, was strongly felt by her at all times, and at a period when no one suspected that it could indicate any morbid state.

132. Experiments with Mlle. Reichel yielded the same results. She perceived the coolness from a wax-light at a considerable distance. Two lights acted at nearly double the distance; an argand lamp still further; and this last furthest of all when a ground glass globe was placed over the flame.

133. I tried, in the way already often described, whether the cause of these sensations could be conducted through other bodies, or communicated to them. I placed two wax candles near a copper plate, connected, as usual, with the hand of Mlle. Maix, but in such a position that she could not see the flames, and their rays, therefore, could not fall on her. She perceived, at the same time, increased warmth in the wire, and the refreshing coolness which she experienced from the sun's rays; but much more feebly than from the sun. This was

often tried, with the same result. I made a similar experiment with Mlle. REICHEL. Eight stearine candles were placed near a large copper plate. The patient was in the next room, and connected with the plate by a copper wire. She perceived the action very strong, and felt the cool aura from the end of the wire at a considerable distance. I then introduced, in a second experiment, a glass plate between the lights and the copper plate; but the effect was very little diminished.

134. These results, according to which the rays of artificial light, on the one hand, directly affected sensitive persons, and, on the other, communicated the force residing in them to other bodies, exciting in them apparent change of temperature, while this force was conductible, and produced luminous phenomena in the dark, visible to sensitives, lead me to the conclusion, that not only the sun and moon, *but also light in general, is a source of the influence under consideration, as found in crystals, &c.; and it is the eighth.*

RETROSPECT.

a. The solar rays possess a power of acting on sensitive patients, coinciding entirely with the force residing in magnets, in crystals, and in the human hand.

b. In the spectrum, the strongest manifestation of this force is found at the outer end of the red and of the violet.

c. The moon's rays are very rich in this peculiar force.

d. Heat is also a source of it.

e. It is excited by friction; and

f. Appears to exist in artificial light.

FIFTH TREATISE.

CHEMICAL ACTION. ELECTRICITY.

135. WHEN we look back to the state of chemistry in the times of AGRICOLA, KUNKEL, and BRAND, we may form some idea of the relation in which that science now stands to the subject of these researches. Scattered fragments of observations lie around; but in what a state! I shall illustrate this by one example. For about seventy years an apparatus has been occasionally used in medicine, called the *magnetic baquet.* I can hardly venture to describe it, because it will prove an abomination to every one accustomed to scientific investigation into natural truths. A tub of wood is filled with a hotch-potch of the most absurd and senseless kind, which is stirred about with magnetised water. A rod of iron is now introduced, and from it woollen threads are carried to various patients, who are thus to receive a health-bringing current of vital magnetism from the *baquet.* Now, this extraordinary mixture consists of iron slags, broken glass, steel filings, roots, iron ores, grains of corn, sulphur, saw-dust, glass plates, wool, old iron, aromatic herbs, mercury; all magnetised and mystically stratified. One naturally, and with reason, asks: Can any thing conducive to health, or any thing intelligible, proceed from such a diabolical cauldron? But how a magnetic effect should be exerted by such a stew, all natural philosophers will certainly agree with me in thinking utterly incomprehen-

sible. And yet all who have occupied themselves with magnetic cures are unanimous in stating, that it is a permanent and ever-flowing fountain of magnetism, which can be conveyed to the patients by the conductors, &c.

136. That the *baquet* is neither a galvanic nor an electric, and far less a magnetic apparatus, is obvious to every one who knows even the elements of these things; and yet it produces an influence which has been known for seventy years, has some analogy with magnetism and electricity, and which must depend on some unexplored cause, whatever that may be, otherwise its turbid existence could not by any possibility have been prolonged to our days. If I asked myself, what could take place in the *baquet*, I could only see my way to one agent, namely, *chemical action.* Space had been blindly given for a confused play of affinities; and combinations and decompositions must go on slowly in the *baquet.* In the preceding treatises I had discovered eight different sources of one and the same power. Now here, according to the statements of physicians, this same force flowed from a mixture of substances, many of which act on and decompose each other. Did chemical action, in itself, then, set in motion the same imponderable agent? Was chemical action another source of the power residing in magnets, in crystals, in living men, in the sun, in heat, &c.?

137. To determine this, I took a glass of water, dissolved in it bicarbonate of soda, introduced the end of a copper wire five feet long into it, gave the other end to Mlle. MAIX, laid a pinch of powdered tartaric acid on the edge of the glass, and, when her hand was accustomed to the wire, sprinkled the acid into the solution. As soon as the decomposition began, the same sensation of warmth, followed by coolness, was perceived as when I touched the end of the wire with my ten fingers, with the point of a large crystal, or with a bar magnet. It became so strong that it produced flushing of the face. It continued uniformly as long as the action lasted, and ceased when it stopped.

138. I had now seized the thread of the new force in a chemical process; but it was necessary to make quite sure.

I had first to meet the objection, that the electricity developed in the process had acted on the sensitive patient. Chemists are not agreed whether, in such a case as the above, any electricity becomes free or not. LAVOISIER and LAPLACE, and more lately GMELIN, thought they had discovered the developement of negative electricity during the decomposition of carbonate of lime by sulphuric acid. PFAFF and others deny this. Without here reckoning up authorities, I thought it safest, in the concrete case, to appeal to direct experiment. I therefore attached a conducting wire to a Bohnenberger's Electroscope, and introduced it into the liquid, isolated in a long stemmed glass, in which I mixed the tartaric acid and bicarbonate of soda. The gold leaf did not move. I then attached the plates of the condensator, and added fresh portions of the salt and acid to the water; but even now, in separating the plates, not a trace of motion in the gold leaf could be detected. If, in so tumultuous an action, free electricity is developed, it is not at all probable that it is so in inappreciably small quantity; at least certainly not from the purely chemical part of the process. I must conclude, therefore, that electricity does not become free here, where it cannot enter into circulation; and that that portion of electricity which is set free by the chemical act, is, according to known laws, again rendered latent at the moment of its production. The sensation in the patient's hand, therefore, cannot be caused by a current of electricity, and belongs obviously to the class of phenomena investigated in this work.

139. I return to the account of my observations. A glass of diluted sulphuric acid and an iron wire were put into the hand of Mlle. MAIX. After a pause, I introduced the wire into the acid, when it began to dissolve with abundant disengagement of hydrogen gas. She instantly felt the wire apparently warm, then hot; while a cool air diffused itself round the glass.

140. A glass of water, and some common salt in a paper, were given to her, and a short time allowed to pass. I now threw the salt into the water, which she gently agitated by moving the glass. She felt the glass, during the solution of

the salt, become more strongly charged with the force, an then it was stationary. The sensation reached up the arm.

141. I tried similar experiments with Mlle. Reichel. First, that with tartaric acid and bicarbonate of soda; next, that with iron and sulphuric acid; and then, that with salt and water. All gave the same results as with the other patient. I also tried with her the following mixtures:—Sulphuric acid and caustic soda ley; iron filings, with an excess, successively, of acetic, tartaric, fumaric, citric, and hippuric acids; carbonate of soda, with excess of sulphuric acid, &c. We had some newly-pressed must of grapes, which was in full fermentation; and this, also, I tried. All these actions yielded abundant developement of the new force.

142. I next tried solutions; as of sugar, alcohol, borax, in crystals, carbonate of soda in crystals, and potash, in water; then effloresced borax, effloresced carbonate of soda, sulphate of calcium and potassium, and quicklime in water. All of them caused partly warmth, partly coolness in the wire, continuing till the solutions were completed, and then the peculiar effect rapidly ceased. *Everywhere, therefore, even where mere solution of water or combination of water of crystallisation occurred, chemical action developed, in an active state, the new force.*

143. I was desirous to see whether a glass of water could be magnetised, as it is called, by chemical action. The mere conduction by a wire could hardly suffice; I required the whole force of the action. To this end, I placed one glass within another. The interior glass contained water, the other, solution of carbonate of soda. I now added to the latter solution tartaric acid, and had the mixture well stirred by female hands, till the action was complete. The inner glass was now taken out and given to Mlle. Maix to drink. It was found as strongly charged as by five minutes, but not so strongly as by twenty minutes of sunshine. After she had tasted it, I subjected the glass once more to the same process; and it was now found nearly twice as strongly charged as before. I made a similar trial with Mlle. Reichel, using carbonate of potash and sulphuric acid, and with the same result. *We are,*

therefore, able to magnetise water by means of chemical action, just as well as by the magnet itself.

144. During these experiments, I caused both patients to hold one end of different copper wires, the other end of which dipped into diluted sulphuric acid. One of these wires was nearly 270 feet long. But even at this great distance the effect was perceived, and Mlle. REICHEL could tell, at her end of the wire, when the other end was taken out of, or put into, the acid; and this, not instantly, but after an interval of fifteen or twenty seconds.

145. It remained yet to examine the luminous phenomena. I made numerous and careful experiments on this point with Mlle. REICHEL. I tried a series of solutions and dissolutions, (as they used to be distinguished) on the one hand, alone, and on the other, with reference to their effect on the further end of a conducting wire, in the dark. Sugar, carbonate of soda, and borax, were separately dissolved in water, and stirred with a glass rod. Even before stirring, the contents of each glass were luminous, giving a red light. A delicate light, flowing upwards, began to play over the liquids. A long luminous bundle rose from the upper end of the glass rod. The pieces of sugar, when dropped into the water, became immediately luminous with red light, and fell to the bottom like red glowing masses. The developement of light, caused by solution, therefore, instantly began when the sugar came in contact with the water. When I stirred it slowly in the dark, I myself saw bright flashes of light whenever the rod rubbed against the sugar. I could hardly look on these as electric, for here the external surface of the sugar was softened and half dissolved. The light produced by rubbing together in the air two pieces of sugar, chalk, &c., is generally regarded as electric, but the proofs of this are very deficient. I placed freshly burnt quicklime in a saucer, and allowed water to fall upon it. As soon as the interior movements connected with the slaking commenced, and vapour began to rise, the whole mass of lime appeared to the observer white hot, and a dull blue flame rose over it to the height of a hand. She supposed the dull appearance to be caused by the steam. These flames

continued about a quarter of an hour after the chemical action was over, in the same degree; they then began to fade, and ceased after half an hour. Sulphuric acid, poured into water, immediately formed red flames in the glass, which played over the water. But when I stirred the mixture, they increased so much, that they rose to the height of a full span above the glass. Here, as also in the last experiment, the effect of heat was added to that of the chemical process. Even the glass rod exhibited fiery bundles at its extremity. These luminous emanations continued nearly an hour. The fermenting must gave continually a turbid yellowish flame.

146. An iron wire, 100 feet long, was passed through the key-hole to Mlle. Reichel on the dark stair, and the outer end was dipped in diluted sulphuric acid. After the lapse of half a minute, she saw a long narrow column of fire, one and a half spans long, rise from the end of the wire; it fell and rose, according as the wire was taken out of or immersed in the acid. The same result happened when the outer end of the wire was immersed in a glass in which sugar was dissolving in water. The flame at the end of the wire appeared to her even larger than that from the iron and sulphuric acid. In the next experiment, a brass wire about fourteen feet long was used, and gave the same result, with this difference, that where the flame from the iron wire was white and reddish blue, that from the brass was white and green. Quicklime stirred with an excess of water, gave at the end of the iron wire a flame of a span in height.

Everywhere, therefore, *where chemical action took place, light and flame, visible to sensitives in the dark, also appeared.*

147. In considering the new force in its relations to chemical action, I cannot pass over the flame of burning bodies, since combustion is a very intense chemical process. But as chemical action is here associated with light and heat, there is, for the present, no prospect of obtaining unmixed results from flame. I might, with equal propriety, have described the experiments which I made on this point, under the heads of heat and light. I brought to Mlle. Reichel a pan of

glowing charcoal. She felt it cold at about forty inches, and perceived the coolness which it diffused, when at the further end of the room. I have already stated the effects of lighted candles on the sensitives. I set fire to a saucer full of spirit of wine, and to another of pure alcohol, and allowed them to burn out near the patient. She found both flames cold, even at small distances. I next burned in her presence various substances, taken from different parts of the electro-chemical series; such as resin, sulphur, and globules of potassium. She found all of them, especially the flames of the two last, very cold. We cannot here easily determine which of the emanations from the flame excited in the sensitive patient the coolness, or its cause, the crystalline force. Light appears to have, comparatively, no chief share in doing so, because the effect of the flames of alcohol and of sulphur, which are very little luminous, was not inferior to that of the bright flame of stearine candles. Again, heat, as a source of the new influence, almost always excites a sensation of apparent warmth, as we have seen in the preceding treatise. Since, therefore, the flame always yielded a predominating sense of coolness, the cause of this must either reside in the products, or, as is most probable, in the chemical action itself, just as in chemical processes not accompanied by flame.

148. Immediately after this investigation of chemical action, we shall come to one on the voltaic pile as a source of the new influence. Since the hydro-electric column or battery is one of the most important seats of chemical action, I shall here venture to anticipate the result of that investigation; namely, that there also, from the process of chemical change, a great fountain of the crystalline force flows, with all the attributes which we have seen it everywhere to possess. I must make this inroad upon the systematic order of my work, because I have to include in the concluding proposition of this section, all kinds of chemical processes.

149. The parallelism between the properties of the so called magnetic phenomena, found associated with chemical action, and those which we have seen to exist in the magnet, in crystals, in the human hand, in heat, in light, &c., is perfect;

and chemical action takes its place in the series as the ninth source of the new influence.

---

150. The field of these researches here opens up to a vast and almost unlimited extent. But this can hardly surprise us; on the contrary, such a result was naturally to be looked for, and was dimly visible on the horizon, when I made the first observations on crystals. Chemical action, in the various forms of solution, decomposition, combination, and changes in the grouping of the elementary molecules, is so closely related to the destruction and reconstruction of crystals, that, as soon as such a force was seen to reside in crystals, it must be necessary connection with those forces and actions, which effect the union and the separation of the molecules of matter. It was therefore to be supposed, *à priori*, that chemical action would here react efficiently. I hope that it will furnish us with a means of concentrating the new force, and of rendering it more manageable than I have hitherto been able to do. I am confined, as yet, to the sensibility of sensitive persons; and it is to be hoped, that above all things, chemical action may yet supply us with what at present is a desideratum; namely, a more convenient reagent for the new force, as well as a sure measure of its relative amount or intensity in different cases.

151. Let us now look back to the magnetic *baquet.* This strange and incomprehensible thing now loses its mysterious character, but only, no doubt, at the cost of becoming still more absurd and ridiculous than before. It is nothing else than a source of chemical action, produced, as it were, accidentally by the throwing together of almost any substances, and going on slowly and continuously. From it the desired force flows, just as from a very slow fire, heat is slowly diffused. It is a slow current, derived from chemical action, of the peculiar influence which flows from magnets, crystals, the human hand, &c.; an influence very improperly called, by the writers on the *baquet*, Animal Magnetism. We can now understand why the *baquet*, when it has gradually become dull in its action, becomes again energetic, when, after some months, it is again

stirred about. Fresh surfaces are then exposed, so that new actions begin. We see also, why every new practitioner could employ a different hotch-potch, and yet obtain the same effect; because it is indifferent what substances act on one another, provided they only ferment, or decompose each other. Lastly we see, why one of them, who filled the *baquet* with glass and water, obtained but little effect, and others who followed him produced no effect at all; because glass and water, no matter how many magnetic charms or spells may be muttered over them, do not chemically act on one another; and so forth. The whole mysterious superstructure of the *magnetic baquet*, which ever since the days of MESMER has largely contributed to render animal magnetism ridiculous, will therefore, in future, be discarded, and replaced by any simple, slow, and continuous chemical process, probably most advantageously by a voltaic battery, of which the circuit is not closed. In this way, we shall have it in our power, to obtain at pleasure a strong or a feeble current of the new force, in numberless different forms, and in the smallest space.

152. But the considerations here brought forward on chemical action, lead us in another and perhaps more interesting direction. They direct us to the source from which, in all probability, the human body itself draws its supplies of the so called magnetic force; to the place where those flames are kindled, which flow from the points of our fingers, and as we shall hereafter see, from other nobler parts of our wonderfully constructed frame. This source is *digestion*. After it has been shown, that in chemical action, in the play of varying affinities, one chief origin of the new force is to be found; and since digestion is nothing else than a process of chemical change, going on, under the influence of the vital force, uninterruptedly during our lives; a constant chemical decomposition and combination;—it necessarily follows, that as we have seen in every, even the feeblest chemical action, the solution of sugar, salt, alcohol, and sulphuric acid in water, the new force must be liberated uninterruptedly along the course of the intestinal canal, and placed at the disposal of the whole organism. But the process does not stop here. The matters

digested in the stomach are further elaborated in the intestines, and there undergo resorption, and chylification, and are mixed with the lymph and the blood by means of innumerable large and small vessels. In the blood, they are subjected to further chemical changes, are carried further and further, and still more and more chemically changed, and are thus continually made use of as inexhaustible stores of crystalline force, in numberless decompositions, till at last they are expelled from the body in various forms.

153. All that I have here said of digestion applies, with a change of terms but with equal force, to *respiration*. It supplies us with oxygen, which is carried by the blood to all parts of the system, there keeps up the general change of matter, and yields, as one of the chief results of the oxidising process, the animal heat. The so called animal magnetism is associated with the animal heat, and comes, goes, flows out, and is dissipated, like it. The chemical action of the body, which yields heat, yields also, as we have seen, crystalline force, magnetism, or whatever we choose to call this peculiar influence. These influences, which constitute the inmost life, whether they be material or immaterial, are perhaps, ultimately, one and the same; and if they here occur to us going hand in hand, this is an additional security for our being on the right track in our pursuit of them.

We have thus traced the origin of the force which flows, in a polar form, from our hands and fingers as from a magnet. We see where it is ever renewed, and whence it, ever burning, blazes out from us. Chemical affinity, which is, to an infinite extent, employed in our bodies, produces, sets free, and supplies this force; and when and wherever we meet with it, we find it exhibiting its innate polarity.

154. By one of the most admirable combinations of profound thoughts to which our age can point, M. Liebig has led us to the conception, that all our motive power is derived from the processes of digestion and nutrition, and all our animal heat from respiration. In other words, both our motive power and heat are derived from chemical action. Although this cannot as yet be expressed, either in an algebraical formula,

or by means of chemical symbols, and although objections may here and there be brought against the form in which this great thought has been expressed, yet the conception of LIEBIG appeals so strongly to our intellects, and finds so powerful an echo in the whole of what we hitherto know of nature, that its final triumph is tolerably certain. And indeed, since nature supplies us, for our existence, only with food and air, we are naturally led to suppose that she has intended us to obtain from food and air every thing which is necessary to our existence. If, for example, we find, on careful calculation, that we daily, on the one hand, assimilate fourteen ounces of carbon from our food, and, on the other hand, give out, in respiration and otherwise, the same quantity of carbon; if we find, further, that we daily absorb forty-seven ounces of oxygen, and give out as much; if we find, lastly, that these quantities of carbon and oxygen are given out in combination, and that their union gives rise to an amount of heat equal to that which we daily require;—such a deduction has so much force, that we may anticipate that doubt will soon give way to conviction.

155. I consider it no trifling guarantee of the essential truth of my researches, that I have been led, as we have seen, by quite a different series of observations and deductions, to the same new field of research on which M. LIEBIG was already occupied. Chemical action yields abundantly the active principle of the crystalline force; the human body overflows with this power. Man digests and breathes; change of matter, combination, and decomposition, occur within him; chemical action is therefore operative at every moment in the human frame. It is then not only plain, but a logically necessary consequence, that man draws from the play of chemical affinities, in other words, from chemical action, that mysterious influence, the existence of which is proved by the present researches. It is possible to doubt whether the sun supplies us with light, because it is often day-light when no sun can be seen. I know this well, and do not contend against such doubts.

156. At the close of this section, I now bring forward a

useful application of the facts already ascertained, which is to me so much the more welcome, as it tears up one of the chief roots of superstition, that mortal enemy to the progress of human enlightenment and liberty. A case which occurred in the garden of the blind poet PFEFFEL, has been widely circulated by the press, and is well known. I shall here mention so much of it as is essential. PFEFFEL had engaged a young Protestant clergyman, of the name of BILLING, as amanuensis. The blind poet, when he took a walk, held BILLING'S arm, and was led by him. One day, as they were walking in the garden, which was at some distance from the town, PFEFFEL observed, that as often as they passed over a certain spot, BILLING'S arm trembled, and the young man became uneasy. He made enquiry as to the cause of this, and BILLING at last unwillingly confessed, that as often as he passed over that spot, he was attacked by certain sensations, over which he had no control, and which he always experienced where human bodies lay buried. He added, that when he came to such places at night, he saw strange (*Scotice*, uncanny) things. PFEFFEL, with the view of curing the young man of his folly, as he supposed it to be, went that night with him to the garden. When they approached the place in the dark, BILLING perceived a feeble light, and when nearer, he saw the delicate appearance of a fiery ghost-like form hovering in the air over the spot. He described it as a female form, with one arm laid across the body, the other hanging down, hovering in an upright posture, but without movement, the feet only a few hands-breadths above the soil. PFEFFEL, as the young man would not follow him, went up alone to the spot, and struck at random all round with his stick. He also ran through the spectre, but it neither moved nor changed to BILLING'S eyes. It was as when we strike with a stick through a flame; the form always appeared again in the same shape. Many experiments were tried during several months; company was brought to the place, but no change occurred; and the ghost-seer adhered to his earnest assertions; and, in consequence of them, to the suspicion that some one lay buried there. At last PFEFFEL had the place dug up. At a considerable depth, they came to a firm layer of white lime, about as long and as broad as a grave, tolerably thick; and on breaking through this, the bones of a human being were discovered. It was thus ascertained that some one had been buried there, and covered with a thick layer of lime, as is usually done in times of pestilence, earthquakes, and similar calamities. The bones were taken out, the grave filled up, the lime mixed up with earth and scattered abroad, and the surface levelled. When BILLING was now again brought to the place, the appearance was no longer visible, and the nocturnal ghost had vanished for ever.—It is hardly necessary to point out to the reader what I think of this story, which caused much discussion in Germany, because it came to us on the authority of the most trustworthy man alive, and received from theologians and psychologists a thousand frightful interpretations. To my eyes, it belonged entirely to the domain of chemistry, and admitted of a simple and clear scientific explanation. A human corpse is a rich field for chemical changes, for fermentation, putrefaction, gasification, and the play of all manner of affinities. A layer of dry quick lime, compressed into a deep pit, adds its own powerful affinities to organic matters, and lays the foundation of a long and slow action of these affinities. Rain water from above is added; the lime first falls to a mealy powder, and afterwards is converted, by the water which trickles down to it, into a tallow-like external mass, through which the external air penetrates but slowly. Such masses of lime have been found buried in old ruined castles, where they had lain for centuries; and yet the lime has been so fresh, that it has been used for the mortar of new buildings. The carbonic acid of the air, indeed, penetrates to the lime, but so slowly, that in such a place a chemical process occurs which may last for many years. The occurrence in PFEFFEL'S garden was therefore quite according to natural laws; and since we know that a continual emanation of the flames of the crystalline force accompanies such processes, the fiery appearance is thus explained. It must have continued until the affinities of the lime for carbonic acid, and for the remains of organic matter in the bones, were satisfied, and finally brought into equilibrium. Whenever, now, a person approached who was, to a certain degree, sensitive, but who might yet be or appear in perfect health; and when such a person came within the sphere of these physical influences, he must necessarily have felt them by day, like Mlle. MAIX, and seen them by night, like Mlle. REICHEL. Ignorance, fear, and superstition, would now give to the luminous appearance the form of a human spectre, and supply it with head, arms, and feet; just as we can fancy, when we will, any cloud in the sky to represent a man or a demon.

158. The desire to inflict a mortal wound on the monster, superstition, which, from a similar origin, a few centuries ago, inflicted on European society so vast an amount of misery; and by whose influence, not hundreds, but thousands of innocent persons died in tortures on the rack and at the stake;—this desire made me wish to make the experiment, if possible, of bringing a highly sensitive person, by night, to a churchyard. I thought it possible that they might see, over graves where mouldering bodies lay, something like that which BILLING had seen. Mlle. REICHEL had the courage, unusual in her sex, to agree to my request. She allowed me, on two very dark nights, to take her from the castle of Reisenberg, where she was residing with my family, to the cemetery of the neighbouring village of Grünzing. The result justified my expectation in the fullest measure. She saw very soon a light, and perceived on one of the grave mounds, along its whole extent, a delicate, fiery, as it were, a breathing flame. The same thing was seen on another grave in a less degree. But she met neither witches nor ghosts; she described the flame as playing over the graves in the form of a luminous vapour, from one to two spans in height. Some time afterwards I took her to two great cemeteries near Vienna, where several interments occur daily, and the grave mounds lie all about in thousands. Here she saw numerous graves, which exhibited the lights above described. Wherever she looked, she saw masses of fire lying about. But it was chiefly seen over all new graves; while there was no appearance of it over very old ones. She described it less as a clear flame,

than as a dense, vaporous mass of fire, holding a middle place between mist and flame. On many graves this light was about four feet high, so that, when she stood on the grave, it reached to her neck. When she thrust her hand into it, it was as if putting it into a dense fiery cloud. She betrayed not the slightest uneasiness, as she was from her childhood accustomed to such emanations, and had seen, in my experiments, similar lights, produced by natural means, and made to assume endless varieties of form. I am convinced that all who are, to a certain degree, sensitive, will see the same phenomena in cemeteries, and very abundantly in the crowded cemeteries of large cities; and that my observations may be easily repeated and confirmed. (Postscript. 1847.—Since these experiments, which were made in 1844, I have taken five other sensitive persons in the dark to cemeteries. Of these two were sickly, three quite healthy. All of them have confirmed literally the statements of Mlle. REICHEL, and have seen the lights over all new graves more or less distinctly; so that the fact can no longer admit of the slightest doubt, and may be everywhere controlled.)

Thousands of ghost stories will now receive a natural explanation, and will thus cease to be marvellous. We shall even see, that it was not so erroneous or absurd as has been supposed, when our old women asserted, as every one knows they did, that not every one was privileged to see the spirits of the departed wandering over their graves. In fact, it was at all times only the sensitive who could see the imponderable emanations from the chemical change going on in corpses, luminous in the dark. And thus I have, I trust, succeeded in tearing down one of the densest veils of darkened ignorance and human error.

---

159. We come now to the province of electricity.

It may have now and then appeared, in the preceding researches, on superficial consideration, as if perhaps electricity, by itself, excited now in one way, now in another, might have the greatest share in producing, if it did not alone produce, the phenomena which have been described. The following investigation will shew what we are to think on this point.

160. The first experiment was intended to ascertain what degree of sensitiveness for galvanism was present in patients, who were very highly sensitive to the influence of the magnet. It was tried with Mlle. NOWOTNY in the presence of M. BAUMGARTNER. I used a pair of zinc and copper plates, of about forty-five square inches surface, between which was placed linen, moistened with solution of salt. The patient held two rods of German silver, connected by short copper wires with the two electrodes. She did not, however, perceive in the least degree the current which was passing through her, although it caused the needle of a multiplicator to deviate. On the tongue she observed the acidulous taste of the positive wire, not stronger, and not otherwise than we who were healthy did. I tried Mlle. STURMANN with a soldered pair of zinc and copper, about fifteen square inches in surface. When this was placed between her moistened fingers, she felt little difference from the effect of single plates of either metal separately. The galvanic pair seemed at most a little stronger, but she did not speak decidedly. The same negative result was obtained from the same experiment by Mlle. ATZMANNSDORFER. Mlle. MAIX could hardly distinguish, and not with certainty, between the zinc of the pair when held in her fingers, moistened with solution of salt, and zinc alone. She felt a larger pair, of forty-five square inches, but little stronger, when tried in the same way. Mlle. REICHEL did not differ from the other sensitives. She could observe no perceptible effect when the zinc and copper were united, whether the surface were large or small, or whether her hand were moistened with plain water or salt and water. The current produced by a single pair, after deducting the loss in overcoming the resistance to its passage, and with the feeble intensity which it has, is no doubt very feeble. But if the reagent on which it has to act be in a high degree sensitive, it is sufficient to produce sensible effects. It excites spasms in the limbs of frogs, causes the needle to deviate, decomposes iodide of potassium, &c.; and

it might therefore be expected to have a perceptible action on sensitive persons.

161. But, on the contrary, it follows from all the experiments, made with the utmost care, and often repeated with all the highly sensitive individuals, and which were answered in the same sense; *that a feeble current of electricity does not observably act more strongly on patients, even on those who have a high degree of sensitiveness for the smallest degree of magnetism, than on healthy persons.*

162. Compound galvanic combinations could not fail to affect them more powerfully than single pairs. I tried on Mlle. NOWOTNY a small pile of copper and zinc plates, of about fifteen square inches. Felt, steeped in solution of salt, was placed between the pairs. The zinc plates were far from bright, and I purposely left them so, to have the means in my power of graduating the discharge. I had put together ten pairs, and the patient, as well as the bystanders, felt nothing. With fifteen pairs she began to perceive a slight effect, but this was the case also with several of the more excitable among those present. With twenty pairs the shock reached through the hands to the elbows, while I could perceive it in the middle of my hand. Some females who were present felt it, to beyond the wrist. Mlle. NOWOTNY was thus the most sensitive among us, but not more sensitive than healthy persons are often found to be. I tried Mlle. MAIX with nine pairs of the same plates. The zinc was this time freshly scoured; and she felt nine pairs nearly as strong as the other patient had felt twenty pairs. But healthy persons, who were present, perceived nearly as powerful an effect. A certain degree of increased sensitiveness must be ascribed to the morbid state in general; but this does not depend on proper sensitiveness, as I have used the word in reference to the new force, but on common sensibility. With Mlle. REICHEL I tried piles of from two to fifty pairs, at different periods of her illness, in July, September, and November. She felt nothing from a few pairs. With fifteen or twenty pairs she considered the shock so trifling, as not even to notice it when she accidentally received it. She perceived vividly the shock from forty or fifty pairs, but made a

joke even of this, taking the shock readily, when others hesitated to do so. No secondary effects were perceived, either at the time, or on the following day. Mlle. ATZMANNSDORFER felt the reaction of a copper and zinc pair, excited by salt and water, no otherwise than when she touched both metals separately. The copper felt warmish, the zinc cooler, but without any other marked effect. Three pairs, and then sixteen pairs, were tried on her. She perceived the current, like any other healthy person. At least she noticed, that, as the number of pairs increased, the zinc became colder, the copper warmer, to her fingers. This was the first trace of the excitation of crystalline force by means of the voltaic pile.

163. When Mlle. REICHEL, by holding a conducting wire in each hand, allowed the current from 50 pairs to pass through her for some time, sensations were gradually developed, which, on the one hand, were felt in the head, and, on the other, extended to the knees. But these were also the first signs in her of galvano-magnetic action, which necessarily appeared after a time, since a continued current is followed by direct magnetic movements in all matter. It was, therefore, most probably not Galvanism, as such, which acted immediately, but rather its effect in converting, according to known laws, the conducting substance more or less into a magnet. The conductor, in this case, was the transverse axis of the patient's body, from one hand to the other, through the chest and arms.

164. When I used the electrical machine, and allowed friction electricity from the conductor to act on her, she amused herself by receiving dozens of sparks; but no special sensation, differing from that produced in healthy persons, could be observed.—Some time afterwards, her physician ordered her electric shocks to enter at the back of the neck, and to be carried from thence through the spinal column. I undertook to administer this severe remedy. According to the prescription, I charged a Leyden phial, having nearly a square foot of metallic surface, to full saturation, and exhibited to the patient, daily, eight such shocks, which were not pleasant to receive. She experienced, however, nothing different from what any of us would have felt.

165. The conclusion, which these experiments seem to me to justify, is this: that an electrical discharge, whether from the pile, the conductor of the electric machine, or the Leyden phial, when passed through the body, is too transient in its action to be able to bring into a state of sensible motion or activity that force, which resides in the body as in crystals; a result quite coinciding with well-known analogous effects in electricity.

166. We shall see, however, that we are in no way entitled to draw the conclusion, that all kinds of electric treatment are equally inert, as are the shock and the discharge of a weak pile.—When I placed in the hand of Mlle. REICHEL a thick copper wire, connected with a weak pile, and after she was accustomed to the wire, closed the circuit with it, so that the whole current should pass through it, without passing through the substance of the hand, she instantly felt the wire to increase very decidedly in apparent heat, that is, in the crystalline force. The hand was here in contact with the wire, which was so thick, (0.08 of an inch,) that actual warming of it by the current was out of the question.

But since the pile is a compound of chemical, magnetical, and electrical action, and consequently could not here yield unmixed results, I tried friction electricity. I caused her to surround with the hand, but without actual contact, a metallic wire connected with the conductor of the machine, so that the wire passed freely through the hollow formed by the hand. As soon as I began to rotate the glass plate, she felt a warm atmosphere round the wire, which sensation she perfectly distinguished from the well-known one of the spider's web.

167. I pass over the experiments on the conduction, charging, and polarisation of this influence, which are, indeed, implied in the following observations, and hasten to the luminous phenomena. I placed Mlle. REICHEL on the darkened stairs, gave her one end of a brass wire twenty feet long, which she held so, that the point was free. It lay on the floor, and after passing through the door, was attached to the conductor. The machine was at that time acting so feebly, that the conductor yielded only sparks of about $\frac{1}{4}$ of an inch. Soon after the

first turns of the plate, a slender flame rose from the end of the wire, such as we have seen to occur in other experiments. It was nearly ten inches high, and had, below, the thickness of a thumb, while above it had a very fine point. When I rotated the plate very rapidly, which the patient could not be aware of, the flame rose higher, and invariably sank, when I stopped. The point of the wire had, as I ascertained, no trace of an electric bundle of light, which indeed, owing to the numerous opportunities of conduction during its course, it was impossible it should exhibit. As often as I ceased to turn the machine, the flame continued uniform for about a minute, and then began slowly to sink. Something, therefore, was going on here, which in no way coincides with the known phenomena of the electrified wire. Every repetition at different times, and with different wires, invariably produced the same appearance.

168. I now detached the wire from the conductor, and attached it to a hollow sphere of polished brass, two inches in diameter. I took hold of the wire, and approached the sphere to the conductor, sideways, at a distance of about $1\frac{3}{4}$ inches, the sparks from which only reached to about the tenth part of that distance. The ball, therefore, was in the electric atmosphere of the conductor when the machine was made to revolve, but it was at the same time unisolated in my hand. A full minute elapsed before the patient, in the dark stair, observed any change at the end of the wire. Then a flame slowly rose to the height of a span, and it was necessary to rotate the plate for four minutes longer, after the first appearance of the flame, before it reached its maximum length, at which it then remained steady. When I ceased turning, and removed the ball from the electric atmosphere, the flame continued unchanged for more than a minute, and then, during several minutes, slowly diminished till it disappeared.

169. I varied these experiments by increasing the distance between the ball and the conductor. I held it now nearly eight inches above the conductor. The patient described the appearances in the same form, time, and order as before; nay, the flame actually became somewhat larger. I now removed the ball to a distance, sideways, of about thirty-nine inches

from the conductor. The result was the same, in every respect, except that the flame was shorter by one fourth.—At last I held the ball at six and a half feet distance from the conductor, but even then, after a pause of two minutes, the flame appeared, increased during four minutes, and when the machine stopped, slowly fell and disappeared as before. But this time it was only half as high as at first.

170. I next placed the conductor and the ball in contact, so that the latter was not in the electric atmosphere, but received directly positive electricity. The same phenomena now appeared, but, instead of doing so after one or two minutes, less than half a minute only elapsed. The duration of its increase and of its decrease remained as before. If any one, during the experiment, touched the conductor with the fingers, or even took hold of the ball, the size of the flame was not perceptibly affected; it appeared, increased, and faded away in equal times, and with the same characters. But when I placed the ball so as to receive a rapid succession of sparks from the conductor, the former appearances did not occur. The patient felt the shock of every spark in her hand, but saw no flame. The rapidity of the true electric action was too great to excite the principle which causes the flame, and which, being more sluggish than electricity, was not set in motion.—I finally repeated these experiments with negative electricity, using the isolated rubber instead of the conductor, but connecting the points with the earth. There was little, or rather no difference in the results, as seen by the patient on the darkened stair. During all these experiments I spoke not a word. Mlle. REICHEL, who was in the dark, separated from me by a wall, and could know nothing of all that I silently carried on, did nothing more than to call out, so that I could hear her through the door, the changes which took place at the end of the wire, as she saw them arise, continue, and disappear. Any kind of deception or delusion was therefore entirely out of the question. On the other hand, the exact coincidence of the appearances, as described by her, with the experiments, and with theory, was a strong guarantee for the truth of the whole.

171. These experiments are so distinct, that I need not

enumerate a multitude of other similar ones, made with the same results. I have only to add, that all these slender flames, visible to the patient in the dark, diffused coolness around.—When Mlle. REICHEL was near the machine, the positive charge of the brass conductor caused warmth; but at some paces distance, she perceived only coolness. The cause of this difference I can only render intelligible in one of the following treatises.

172. I held a tin electrophorus plate, about one foot in diameter, by its wooden handle above the conductor, so as to be somewhat within its atmosphere, for one minute. When I now brought it, after touching it with a moist finger, near Mlle. REICHEL'S face, she felt a strong impression of coolness flowing from it for several minutes. This confirms, in a different way, the preceding experiments. I did the same thing with an isolated body. The rod of German silver before mentioned, was suspended by a silk ribbon over the conductor. Its effect, when brought near the patient's face, was, as might have been expected, the same as that of the unisolated tin plate. We do not possess, as yet, any means of isolating the crystalline force. Similar experiments were tried, with the same results, with Mlle. ATZMANNSDORFER. The tin electrophorus plate was first held near her face, before being electrified. She found it, like metals in general, to diffuse warmth. It was now held a handbreadth above the weakly-charged conductor, and then, for the possible event of its retaining traces of electricity, touched with my moistened hand, and entirely discharged. When again held near her face, she felt it strikingly cool, diffusing coolness around it.

173. Everywhere, therefore, where electricity is excited so as to continue for some time, the peculiar force which I have endeavoured in this work to investigate, appears; *and electricity is thus the tenth source of it.*

RETROSPECT.

a. Chemical action is a wide and comprehensive fountain of this peculiar force; whether it be simple chemical action, or that of combustion, or of the voltaic pile.

b. The feeblest chemical action suffices to develope the new force abundantly ; to charge other bodies with it ; to excite polarity ; and to produce light, visible in the dark to sensitive persons.

c. The magnetic *baquet* is nothing but a source of slow chemical action.

d. Digestion and respiration, as well as the change of matter, being chemical processes, are the sources of the magnetic power, as it is called, which resides and acts in the human frame.

e. The spectral or ghost-like luminous appearances, seen over graves, which have been ridiculed and denied by the healthy, are of purely chemical and physical nature, but can only be seen by the eyes of the highly sensitive.

f. Electricity is also a source of the new influence found in crystals, &c. This is true both of friction electricity, positive and negative, and of contact or voltaic electricity.

g. Even the electrical atmosphere can, at very considerable distances, set in motion the new force.

SIXTH TREATISE.

THE MATERIAL UNIVERSE.

174. THE force which we are here investigating has now been seen proceeding from ten different, but always *special* sources. We shall now find it flowing from a *more general* source; we shall seek it in the material universe, and endeavour to ascertain something of the great and important part which it has to perform in nature.

175. Every one knows that there are many persons on whom certain substances produce a peculiar effect, generally disagreeable, the results of which are often ludicrous. I do not here allude to the strange fancies, likings or dislikings, instinctively felt by pregnant females. But when we see people who cannot touch fur; others who cannot look at feathers; nay, some who cannot endure the sight of butter, &c., &c., without feeling, when they are forced to do these things, ill and faint; we might, if we considered a single case, ascribe it to defective education. But experience has shewn that such special antipathies always return in the same form, and with reference, in each case, to the same substances, in the most widely different persons, and in the most widely distant countries. This proves that they are far from being always manifestations of ill-breeding or bad manners; and that they must proceed from some special cause, whether objective or subjective; further, that when they occur, they

ought not to be treated with anger or scolding, but often deserve to be attended to, and carefully investigated.

Now, accurate observation teaches us, that these strange and often violent antipathies chiefly occur in persons, who, although apparently healthy, are more or less sensitive; and that they increase in force and in variety as the persons who experience them are more sickly, and subject to nervous disorders, spasms, and similar affections. This went so far in Mlle. STURMANN, for example, that she sometimes could not touch a key or a lock, without her fingers and hands being paralysed by spasm, although, at the time, she was able to go about, as a healthy person does, in the house, in the garden, or through the streets.

In my multifarious researches with very sensitive persons, I soon observed that there was, in such antipathies, something in common, and some agreement, which, if pursued and compared, would certainly give a hope of reducing the phenomena to a relation of cause and effect, and thus enable us to approach nearer to the common but more hidden natural cause which produced them. I found that certain special sensations always presented themselves, and that, when I only took care clearly to ascertain the feelings of the patient, and reduced them to a common nomenclature, the apparent variety might be reduced to a few which were constantly observed. And I soon found that these few might be brought under certain rules. They consisted of sensations of apparent heat or cold in different substances of the same temperature; in decidedly agreeable or unpleasant feelings, the latter reaching to the point of exciting spasm; in sensations of pulsative pricking, and of dragging; and in painless tonic spasms.

In the second of these treatises, which treats of the force residing in crystals, I have shown, that in Mlle. NOWOTNY, the last-mentioned effect, namely, painless tonic spasms, was produced by the emanation from the poles of the axes of crystalline bodies, and that the power of exciting the spasms was found in different degrees of energy in different substances, but was never absent in those which appear in detached crystals, whether they were simple bodies or of highly complex composition. This kind of sensation has therefore been, to a certain extent, discussed, and, for the present, settled. We have still, therefore, to examine those of *apparent difference of temperature*; those which are *simply disagreeable or offensive*; and the apparently mechanical agitations of *pulsative pricking*, &c. Some of these we shall now proceed to test.

Among these sensations, some were experienced by healthy persons; but the highly sensitive perceived them all, more or less strongly according to the nature of their disease, and its actual intensity at the time.

I was first induced to make researches in this direction by observing, in Mlle. NOWOTNY, that all amorphous bodies, which, as we know, do not possess the peculiar action of crystals, yet generally caused a disagreeable sensation, and sometimes also one of heat or cold. This property adhered pretty uniformly to certain substances, and was of different degrees of strength in different bodies. Where I formerly saw the crystalline force arising from the nature of the mode of aggregation of the molecules, I now saw something dynamic, of another kind, proceeding from matter, as such. The nature and the form, therefore, of substances, exhibited strongly marked differences in their action on other bodies.

176. Since something obviously lay here concealed, which could not but be of interest, whether in reference to physiology or to physics, I took the trouble, which was certainly not small, to test, as to the power of exciting a disagreeable or offensive sensation, more than 600 substances; in fact, the greater number of the chemical preparations in my collection. It appeared that the power had very distinct degrees of energy, and that the patient could distinguish these so acutely, that she was able to assign to each substance its proper place in the series, between two others. She did this with so much precision and certainty, that when I showed her the substance which she had arranged, three or four days afterwards, she assigned to them exactly the same relative places as on the first occasion. It is obvious that she could neither recognise nor understand the nature of so many bodies, most of which were white powders. But these sensations were to her quite

as distinct as ours can be in distinguishing shades of colour by our eye, or harmonious and discordant sounds by the ear. If we further consider that in Mlle. NOWOTNY this sense was deficient in that exercise by which, in their daily use, sight, hearing, and the other senses acquire, during a long life and by degrees, their perfect developement and sensibility; which exercise we know, from physiology, to be of preponderating influence on the clearness and distinctness of our sensual perceptions; we may form some idea of the extraordinary internal acuteness which must belong to this peculiar and probably morbid sensitiveness. These considerations will, in the sequel, serve to explain many things as yet involved in complete obscurity.

177. The investigation had hardly reached so far as to enable me to classify a dozen substances among themselves, when I could perceive a law developing itself. The bodies arranged themselves according to their electro-chemical order, but in such a way, that the most powerfully electric stood at the top of the list, the indifferent bodies at the bottom, without distinction of their polar opposition. I cannot inflict on the reader the whole list of more than 600 substances, but I shall extract a few. The most active substances, which stood highest, were oxygen, sulphur, caffeine, sulphuric acid, potassium, phosphorus, sodium, selenium, iodine, cinnabar, lead, potash. Between the numbers 30 and 110 stood bismuth, arsenic, mercury, morphine, zinc, iodide of potassium, tellurium, bleaching powder, chromium, lithium, purple of cassius, oxide of nickel, tin, iridium, nickel, alcohol, chlorine gas. Between 200 and 400 were found paraffine, rhodium, acroleine, piperine, kreosote, sea salt, quinine, brucine, cantharidine, strychnine, crystallisable acetic acid. From 500 to 600 stood, among others, cinchonine, quartz, hippuric acid, mastic, chalk, gum, almost all vegetable acids, sugar, sugar of milk, mannite. At the lower end of the series were palladium, platinum, copper, iron, gold, amber, water. These last were almost indifferent to Mlle. NOWOTNY. With a few exceptions among the rarer metals, which probably were not quite pure, all the highly polar substances appeared at one end, the indifferent at the

other. Among the latter, iron alone, with its magnetic susceptibility, was an exception. But, strangely enough, the patient could not discover any distinction between the positive and negative substances, although at last I tried to explain to her the important difference, and thus purposely directed her attention to it. Oxygen, sulphur, and phosphorus, ranked along with potassium and sodium.

178. This comparison, which cost a whole week's labour, was carried out while the patient was perfectly conscious and clear. I now tried to control it by observations made on her in the insensible cataleptic state. When I brought near to her in this state any substance, it acted on her. The indifferent bodies, if not crystallised, affected her but slightly; those standing towards the middle of the series caused restlessness in the hand; those lying higher excited trembling and spasmodic contraction of the hand; and with substances at the head of the list, such as sulphur, caffeine, galena, iodine, cinnabar, and minerals such as heavy spar, fluor spar, iron pyrites or gypsum, which also, although not in detached crystals, stood at nearly the same point of the series, the whole arm was seized with such violent spasm, that it was raised mechanically, like the galvanised limb of a frog, and flung away the body lying on the hand. It then, as in all cataleptic cases, continued rigidly fixed in this constrained position. These testing experiments prove, that the unknown action of these bodies on the cataleptic patient was, qualitatively and relatively to the power of different bodies, exactly the same as when she was in her ordinary state, but that quantitatively the effect was considerably increased. That which, when she was conscious, produced only a very unpleasant sensation, excited the most violent spasm when she was in the cataleptic state. When she awoke, after some time, from this state, and became conscious, she regularly complained of pain and weariness in one arm; and this was always the arm with which the experiments had been tried, although she retained no consciousness of them, and was never informed that we had made such experiments. The spasms were therefore accompanied with violent excitement and exertion, which left a feeling of

exhaustion and fatigue. From all this we may confidently conclude, *that all solid bodies in contact with this sensitive patient, excite feelings of a peculiar kind, differing in degrees of intensity according to the chemical nature of the substances.*

179. I frequently noticed the remarkable fact, that some of these bodies, even before they actually touched her hand, and while I still held them over it, began already to excite restlessness. On attending more closely to this, I found that many substances, even when I only laid them on the bed near her hand, were yet able to produce an effect on the patient. Such bodies were sulphur, galena, fluor spar, rock salt, cinnabar, tin in grains, gypsum, arsenic, sal ammoniac, prussiate of potash, antimony, telluric acid, wolfram, apatite, celestine, white lead ore, cyanide of potassium, sulpho-cyanide of potassium, orpiment. The hand near which they lay began to tremble, the motion increased, and frequently became so great, that the hand touched the substance, when it flung the body away, or became violently affected with spasm. —It was therefore obvious, *that even amorphous bodies, if only of a decided electro-chemical character, acted at sensible distances.*

180. In order to ascertain how these bodies would act on the patient in her ordinary conscious state, I repeated the experiments on the following morning, at an hour when she usually felt better than at any other time. But she felt nothing, in often-repeated trials, from these bodies when merely laid near her hand. She closed her eyes, and covered them up herself, to be quite sure that she really felt nothing when those substances were brought very near to her hand, which the evening before had caused contraction and spasms, but in vain; she could never tell whether they were near or more distant; and only after contact did the unpleasant sensations arise, which she perceived from these bodies in her conscious state. *The cataleptic state is therefore a condition which exalts to a very great degree the sensitiveness of patients to certain unknown properties of substances;* and these substances must possess some unknown power, in virtue of which they affect cataleptics, even at some distance, in a manner analogous to

that in which they act on the same patients in the conscious state, when in actual contact with them.

180. I soon afterwards extended these observations further, by means of Mlle. MAIX. She found most of the bodies which I placed in her hand warm or cold, as I have already mentioned. But along with these sensations, which she only felt on contact, and at the surface of the substances, she perceived another, namely, coolness in the form of a cool aura, (analogous to the effect of the positive electric discharge from a point on healthy persons,) which was diffused by many bodies. Such a body was sulphur. When I placed a bit of sulphur in her hand, she perceived, besides the warmth felt at the points of contact, a coolness which diffused itself over her whole person, like a cool zephyr. This soon extended from the hand over the arm to the face, and was felt in the other hand, from which it passed to the other arm. It then forced its way through her dress to the chest; at last it flowed through the bed-clothes, and the whole abdomen; and, finally, the feet were affected by the cool emanation derived from the sulphur.

182. I put the sulphur in an open tumbler with thin bottom, which I laid on the patient's hand. She thus held the sulphur without touching it, and was, at the same time, as I hoped, protected from direct radiation from it. The tumbler alone felt warm, without any further emanation. As soon, however, as the sulphur was introduced, while the surface of the glass in contact with the hand continued warm, as before, a cool aura flowed from the glass on all sides, and spread over her hand. It seemed also to flow from all those parts of the under surface of the glass which were not in contact with the hand; to fall down from the sides of the tumbler, and to overflow its upper edge, falling down on the hand. But this coolness, as it appeared to pass through the glass, was now far more grateful to the patient than from the sulphur by itself. It seemed, she said, finer, more transparent, purer, more etherial. It soon penetrated through the whole hand, which it rendered cold and stiff; and this effect lasted a long time after the glass was removed. It was then sensible in the face, the

other hand, and elsewhere. Gypsum produced the same effect as sulphur.

183. After I had removed the sulphur from the glass, and when I had laid it for a moment on a table near the bed, while I prepared another experiment, Mlle. MAIX soon after told me, that she perceived the effect of the sulphur from where it lay. I now removed it to more than three feet distance, but she still felt it. Even at six and a half feet she still observed traces of cooling influence from a piece of sulphur about as large as a finger. I next prepared a surface of sulphur about eight inches square, or nearly half a square foot, by attaching closely on a board six rods of sulphur, such as are here used for burning in wine casks, for the purpose of consuming the oxygen of the air. With this extemporised apparatus, I could go gradually as far as the room permitted me, about twenty-four feet from the patient, who still felt, even at that distance, a feeble though distinct cool aura flowing upon her from the surface of sulphur. At this distance, the sensation was exactly like that caused by the point of a large rock crystal, or by an open magnet, although the influence of the two last was more powerful.

184. As sulphur, when in contact, caused a sensation of warmth, I next tried a substance which, on contact, appeared cold. Such a body was oil of vitriol. When it was given to her in the glass, she felt intense cold. But also, when removed from her, even to the distance of several paces, it diffused a cool aura. Nitric acid felt, through the glass, equally cold to her hand; but its action, at a distance, extended half as far again as that of the sulphuric acid.

185. I had now to try bodies which, on the hand, were neither warm nor cold, but indifferent. Such were paraffine and cane sugar. But both of these produced a cool aura when removed, the former to six and a half feet, the latter to nearly forty inches.

186. I now went through, with Mlle. MAIX, a number of the most different bodies, to obtain confirmation and extension of what I had observed of this peculiar phenomenon. A jar of oxygen gas was felt by her peculiarly hot (boiling hot, as

she said), soon rendered her hand and arm rigid and contracted by spasm, was highly offensive to her, and diffused a cool aura from a distance of about twenty inches. So small a quantity (by weight) of an energetic substance, in which, though in a more diluted form, we constantly live, sufficed in this case, as in that of Mlle. NOWOTNY, to produce a very strongly-marked effect! I had, in a glass tube 0.8 of an inch wide, and closed hermetically at both ends, some chromic acid. Mlle. MAIX felt it, through the glass, burning hot, but diffusing a cool aura even at a distance of ten or more feet. Phosphorus, under water in a bottle, felt like sulphur, only not so strongly, cooling, and diffusing a cool aura at ten feet. Selenium was like sulphur, and some small pieces diffused a cool aura at twenty feet. Tellurium acted very like sulphur, and the cool emanation was felt as far off as the room allowed me to remove it. Charcoal had like effects, but in a very much weaker degree. An empty glass on the hand felt warm without aura. But when I closed it with a watch-glass, a cool breath, as it were, descended on the hand. When I removed the watch-glass the coolness ceased. It was, therefore, the inclosed air which produced the cool sensation. This experiment, repeated in many various forms, always gave the same result. The oxygen of the air, when shut up, and warmed by the hand, had, no doubt, acquired a preponderance in its action on the patient, over the external colder air, which was constantly descending; and as the oxygen was here the most active substance, nay, is probably of all substances, in this respect, the most energetically active, a very slight rise of temperature, the effect of which we have already seen in § 177, was sufficient to produce a sensible effect.

187. Almost all metals felt warm to the hand, but all also yielded the emanations which the patient called cool air. In the order of their energy they were nearly thus: chromium, osmium, nickel, iridium, lead, tin, cadmium, zinc, titanium,* mercury, palladium, copper, silver, gold, iron, platinum. A

* If this was the so-called metallic titanium, occurring in copper-like cubes in certain slags, it has lately been shown to contain also carbon and nitrogen. —W. G.

thin copper plate, of nearly eight hundred square inches, placed near and opposite to the bed of the patient, caused the sensation of a lively current of fresh cool air, which by degrees seemed to penetrate the whole bed, and was very agreeable to the patient. A zinc plate of the same size produced a similar effect, but not so powerfully. Plates of lead and iron were still weaker. On the other hand, when I held towards the patient a mirror, with the back, that is, the metallic surface nearest to her, the action on her was very powerful. But it was far more powerful when the glass surface was turned towards her. The radiation from the polished metal, through the glass, diffused that etherial and delightful coolness described in § 182 as proceeding from sulphur and gypsum, also through glass. She felt her whole person, from head to foot, pervaded by a pleasurable sense of comfort.

188. I made some less-extended experiments with Mlle. STURMANN, but the results were sufficiently decisive to yield positive confirmation of the former. She felt oxygen gas, on contact with the containing bottle, very hot; sulphur, selenium, iodine, bismuth, chloride of gold, iridium, purple of Cassius, and morphia, on the hand, all warm; antimony, mercury, zinc, copper, tellurium, lunar caustic, gold, lead, tin, iron, were all, in different degrees, cold. Potash yielded uncertain results. She found crystals of calcareous spar, doubly refracting Iceland spar, arragonite, tourmaline, and rock crystal, at one pole warm, at the other cold. I tried her with surfaces of half a square foot of sulphur, lead, zinc, iron, tin, copper, silver, and gold; all of which gave out, at forty inches distance, emanations varying from hot to warm, tepid, and cool. Palladium diffused a delicate cool aura, flowing from it on all sides. When I found her in the cataleptic state, and placed sulphur, selenium, tellurium, mercury, in a bottle, antimony or zinc in her hand, she struck out, like Mlle. NOWOTNY (§ 178), and hurled the objects to a distance. But when I placed them near her hand, it began to tremble, drew itself slowly away, and was partially convulsed, exactly as with Mlle. NOWOTNY (§ 179).

189. Mlle. ATZMANNSDORFER felt, on contact, sulphur,

selenium, iron pyrites, antimony, zinc, lead, Egyptian jasper, common salt, alum, potash, and brucine cold; but platinum, silver, bar iron, copper, gold, and mercury warm; gypsum, fluor spar, iron pyrites, alum, tellurium, lead, common garnet, and gallic acid, yielded a cool aura from some distance. This aura appeared to flow upon her from all sides.

190. M. SCHUH felt, on contact, roll sulphur and the powder of sulphur equally warm. Oxygen gas, iodine, bromine, coppernickel, cyanide of gold, cyanide of potassium, felt warm, and soon caused headache, which by degrees became intolerable. He arranged in a series, proceeding from cold to warm, according to his sensations, the following minerals: iron pyrites, fluor spar, calcareous spar, specular iron ore, staurolite, rock crystal, tungstate of lime, shorl, sandy calcareous spar from Fontainebleau, heavy spar, topaz, rock salt, analcime, felspar. In blue vitriol and carbonate of soda he perceived points which diffused coolness. In order to prevent self-deception, he wrapped them in paper, and tried them again. To his satisfaction he found, on uncovering them, that he had fixed on precisely the same points. He perceived distinct coolness at ten inches from slices of sulphur and from a thin layer of oxalic acid; and warm emanations at the same distance from leaves half a square foot in surface of tin, lead, copper, silver, and gold. He also found silver and gold coins, and steel tools, warm, when he held his hand over them at a certain distance. He could not long endure the effect of standing before a large cheval glass. Its emanations caused headache, stupefaction, and pain of stomach; and if he stood with his back towards the glass, these painful sensations attacked him much more rapidly. He tried, some hours afterwards, the same experiment with a large mirror, fixed on the wall of the room; the same sensations now returned sooner, and were more powerful.

191. M. STUDER grasped in the hollow of his hand a number of substances, or held them near his eyes, with which he could even more sensitively distinguish the cool or warm sensation produced. He found, in this way, without contact, that sulphur, iron pyrites, gypsum, tellurium, bleaching powder, persulphate of iron, sulphuret of potassium, binoxalate of

potash, seignette (rochelle) salt, rock crystal, sugar, diffused coolness; while gold, silver, copper, tin, lead, zinc, potassium, solution of potash, diffused warmth.

The carpenter KLAIBER felt, at ten inches, a cool aura from sulphur, sulphuric acid, gypsum, rock crystal, &c.; and a warm aura from gold, silver, copper, tin, lead, zinc, solution of potash, &c.

Many other persons, who came to visit me, were tried with these substances; and I went through, with these persons, a comparison of some bodies. At last I found reason to regard two selected substances, as representatives of all the rest, in their two chief divisions of cool and warm. These were sulphur and gold. I had a surface of half a square foot of each always at hand. Almost all the persons tried declared unanimously, when I caused them to hold the hand near those surfaces, at a distance of an inch or thereabouts, that the sulphur diffused coolness, the gold warmth. I am permitted to name some of these persons. M. KOTSCHY, the celebrated oriental traveller, perceived both sensations vividly. Without having heard any thing of the matter, he complained of a peculiar pricking sensation from the sulphur, even without contact, like that of little star-shaped points over his hand. Dr. FENZL, the distinguished botanist, perceived the difference of cold and warmth between sulphur and gold very plainly. Mr. INCLEDON, an English gentleman residing in Vienna, not only felt this most vividly, but also described the pricking caused by the sulphur as much resembling that experienced when the hand, has been asleep, as it is called, and when the restoration of its sense of feeling is so nearly completed that only a few scattered prickings are felt.

192. These phenomena present, hitherto, a variable mixture of results; but in the midst of this apparent inconstancy, we can perceive something which is very constant. The variation occurs in the nature of the sensation, whether cool or warm, which the same substance excites in different individuals. This part of the subject requires a separate examination. I therefore exclude it from my concluding proposition, reserving it for an early opportunity, and here only lay down with con-

fidence the general and constant *fact, that all the bodies tried act on sensitive persons, so as to excite feelings of an apparent change of temperature, whether the sensation be cool or warm.*

But I was enabled, more clearly than with any of the persons yet mentioned in this section, to throw light on these phenomena, with the help of Mlle. Reichel; and to bring them to such a degree of distinctness, that I might pass over all that I made out by means of other sensitives, were it not that such observations, from their very nature, can only acquire stability by a large number of the most varied repetitions. Mlle. Reichel was so kind as to spend some time, as a guest, at my country seat, Schloss Reisenberg, near Vienna; and this gave me the opportunity to carry out, much more perfectly and regularly than I could in the houses of others, a large number of experiments, with the help of physical apparatus.

193. To render the following experiments intelligible, I must prefix some account of the localities of my house, in which they were made. The so-called castle is built in such a way, that in the front two parallel rows of nine rooms are connected by doors so as to yield, when the doors are opened, in each row a continuous straight line of 160 feet. As there is at each end a balcony, ten feet wide, and the doors of the balconies are in the same line as the rest, I had a straight line of 180 feet from one balcony to the other, in which the air was tranquil. When it was necessary to use long wires, they could be carried through the second row of rooms, so as to give conveniently a length of about 365 feet. I placed Mlle. Reichel at the end of one of the exterior rooms, and tried experiments on the distance at which different bodies acted on her. And first, in order to have some kind of measure, I used magnets. With a small horse-shoe, having limbs of two and a half inches, it was necessary to go to the distance of nearly eight and a half feet before the effect disappeared. Another, with limbs of about eight inches long and one and a half broad, I had to carry to the distance of 60 feet. A heavy nine-bar horse-shoe, supporting at that time more than forty pounds, acted on her until I reached the distance of one hundred feet, the length of six rooms. Comparative experiments, with a bar mag-

net eighteen inches long, yielded, as the utmost acting distance—

For the positive or southward pole,	76.5 feet.
For the negative or northward pole,	60 ...

194. I next examined the soft iron bars, which served, arranged in a parallelogram, as armature for this bar magnet when in its case. The long one was exactly of the same form and size as the magnet. The two shorter ones had the same section, but only one-fifth of the length. They did not attract iron filings, when separated from the magnet, and were, in short, perfectly unmagnetic. Notwithstanding, the patient felt the action

Of one end of the long bar at	76.5 feet.
Of the other end at - -	65 ...
And that of the cross bars	
Of one end of each at -	26.5 ...
Of the same end of both together at	50 ...
Of the other end of each at -	36.5 ...
Of the other end of both together at	70 ...

The separate and unmagnetic soft iron bars, therefore, acted nearly as powerfully on the patient as the magnet, and must consequently have come, in regard to crystalline force, nearly into equilibrium with it. It was here impossible that the patient could be deceiving herself, as one might at first sight imagine; for at such distances she could not possibly tell which bar I held towards her, nor whether it was a few feet nearer or further off, and all her statements harmonised perfectly among themselves, and thus served mutually to control each other.

195. Three days afterwards, during menstruation, her sensitiveness had much increased. I repeated the measurements of the distances at which magnets affected her. But this time I was compelled to use all the rooms in one straight line, and both balconies; and when I had thus reached a distance of 160 feet, she felt the action of the great nine-bar magnet so powerfully, that, according to her estimation, twice the distance

would not have sufficed to attain the limit of the influence. I now tried half a square foot of surface of slices of sulphur. Mlle. Maix had felt their action at about 24 feet, but the size of her room did not permit me to try a greater distance. But here Mlle. Reichel felt the sulphur to diffuse coolness at 124 feet. Astonished at this, I tried a copper plate of more than 4 square feet. It diffused warmth to the distance of 94 feet.

A plate of iron, 6 square feet, was felt warm at	146	...
Thin lead foil, of the same size, at	75	...
Tin foil, ... at	70	...
Zinc plate, ... at	64	...
Silver paper (genuine) of one square foot, at	24	...
Gold paper (genuine) of 3 square feet, at	67.5	...
An electrophorus plate, 16 inches in diameter, at	98	...
A mirror of about 10.5 square feet, at	106	...
A small bottle of oxygen gas, at	19	...

A number of other substances, such as brass utensils, porcelain vessels, glass, surfaces of stone, coloured paper, boards of wood, linen, open or shut doors, lustres suspended to the roof, trees, human beings, horses, dogs, cats, approaching her, pools of water, especially after having been long exposed to sunshine;—in short, all and every thing of a material nature, aeted on her, diffusing in some cases warmth, in others coolness; and many things acted so strongly as to attract her attention and annoy her; others so feebly, that, becoming accustomed to them, she no longer regarded them.

196. This astonishing phenomenon, namely, that a human being distinctly perceived a metallic plate, gold paper or tinfoil, without seeing them, at the distance of 100 feet or more, surprised me so much, that I could not suppress expressions of wonder, at which Mlle. Reichel only smiled, as she had all her life been accustomed to it. All my sensitives, healthy or otherwise, had, without exception, observed the same thing, under favourable circumstances, more or less strongly and distinctly, and in a more or less extended measure, according to their individual sensitiveness. It was obviously nothing else than a manifestation of the often mentioned crystalline

force, in a more general and perhaps altered form. The concluding proposition, deduced in § 178 from observations on Mlle. NOWOTNY alone, is now valid for all the persons tried; and I may now extend it as follows:—

ALL *solid bodies in contact with persons sufficiently sensitive, excite peculiar feelings, differing in degree according to their chemical nature; these sensations are chiefly those of an apparent change of temperature, such as cool, tepid, or warm, with which a pleasant or a disagreeable sensation keeps pace, more or less uniformly. Lastly, these re-actions are in all respects similar to those produced by the force of magnets, crystals, the human hand, &c.*

It is now necessary, in order to demonstrate the identity of the cause of the phenomena just described with the crystalline force, to study the other characters of that cause.

197. Is this universally diffused influence communicable to other bodies, or can they be charged with it, as I have repeatedly expressed it, without however thereby meaning to assert that it is material? A singular observation, which I made on Mlle. MAIX, leads to the answer to this question. Her sister, Mlle. BARBARA MAIX, came to see her. This young woman was apparently healthy, yet suffered from various nervous affections. She took hold of her sister's hand, and had hardly done so when she started back trembling, and let it go. "What in the world have you got in your hand that pricks me so?" cried she. Now, there was nothing in the patient's hand; but she had held in it, immediately before, a piece of sulphur. And when this effect had gradually disappeared, it could be renewed at pleasure, as often as she, Mlle. MAIX, held a piece of sulphur for a short time in her hand. This clearly shewed, that the power of exciting a pricking sensation is simply the result of a transference to the hand of certain unknown properties of sulphur, and which was capable of acting on a second hand, similarly sensitive to sulphur and other like substances. When I held sulphur for some time in my own hand, and then took hold of that of Mlle. MAIX, she felt the same sensation, and thus discovered what it was I had held.

198. Direct experiments proved this more decidedly. I gave to Mlle. MAIX the German silver rod, and when she was accustomed to it, I caused her to lay it down, and then placed on it a piece of sulphur. After a few minutes, I removed the sulphur, and she again took hold of the rod. She instantly recognised a strong impression, the same as was always caused by sulphur; and therefore something from the sulphur must have remained in the rod, which adhered to it, and could be perceived by the hand of the patient.

199. Experiments with Mlle. REICHEL gave similar results. I first repeated that experiment made by the sisters MAIX, without telling her of it. I held in my hand for five minutes, a roll of sulphur. I then laid it down, and with the same hand took hold of that of Mlle. REICHEL. I had not held it long, when she tore her hand away, and said that my hand was full of needles. On the whole surface of contact she felt innumerable small punctures; an exact confirmation of what Mlle. MAIX had observed, in the hands of two different persons, and a striking indication of how much objective reality lies at the root of these phenomena. My hand, although I purposely refrained from touching any thing, was not in its natural state at the end of a quarter of an hour; but still felt, although in a less degree, to the patient, as before. It had therefore received from the sulphur a charge, which it retained for a considerable time, and which was only slowly dissipated.

A pair of steel scissors, which by itself felt warm, was laid for a time on sulphur, and then taken up by her. The scissors were now cool, and caused the pricking sensation, as above.

A glass tube, forty inches long, by itself feebly warm, was covered with sulphur at one end for a minute, and the other end given to her. She found it now very cool. Five minutes afterwards it was still cool, more feebly, but quite perceptibly. At the end of half an hour, it had recovered its original slight warmth.—Another glass tube, placed in contact with the outside of a bottle of sulphuric acid, produced the same effect.

A few milligrammes (a milligramme = one sixty-fifth of

a grain) of caffeine in a thin glass vessel, laid in the tube for a time, and then removed, left the tube considerably warmer than before.

My younger daughter, OTTONE, laid her hand, which the patient had previously tried, on several folds of gold paper. She then took hold of the hand of the patient, who found the hand much warmer. After three minutes, it appeared as if half the effect still remained; after seven minutes the hand had acquired its natural state.

A glass bottle, full of powdered gum, by itself slightly cool, was placed close to a bottle of potassium in naphtha, the potassium being, by itself, one of the very warm bodies. After a few minutes, it yielded a remarkable sensation of cool and warm, and it appeared as if a part of the gum had been overpowered by the potassium, while another part, possibly the interior, had remained unaffected in the given time.

Pure gold leaf always caused a strong feeling of warmth in this patient. An empty glass bottle, by itself slightly cool, was placed on gold leaf, and turned about on it in all directions, to obtain as full an effect as possible. She now felt the bottle, not cool, but strikingly warm. The warming power of the gold had been communicated to the glass.

Gypsum, cold by itself, was held for some time in my right hand. It was not perceptibly altered. In my left hand it became warm to her. The same thing happened with oxide of copper, cold by itself, when held in my left hand.

I covered with many sticks of sulphur a bottle of potassium, in naphtha, itself very warm. After a few minutes it was felt as cold as it had been formerly warm. The sulphur had not merely destroyed the apparent warmth of one of the warmest bodies, but had also communicated to it its own cold. This effect lasted for several minutes after the separation.

The German silver rod, by itself feebly warm, was covered with common salt. After a short time it was found, when taken out, quite cold. Fluor spar was now laid on it, which made it much colder. It was now covered for a minute with sticks of sulphur, and again the cold sensation was found more intense.

A rod of sulphur was laid so as to lean against the potassium. When removed, it was much less cold. If left longer in contact, it became warm in some parts, cool in others. When now placed for a time in my left hand, it lost all coolness; and when I had lastly wrapped it up for some time in gold leaf, the sulphur had actually become warm to the patient.

200. Mlle. ATZMANNSDORFER had felt copper always tepid. It happened one day, that when I tried it, she felt it cold. As this did not agree with the previous observations, I was surprised, and endeavoured to discover the reason of it. It was a smooth thin plate, and lay on a polished table of walnut wood. I made her feel the table, which, in all parts, she found cool. I now laid the copper for some time on a seat covered with silk. When she now touched it, it was tepid as usual. I then laid it again on the table. In five minutes it again felt cold. I fastened it next between the cheeks of a small vice; it was very soon again warm. But as soon as it was left for a little on the table, it again became cold. The great mass of the colder table had obviously overpowered the action of the slight copper plate, and charged it with the cooling power, whenever it remained a short time in the sphere of action of the walnut wood.—It was therefore established, by numerous experiments, *that the peculiar property here treated of may be communicated from one body to another by contact, exactly like the crystalline force.*

201. We have seen effects produced at a distance by magnets, crystals, fingers, heavenly bodies, &c., in virtue of the new force. We have also seen that bodies in general acted at a distance on the cataleptic patients, Mlles. NOWOTNY, STURMANN, and ATZMANNSDORFER, and in the natural state also on Mlles. MAIX and REICHEL. The question now arose, whether this property of matter in general could be communicated to other bodies at a distance, that is, without contact? I placed beside the German silver rod, at the distance of three quarters of of an inch, a rod of sulphur of equal length. After a few minutes, the rod was tried by Mlle. REICHEL. Naturally causing warmth by itself, it now felt quite cold, as much so as if it had been in actual contact with the sulphur. After

four minutes, the coolness was only half dissipated, and did not disappear under a quarter of an hour.

Blue vitriol, bruised small and wrapped in paper, was laid at the distance of ten inches from the forty inch glass tube, and contact avoided. After five minutes the patient examined both ends of the glass rod, which, in spite of the considerable distance, had become strikingly cool, an effect which lasted for several minutes.

I learned from these observations, that I ought not, in these delicate experiments, to use my own hands, on account of their magnetic force, if I wished to avoid complication in the results. I also obtained hints towards the explanation of many anomalies in my earlier experiments, where the sensitives often did not agree in the sensations of heat or cold. It might often have been my hands, which had by their own force altered the natural state of the substances. I therefore caused my daughter HERMINE to place her hand in that of Mlle. REICHEL, and after she was accustomed to it, hold it over a surface of sulphur without touching it, for two minutes. She then again took the patient's hand. Very soon, besides coolness, the patient perceived the pricking sensation, which, in the former experiments, had always been caused by contact with sulphur. My daughter OTTONE repeated this experiment with the same result. The pricking was even felt from the hand held over sulphur for seven or eight minutes. I pass over other similar experiments.

202. It is therefore demonstrated, *that the charge of this peculiar form of the force may be effected, without contact, by the mere approach of the substances to each other.*

203. Its conductibility is sufficiently shewn by many of the above experiments; but I shall add here a few remarkable confirmatory examples. I connected with Mlle. REICHEL, by means of an iron wire 100 feet long and one-twelfth of an inch thick, a copper plate, and laid on the plate, successively, zinc, tin, lead, gold, mercury, potassium, potash ley, minium. All of them, after the lapse of half a minute, caused, through the wire, warmth in her hand. Sulphur, carbon, oxalic acid, aqua regia, green vitriol, and sea salt, after the same time,

caused coolness. After removing the objects, the sensation began, also after half a minute, to fade, and always required several minutes to disappear.—Sulphur, when only brought near the plate, without contact, caused coolness.—The German silver rod, when laid on the plate, yielded warmth, but after having been for some minutes in contact with sulphur, and again, with my right hand, laid on the plate, it gave durable cold.—An empty bottle, which by itself felt slightly cold, (different glasses varied in this respect,) was wrapped in gold leaf for some minutes. When now laid on the copper plate with my left hand, it caused lively warmth. I next placed, successively, in a wide glass tube, a number of the most widely different substances in their bottles; some of these solid, others liquid. The patient, holding the other end of the tube, described to me exactly the same sensations as had been already caused by contact with the same substances.—I also gave her a long thin glass rod, like a thermometer tube, but stouter, one end of which she held, while I placed the other, successively, in the contents of many bottles, containing amorphous preparations of all kinds. She now described the sensations exactly as when the wide tube was used, so that finally, a glass rod was found the most convenient of all means for trying the magnetic force of dry, fluid, corrosive or offensive preparations.

Mlle. Maix gave me some additional proofs. When I tried her with the copper plate and wire, and laid sulphur, sulphuric acid, selenium, sugar, silk, moist linen, &c., on the plate, some directly, others in bottles, she named for all the same sensations as they caused when placed in her hand.

My daughter Ottone gave one hand to Mlle. Reichel, and held the other over sulphur, without contact. At the end of half a minute, the patient felt her hand cold; and after one minute the pricking sensation was felt.—An hour afterwards, the experiment was repeated, only that my daughter's hand was held this time over gold leaf. In half a minute warmth was felt by the patient. It increased during a minute, and then remained stationary.

204. All this proves, *that the force which emanates from all*

amorphous bodies is conducted, and its effects conveyed, through bodies of all kinds, even through living persons ; and that this occurs by merely approaching the body, without contact, to the further end of the conducting mass.

205. It is hardly necessary to speak of the capacity for receiving or retaining the charge, &c., which are all implied in the preceding experiments.

206. We have now to study the luminous phenomena. These are, indeed, surprising enough. When I made experiments with Mlle. REICHEL, in the dark, as to the light from crystals, she spoke to me of locks, window frames, and other metallic objects, all of which she said she saw. When I showed her a freshly-scoured copper vessel, she said that it shone over its whole length, and that a thin, vaporous green flame played immediately over its surface, but flowed chiefly from its edges. On this, I undertook a long investigation, which continued at different periods in Vienna and at Schloss Reisenberg, and confirmed the results by frequent repetition. From this it appeared, that all metals, and in general all elementary bodies, without any trace of crystalline structure, appeared, in sufficiently dark rooms, luminous to highly sensitive persons. Compounds appeared also luminous, but more feebly, and the light was feebler in proportion to the complexity of their composition. To study this as fully as possible, I brought into the dark chamber, one after the other, a large number of different substances. Mlle. REICHEL saw most metals red, almost as if red hot ; some of them gave a white light, some a yellow. (Copper, as we have seen, gave a green light.) A delicate vaporous flame played over all, undulating backwards and forwards. It had different colours in different metals, but the same colour constantly in the same, and could be moved like ordinary flames by the motion of the air, by the breath, or by the hand. More complex substances only showed flames at their points, when crystallised. Otherwise they were either surrounded by luminous vapour, or luminous in their mass as if red hot. The darkness enabled me perfectly to control these statements. I brought to her, at different times, in the dark, different substances, and frequently

the same ones, which no one could discover in the dark, and it was thus easy to see if her later statements were consistent with her earlier ones. This turned out to be perfectly the case in regard to the luminous appearances generally, their intensity and form, and in the simple bodies; their colour also. But in regard to the colour of the more complex compounds, it was not exactly so. The colour of the lights appears, like that of ordinary flame, to depend on the nature of the substances, and of any impurity, however trifling in amount. Thus Mlle. REICHEL saw the following bodies, however often I tried them, uniformly the same in this respect:—

Copper plate; a red glow, with a green flame over its whole surface, especially at the edges, flowing over them with a width of three-fourths of an inch to one and a half inch; undulating from the middle towards both sides.

Iron plate; a red glow; the flame of the whole surface undulating towards the middle, and there a little raised; about half a hand high; with a play of red, blue, and white colours.

Bismuth; a red glow; flame of surface and of edges bluish red.

Zinc plate; a pale red glow; flame white and turbid, at the points bluish red, jagged at the edges, forming bundles at the corners.

Tin, tinfoil; a dirty bluish white flame; flowing feebly over the edge, without corner bundles.

Lead; a blue glow; flame dull blue, at the edge feebler than tin, without corner bundles.

Cadmium; a white, somewhat bluish flame.

Cobalt; feebly blue.

Silver paper, highly polished; a white glow; the flame white, a finger-length high, not undulating, the same at the edge as in the middle, no corner bundles.

Gold paper, polished; a white glow; flame white, two fingers high, not undulating; the edge like the middle behind the paper, a white luminous mist.

Palladium; a strong blue glow, with pale blue vapour.

Platinum; a white glow, with bluish flame.
Antimony; a white glow, with bluish flame.
Rhodium; red (glow?) with pale yellow (flame or vapour?)
Tellurium; a red glow, white at the edges, no flame.
Osmium; a red glow, with greyish red vapour.
Mercury; a red glow; strongly flaming white flame, with a white vapour.
Chromium; a green glow, with a tinge of yellow.
Nickel; a red glow, with a tinge of greenish yellow.
Titanium; a lively red glow, with a tinge of violet.
Arsenic; a bluish red glow, with a pale red vapour.
Iodium; a blue glow; the flame reddish blue.
Sodium; a red glow; the flame turbid white with a tinge of lilac.
Potassium; a red glow; on the fresh cut surface, a large yellowish red flame.
Carbon; a red glow; red flame at the edges.
Diamond; a white flame, blue in the middle, red at the upper end.
Iodine; a red glow; when shaken, exhibiting a tinge of green.
Selenium; a bluish red glow; the flame blue.
Sulphur; a blue flame, with a turbid white vapour.

A glass box, filled with objects of silver, appeared to the patient at night full of fire; all the objects covered over their whole surface with white flames.

I found all these statements, after many trials, to be invariably the same, provided the darkness was complete. But when it was less perfect, slight variations occurred; bluish red became blue, and so forth. On the other hand, the colours of the lights from compound bodies were less consistently described, and even varied considerably. I cannot, therefore, here give them, but must delay doing so till I have again and thoroughly investigated the subject. In particular, the same alkaloid, as prepared by different makers, often yielded lights of essentially different colours, which depended, no doubt, on its different degrees of purity.

In general, these lights had somewhat of the character of

electric light, so that their colour, which, like that of electric light, varied from red to blue, green, and yellow, backwards and forwards, was frequently not to be determined with precision. In aspect, they had the greatest resemblance to the flames of magnets and crystals, and were regarded by the patient as nothing else than a lower degree of the, to her, familiar appearances which she had known from her childhood, and of which, in compliance with the warning given by her mother, now dead, she had hitherto said nothing, from the fear of being regarded by other persons as one possessed of supernatural and forbidden powers; *Scoticé*, as an *uncanny* person.

207. It follows, then, from these observations, that *all liquid and solid bodies, that is, all bodies of a certain density, yield luminous emanations in the form of glow, flame, and nebulous or vaporous lights, just as magnets, crystals, &c. do.*

208. In order to complete these considerations on the material universe, we have still to cast a glance towards the starry heavens. In the fourth treatise, we have seen the powerful influence of sun and moon; and this naturally led me to examine whether the starry host of heaven were without action on the sensitive nerve, or whether, on the contrary, they might not present an influence, feebler, in consequence of the greater distance, than that of the two great and near heavenly bodies, yet possibly sensible in a small degree.

209. Even from the windows of my house, which commands an uninterrupted view towards east and south, to the distance of 190 to 140 miles, and before which I placed Mlle. Reichel in a clear night, she perceived an indubitable effect, such as I had suspected. In the middle of October, at 8 P. M., I went out with her to the neighbouring isolated hills, from which there is a wide view all round. The moon was not visible, and the air was perfectly calm. She perceived coolness from some quarters of the heavens, and warmth from others. This was repeated in different nights, and at different hours; soon after sunset; then at 9 P.M.; twice at midnight; once at 4 A.M.; and shortly before sunrise. The general result was, that soon after sunset, namely, at 6 P.M., the most vivid coolness proceeded from the west; but shortly before sunrise,

at 6 A.M., it proceeded from the east. That long after sunset, at 9 P.M., north and north-west were the coolest, south and south-east the warmest quarters; while long before sunrise, at 4 A.M., north and north-east were the coolest, and south and south-west the warmest directions. Lastly, at midnight, north was cool, south warm, and east and west so nearly alike in apparent temperature, that she could only say that east was rather warmer than west. Had an observation been made between 2 and 3 A.M., it is probable that east and west would have been found exactly equal.

210. This variation of the results was obviously the consequence of the position of the sun. We have seen that the sun's rays cause a cold sensation. Now, at the quarter of the heavens to which he was, at any given time, nearest, as west after sunset or east before sunrise, the greatest coolness was always perceived. At midnight, when the sun had reached its lowest point, the equilibrium between west and east was nearly restored, but west still retained a slight excess of power, the sun having so recently left it, and the perfect equilibrium could not take place till 2 or 3 A.M. That the variations depended on this cause, is proved by the fact, that it was the same when the sky was cloudy. But in all these cases, north was always cool and south warm; and when I desired her accurately to point out the direction of the middle of both, she always pointed in the line of the magnetic, never in that of the astronomical meridian. More particularly in the south, she affirmed that she was sensible of a sharply defined stripe of the greatest warmth, as it were projecting from the rest of the sky. The warmest direction was always that of the magnetic south, and the coolest that of the magnetic north, when compared with other points of the compass. This clearly indicates the true interpretation of these statements.

211. But these half terrestrial, half solar phenomena, must not be confounded with the stellar, with which, in our sensations, they are complicated. When Mlle. REICHEL was out in a clear night, she always pointed out the milky-way as decidedly cool; as also the Pleiades, the Great Bear, and

others; and, in general, the starry expanse was felt cool, and only individual stars caused a sensation of warmth. These were invariably stars of the first magnitude, and when I examined them with the dyalite, I found them to be Saturn with his rings; Jupiter with his four satellites; Venus; in short, always a planet. It appeared, therefore, that stars shining with borrowed light, appeared to the patient warm, and all others, shining with their own light, appeared cool. This coincided very beautifully with the previous observations, that the moon yielded warmth, the sun, (and, therefore, the fixed stars,) coolness.

212. I was even able to establish this further. She was unable to bear long the aspect of Jupiter, as of all bright light. She said that all the stars together acted on her like a weak magnet, not only in front, but from behind, on the spinal column, but especially on the head, where she was most sensitive to magnetic influences. I connected a copper plate of about one square foot with a long brass wire, the other end of which reached to the patient, who was shut up in the darkened stair. The wire gave, by itself and with the plate, a small flame at its end; but when I, without her knowledge, allowed the star-light to fall on it, after a short pause she announced the rise of a slender flame, more than a span in height, which fell and rose, according as I moved the plate into or out of the star-light. A zinc plate gave the same results, somewhat more feebly. The sensations corresponded; the wire became cooler, when the star-light fell on the plate, and that more distinctly, when no great planet shone on the plate; less so, when such a planet partly neutralised the combined effect of the stars.

There is nothing in these observations, which, after the contents of the preceding treatises, can much surprise us; but they are certainly a fine additional confirmation of what has been stated in regard to the sun and moon, and also of the fact, that the whole material universe, even beyond our earth, acts on us with the very same kind of influence which resides in all terrestrial objects; and lastly, it shows, *that we stand in a connexion of mutual influence, hitherto unsuspected, with the universe ;* so that, in fact, the stars are not altogether devoid of action on our sublunary, perhaps even on our practical world, and on the mental processes in some heads.

213. We thus reach the concluding proposition of this treatise, namely, that, just as magnets, crystals, organised beings, the solar and lunar rays, heat, electricity, &c., possess the power of exerting a peculiar force, common to them all; *so this force resides also in all, even the most widely different bodies yet examined, and consequently, without doubt, in all amorphous bodies, even when gaseous, including the stars ; and, therefore, takes its place as a general and universally diffused natural force.* In the first ten sources of it, we have seen it appearing *concentrated in certain points of the material universe ;* but here we find it to be *a universal property of all matter, in variable and unequal distribution.*

214. Whether this *natural force, embracing the universe,* be an entirely new one; or a hitherto unknown modification of a force already known; or a complex force, derived from several known forces in a form of combination hitherto unknown;—these, and many other important questions, I leave, for the present, undecided. I have now collected and arranged all the sources from which I have seen it flow. In future treatises I shall compare these together, and endeavour further to develope them, in many points of view, some of which I have only been able to indicate. Higher judges than myself will then perhaps undertake to pronounce a verdict on the whole investigation.

---

215. In conclusion, I shall here venture to make a proposal for removing the difficulties in language and nomenclature which are connected with such a subject, and with which, as the reader must have observed, I have had to struggle throughout the preceding pages. Wherever, for the last seventy years, this force has been observed in isolated manifestations, it has received a number of different names, almost all of which are derived from analogies to, or connections with, magnetism. It has always been viewed as more or less identical with magnetism; but we have seen that it has no greater resemblance

to that force than magnetism has to crystallisation, crystallisation to electricity, electricity to chemical attraction, heat to light, &c. We no doubt have a presentiment of the final unity of these imponderables in a higher form; but we are still far removed from this much desired goal of natural science. We cannot yet even fill up the gap between magnetism and electricity, which appears so narrow that we might almost expect to reach with our hands from one bank to the other. But as long as we can, in reference to the force under consideration, effect almost as much with a soft iron bar, which cannot attract one particle of iron, as with a powerful steel magnet, (§ 194); as long as there are magnets and crystals, which act with equal energy on the sensitive nerve, while the former supports a mass of iron, and the latter cannot move one filing of iron, (§§ 37, 42, 43, 44); and as long as no scientific explanation of this vast difference can be given; so long must the two forces, magnetism proper and the new force, remain essentially distinct, in so far as we are unable to regard them from a common point of view. And thus, for the present, a fit and proper name for the new influence is necessarily demanded. Reserving for another opportunity the etymological justification of the term, I shall here venture to suggest, for the force here treated of, the short name of ODYLE. Every one will think it desirable, that for an object, universally occurring in an infinite variety of relations in the material world, we ought, with a view to the numerous compound terms which are required, to select, if possible, a short word, and for convenience, one commencing with a vowel. The words magnetism, electricity, &c., are far too long for convenient use in science. When they are lengthened by additions, as vital magnetism, animal magnetism, &c., such terms are as troublesome as they are in themselves fallacious, for these things do not exclusively, or even chiefly, belong to vitality or living beings, and are still less identical with magnetism. To the force, which carries iron, and gives to the compass needle its direction, we leave its old name, with the original idea of attracting or lifting iron, which belongs to it. If, now, the term ODYLE should be found, in general, advisable for that force which attracts no iron, and for which we require and are seeking for a name, the nomenclature for the different sources or origins from which it proceeds can easily be compounded. Instead of all circumlocution, and instead of saying "the odyle derived from crystallisation," this form of it might be called crystallodyle; that from animal life, vital odyle; that from heat, thermodyle; that from electricity, electrodyle, or shorter, elodyle; from light, photodyle; and magnetodyle; chemodyle; heliodyle; artemodyle; tribodyle; and for the whole material universe, pantodyle; &c. &c. I know that objections may be taken on the score of grammatical correctness. But when it is possible that new words may become important in practical life, custom and convenience alike demand, that something of grammatical accuracy should be sacrificed to euphony. It is possible, nay even very probable, that we may one day succeed in bringing under a common denominator, the hitherto incommensurable fractions called magnetism, electricity, crystallisation, light, heat, chemical affinity, &c.; but the numerators must always remain unequal and distinct; and we shall always have groups of phenomena, collected and kept separate from other groups, under the names of magnetism, electricity, &c. And thus, whatever may be the fate, as a matter of science, of what I have found it necessary to unite under a new name, yet we must always require a word, like odyle, or any other of the same value, to enable us conveniently to group and describe these phenomena.

Retrospect.

a. Not only magnets, crystals, the human hand, chemical action, &c., but all solid and liquid bodies, without exception, excite sensations of coolness or warmth, which are here synonymous with pleasant and disagreeable.

b. The acting force belongs not only to certain regular forms, or to peculiar conditions of matter, but resides in matter as such.

c. This force acts, not alone on contact, but also at a distance, as from the sun, moon, and stars, so from all matter.

d. In regard to this force, bodies arrange themselves in the electro-chemical order.

e. At one end stand the electro-positive bodies, with potassium at their head; at the other the electro-negative bodies, headed by oxygen. All the electro-positive metals are found on the side yielding warmth and a disagreeable sensation; all the metalloids on that yielding cold and a pleasureable feeling.

f. This force is conductible; bodies may be charged with it; they shine with a luminous glow, visible to sensitives in the dark, with flame and luminous vapour.

g. This force is one which embraces the whole universe.

h. Nomenclature. The word odyle; odylic; with inflections and compounded terms derived from it.

SEVENTH TREATISE.

DUALISM IN THE PHENOMENA OF ODYLE.

216. The polar opposition in the magnet; the dualism in every crystalline form; the symmetrical and sexual opposition in all living organisms; naturally led me to suspect that something analogous might here be found to prevail. The first indications of this which presented themselves were, the constantly recurring sensations of warm and cool, of agreeable and disagreeable, which all sensitive persons, healthy and diseased, perceived in all material substances. I found, indeed, that these persons were not always unanimous in regard to the sensations caused by the same body or substratum; but when they had once given to any substance its place among the warm or cold substances, they were, in subsequent trials, almost invariably consistent with themselves. There must therefore have been in operation here, objective differences in the bodies tried, and subjective differences in the observers, connected probably with the form of their disease, which determined on the one hand inconstant, on the other, constant results. The attempt, therefore, to reach, in pursuing the sensations of warmth and coolness, a path that might lead me to a sure scientific truth, was here surrounded by manifold and peculiar difficulties. These could only be overcome by patient perseverance.

217. The first question was this: What does the expression warm mean, when used by sensitives? What does cold mean? The objects examined by them were all of equal temperature; these words, consequently, could not mean a real, but only an apparent temperature, or difference of temperature; and the expressions must be taken in a peculiar sense, different from their usual one. They imply *effects on the sense of touch*, proceeding from an unknown cause, *which are analogous to those of heat and cold.*

218. Mlle. STURMANN found a bottle of oxygen gas and a piece of sulphur both hot; Mlle. REICHEL found them both cold; and Mlle. MAIX felt them both, when on the hand, hot, but diffusing in every direction a cool aura. They all agreed in this, that they perceived a variation from the temperature of the air; but in determining the degree of this, they gave me different accounts; three observers made three different statements; and all three continued at all times, and on every repetition, to be consistent, each with herself.

219. From this it plainly appeared, that not only the objective cause residing in matter, was present in different degrees in different substances; but that also in different diseases, different degrees of sensitiveness occurred. These last, again, might cause merely quantitative differences of sensation, so that a substance acted more energetically on one patient than on another, perhaps over-exciting one of them; or else they might lead to qualitative differences of result; so that, in one case, a certain body might appear always warm, in another always cold.

220. In order to make a nearer approach to the natural laws which here lay concealed, it was necessary to state the question more simply; to begin, not with substances which had different effects, but with such as produced the same sensation. I therefore again had recourse to my rock crystal and gypsum, as with them I might hope to be able to observe, in one and the same specimen, the different sensations, which, with sensitives, seem to run through all nature; and from this point to pursue the investigation further. I therefore passed a crystal of gypsum four inches long, with a natural acumina-

tion at both ends, over the inner surface of Mlle. NOWOTNY's hand, from the wrist to beyond the point of the middle finger, with the point next her hand, but without contact, while she was in the position north to south. She felt the cool aura as if she was blown upon through a straw, as already stated, § 33. I then turned the crystal round, and proceeded as before, with the opposite end nearest her hand. She now felt no coolness, but a tepid warmth, which was disagreeable. A rock crystal, somewhat longer, applied in the same way, gave the same results. Mlle. STURMANN felt the pass with a crystal of tourmaline warm, when the one pole was drawn downwards over the hand; cool, when the opposite pole was used. A crystal of Iceland spar had the same effect on her.

221. Mlle. ATZMANNSDORFER expressed herself in the same way. The crystal of gypsum, passed downwards over her right hand, caused coolness with the same end as had been felt cool by Mlle. NOWOTNY. The same end was felt still cooler over her left hand. The opposite end was felt warm on the right hand, warm and disagreeable on the left. I obtained exactly the same results with Mlle. REICHEL. The same end of the crystal of gypsum, which, in the two cases above mentioned, had caused coolness, was felt by her right hand cool in the downward pass; by her left, still cooler and pleasanter. The opposite end, on the downward pass, caused hardly a trace of coolness in the right hand, but warmth on the left. She said, that the downward pass felt as if something were taken from her, the upward pass as if something were communicated to her. Mlle. MAIX, in similar experiments, expressed herself exactly in the same way.

222. I have already reported similar results in the cases of M. SCHUH and Professor ENDLICHER, in §§ 33, 34, and 35. Other persons, who experience the same sensations, have also permitted me here to name them. M. KOTSCHY, Dr. FENZL, M. VOIGTLANDER, optician, Mr. INCLEDON, M. STUDER, and the carpenter KLAIBER, are all of them sensitive in this particular, and all healthy. M. KOTSCHY and Mr. INCLEDON could not endure the pass with the cold end of a large rock crystal, from the head downwards, and over the body, more

than a few times, without being so strongly affected in the stomach that I was compelled to stop.

223. All the experiments and all the witnesses, then, agree generally in this, that *in crystals, on the downward pass, one pole causes coolness, the other warmth.* I say expressly they agree generally; for there are always, here and there, some persons, who are not quite clear as to their sensations of cold or warmth, and describe the same pass as at one time cool, at another warm, and vary in different degrees between these feelings; or who can only become certain as to the sensation, and consistent in their account of it, after several repetitions of the pass. These are, however, always either healthy or merely delicate persons. The highly sensitive are hardly ever in doubt. And especially it is those less sensitive persons who are unaccustomed to the experiments, that, at first, are more or less uncertain in their statements. But I shall hereafter describe more particularly the cases in which this uncertainty occurs, even in such individuals, and reduce them to a few clear and decided instances.

224. After these experiments, along with those described in the second treatise, had established the existence of dualism in the phenomena, as presented by crystals; which dualism exhibits an unmistakeable parallelism with that of crystallisation itself;—the questions now arose: Of what nature is this dualism? Does it proceed from a duplex influence? or does it depend on the actual presence and absence of a certain influence? or, finally, does it correspond to positivity and negativity?—I confess, that to these questions I can give no more a distinct answer than I am able with certainty to answer similar questions as to cold and heat; as to + E and — E; as to + M and — M, &c. I was obliged to content myself, for the present, with endeavouring to establish the parallelism which I hoped might be traced between odyle and crystallisation, magnetism and electricity.

225. We already know that odyle has many points of coincidence with magnetism, when we abstract from the phenomena of the attraction of iron, and of giving direction to the needle, &c.; and particularly that of acting on sensitive

persons in exactly the same way. A pass with a magnetic needle over the hand of a sensitive, produces, as we already know, the very same sensations as when a crystal of gypsum, calcareous spar, or topaz, &c., is used. The northward or negative pole of the needle generally caused coolness; the southward or positive pole, warmth. If now we should succeed in proving a perfect coincidence between certain poles of crystals, and certain poles of magnets, we should be entitled to conclude that the cause, in both, was similar, and to apply to those crystalline poles which produced the same physiological effects as certain magnetic poles, and therefore had similar properties, similar names; so that where odyle agrees with + M, we should call it + odyle; and on the same principle we should speak of — odyle.

226. In order to ascertain this, I studied more accurately the relation between magnetic poles and the sensitive nerve. I caused Mlle. MAIX to take a small bar magnet in both hands. It was about four times as long as the breadth of her hand. I made her first place both hands together on the middle of the bar, the northward pole being towards her left side. She perceived in this way a moderate degree of disturbance or uneasiness. She now moved her hands towards the opposite poles, till her left hand grasped the northward pole, and the right hand the southward pole. The effect of this change was very sensible. She now felt a vivid uneasiness through her arms, chest, and head. When she took away one hand, the uneasiness instantly ceased, and it returned and disappeared as often as she applied or removed her hand to or from one pole, holding the other all the time. The same result occurred whichever pole she let go. There was here, obviously, something of the nature of a current or circulation, such as was formerly described as being caused by my hands, in the third treatise, § 86. To control this, I next used a large horseshoe magnet, placing one of her hands round each pole, the left hand round the northward pole. She instantly felt uneasiness and oppression of the chest from the current, which passed through the chest from the arms. The head became oppressed, and soon stupified; and she once more made use of

the simile of the carrousel, or riding round the circus, to describe the current formerly mentioned. When I caused her to remove one hand, she instantly felt the current interrupted, and she breathed deeply, with apparent relief and refreshment. On every repetition the result was the same. In both experiments, especially in that with the large horse-shoe, it was necessary to have the northward pole in her left hand, the other in her right, in order that the sensation should be in some degree endurable. When the poles were changed she could not hold out; but perceived the peculiar strife or contest in herself formerly described, which caused such severe distress, that I was instantly compelled to put an end to the experiment. If I may now venture to assume the existence of a current in this experiment, analogous to the galvanic current, I must conclude that it proceeded from the positive southward pole of the magnet, through her right arm and body to the left side, and flowed down the left arm to the negative or northward pole of the magnet. Therefore her left hand corresponded to the southward, and her right hand to the northward pole of the magnetic needle; in other words, her left hand was positive, her right hand negative, in relation to the poles of magnets (which poles apply to the established use of the word magnetism); and the state of the left hand must of course be named + odyle, that of the right hand — odyle.

227. We have seen (§ 86), that when I held her right hand in my left, and her left hand in my right, a similar current was perceived, which was endurable; but that, when I crossed my hands, and thus placed right in right, and left in left, the singular internal strife arose, which she could not endure, its effect on her being in the highest degree distressing. It follows that my hands also, in the quality of their magnetic state, correspond perfectly with those of Mlle. Maix; that my right hand represents the negative, or northward pole, and my left the positive, or southward pole, of the magnetic needle The order of these properties is in me, then, the same as in her; and, consequently, *males and females possess the same polar organisation in this particular.*

228. After this had been cleared up, I gave to Mlle. Maix,

in her left hand, the bar magnet, in such a way that it lay extended from the point of the middle finger, along the hand, and up a part of the arm. The northward pole lay on the arm, and the other pole on the point of the finger, and thus all was in the natural order. When I turned the bar round, disagreeable sensations were perceived, and the before-mentioned strife began from the point of the finger to the root of the hand. I now drew the magnet up, so that it rested on the fore arm. When the northward pole was at the elbow, the sensations were those of the natural order; but when the southward pole lay at the elbow, the painful strife instantly appeared.

I repeat here what was formerly stated, and must not be overlooked in judging of these experiments, that they were all made with the patient in the position of north to south; the head towards north, the feet towards south, with the face turned to the south.

229. If now, as all hitherto had indicated, the same power of acting on the sensitive organism resides in crystals as is found in magnets, it is obvious that a crystal, applied in the same way, must produce the same or similar effects. I placed a crystal of gypsum in the hands of the patient. It soon appeared that it was by no means indifferent to her in what way it was held. She soon discovered that the two opposite and most distant points of the rhombohedron were indeed the opposite foci of an internal influence; but they were not the strongest. There were two others in the line of the shorter diagonal, which were much more powerful, and lay in the polar principal axis. The poles of this axis were not equal, but one was distinctly warmer, the other cooler, as she had, in the same way, so uniformly observed. When she now held the crystal, so that the cooler end lay on the point of her left middle finger, and the other on the point of her right middle finger, the sensation was, in a manner, normal. But when the crystal was reversed, the disagreeable sensation so often mentioned was instantly observed. The cool pole of the crystal corresponded, therefore, to the northward, the warm pole to the southward pole of the needle. When I placed a crystal in her left hand, it was by no means indifferent to her, in which

direction the axis lay, however short the crystal might be, (here the principal axis was four inches long.) When the cool pole lay near the wrist, the other towards the middle finger, the effect was agreeable; but when the warm pole lay near the wrist, the uneasy feeling of internal strife instantly set in, although only over a short surface. Similar experiments with garnet, staurolite, and heavy spar, gave exactly the same results.

230. I would hardly venture to attach so much importance to these observations, had I made them only on Mlle. MAIX. They might depend on a peculiar, and possibly transient or variable morbid state. But on repeating them, in cases of quite different diseases, I obtained the same results. When I gave to Mlle. NOWOTNY, already far advanced in convalescence, my hands, she felt each of them singly, as Mlle. MAIX had done; and when she held both in the natural order, the feeling of a current through her became so strong that she could not long endure it. Mlle. ATZMANNSDORFER felt my right hand in her left tepid, my left in her left hot. With both hands she instantly felt the current, which affected her whole body, and stupified her head. But when my hands were crossed, I could not proceed far, for the effect was so violent, that in the course of a few seconds she began to lose consciousness, and I was obliged to stop. Mlle. REICHEL felt nothing disagreeable when I placed my right hand in her left, but when I placed it in her right hand, the effect was painfully disagreeable. When both hands were given to her in the natural way, she felt the current up the arm into the head, but far less strong and intolerably distressing than when my hands were crossed. All this agreed perfectly with what I had observed in Mlle. MAIX.

231. We thus obtain the law, that certain poles of crystals and of living beings correspond, in regard to odyle, with certain poles of the magnet; *that crystals, in this respect, have well-marked northward and southward poles, the cooler always corresponding to the northward, the warmer to the southward pole of the magnet; and lastly, that the right hand, both in males and females, corresponds to the northward, the left hand to*

the southward pole of the magnet. Equivalent to, and parallel with, the + M of the north pole of the earth (or the southward pole of the needle), we have here + odyle; in this case crystallodyle and vitalodyle; and corresponding to — M, we have — O, &c.

232. To confirm further the proposition here laid down, I shall now describe similar results obtained with healthy subjects, and particularly with M. SCHUH, a cultivator of physical science. In appearance, a healthy vigorous man of little more than thirty years of age, but of lively nervous temperament, he has yet far more sensitiveness for odyle than many other persons, so that he occupies a sort of middle position between the highly sensitive patients, and insensitive healthy persons. He had hardly ever been ill, but occasionally suffered from headache, when he over-exerted himself in his studies. He is very sensitive to the action of all crystals; large magnets affect him distinctly at the distance of nearly three and a half feet. When I placed my right hand in his left, he felt, after a few seconds, an unpleasant sensation in his head; when I, in addition, placed my left hand in his right, this sensation rapidly increased, and in half a minute caused an undulating headache, that soon became almost intolerable, and after I had removed my hands, continued for seven or eight minutes, and then slowly disappeared. When I crossed my hands, and took hold of his, he felt it most disagreeable. Proceeding from the observation, that right and left are to each other as negative to positive, I begged him to place his own right hand in his left, leaving out my hands. To the no small surprise of himself, as well as of those present, he found that his headache instantly set in, and almost as severely as before; and that it ceased and returned, gradually, as he separated or joined his hands. His negative right hand formed with his positive left a kind of pair or simple arrangement, (if I may use the terms of galvanism,) and the circuit was closed by his arms and body, so that polarisation, or, if we may for the present adopt the word, the current began, and soon acted on the brain. Several months afterwards, he told me, that he dared not, at any time, leave his hands together.

Since he had learned the effect of doing so, the sensations always reminded him, if, during the night, he accidentally brought his hands in contact, to separate them immediately. Other healthy sensitives exhibited similar effects. M. KOTSCHY was immediately affected by my hands; and when I held both, described the effect as a kind of circulation from me through him, and back to me on the other side. When my hands were crossed, he described his sensations as of a disagreeable distressing undulation in the arms and head, almost in the very words of Mlle. MAIX. Now M. KOTSCHY never suffered from headache in his life. When I gave my hands to Mr. INCLEDON, and above all, when I crossed them, he felt an absolutely intolerable headache.

233. We may now look back to the very remarkable phenomenon in all sensitives, that when lying in bed, or reclining in a chair, they were least able to endure the position from west to east. Here the head is towards the west, the feet are towards the east, and the face looks eastward. But in this case their whole right side is towards the south, their left side towards the north; or, in other words, the positive side of their body is towards the positive pole of the earth, the negative side towards the negative pole. Like named poles, that is, mutually repellent poles, are thus turned towards each other, and we can thus comprehend how this position should be so intolerable and injurious to the patients. In July, when Mlle. NOWOTNY was beginning to go out, she could not at all endure, even in the open air, continued walking from west to east. It is impossible to imagine more beautiful confirmations of my observations; and M. SCHUH is no bed-ridden and concealed patient, but a vigorous man, known and seen over the half of Vienna and Berlin.

234. I undertook a minute control of the law thus indicated, with the help of Mlle. REICHEL. The reader is aware that I had series of elementary and compound bodies arranged by Mlles. NOWOTNY and MAIX, according to the strength of the disagreeable sensation they caused. But in these arrangements, although they proceeded from the most powerful to the least powerful or the indifferent, no attention was paid to their

position in the electro-chemical series as positive or negative. I had only attended to the quantity, not the quality of their effect on the patients. If, then, as appeared probable, the difference between cold and warm was founded on a difference between electro-negative and electro-positive, as between the poles of magnets and crystals, these mixed series must be capable of division into two, one of warm, the other of cold bodies, one of which would then include the negative, the other the positive bodies. In this experiment, I took the series of Mlle. MAIX, and caused Mlle. REICHEL to go through it, and divide it into two groups of cold and warm substances, proceeding from the strongest to the weakest. I give here the result. The numbers are those of the series of Mlle. MAIX, before it was divided into two by Mlle. REICHEL.

WARM.	COLD.
2. Potassium.	1. Oxygen.
3. Caffeine.	4. Sulphuric acid.
5. Purple of cassius.	6. Iodide of gold.
7. Brucine.	8. Diamond.
15. Chromic acid.	9. Chloride of gold.
19. Picamar + lime.	10. Sulphur.
21. Bromide of silver.	11. Bromine.
22. Iodide of silver.	12. Tellurium.
23. Iodide of bismuth.	13. Osmic acid.
26. Picamar.	14. Gypsum.
27. Atropine.	16. Lunar caustic.
28. Acroleine.	17. Orpiment.
31. Rhodium.	18. Corrosive sublimate.
32. Narcotine.	20. Oxide of platinum.
35. Strychnine.	24. Iodide of carbon.
37. Sesquioxide of lead.	25. Periodide of mercury.
38. Alloxan.	29. Iodine.
40. Picrotoxine.	30. Telluric acid.
41. Ultramarine.	33. Bicyanide of mercury.
43. Mesite.	34. Selenium.
46. Citronyle.	36. Paracyanogen.
47. Draconine.	39. Tungstic acid.

WARM.	COLD.
48. Bismuth.	42. Sulphuret of potassium.
52. Creosote.	44. Arsenic.
53. Potash.	45. Peroxide of mercury.
57. Lithium.	49. Iodide of lead.
58. Cantharidine.	50. Chloride of cyanogen.
61. Cetine.	51. Bleaching powder.
64. Æsculine.	54. Oxide of copper.
66. Baryta.	55. Cyanide of potassium.
70. Melamine.	56. Sulphuret of calcium.
74. Grey cast-iron.	59. Sulphate of morphine.
75. Murexide.	60. Bromide of potassium.
76. Protoxide of manganese.	62. Cyanic acid.
78. Hydrated oil of turpentine.	63. Antimonic acid.
79. Cholesterine.	65. Sulphuret of cyanogen.
80. Asparagine.	66. Hydrate of baryta.
82. Hyoscyamine.	68. Parabanic acid.
85. Alloxantine.	69. Borax.
88. Caryophylline.	71. Acetate of morphine.
89. Allantoine.	72. Hydrochlorate of citronyle.
90. Sulphuret of ammonium.	73. Phosphuret of nitrogen.
91. Lime.	77. Oxide of cobalt.
94. Gold.	81. Titanic acid.
97. Zinc.	83. Uric acid.
98. Stearine.	84. Neutral phosphate of lime.
99. Chromium.	86. Chloride of carbon.
101. Osmium.	87. Carbazotic (picric) acid.
104. Palladium.	92. Phosphorus.
107. Mercury.	93. Bichromate of potash.
108. Delphinine.	95. Oxide of nickel.
109. Daturine.	96. Alcohol.
110. Lead.	100. Chloride of chromium.
113. Oleic acid.	102. Albumen.
116. Cadmium.	103. Ammonio-chloride of Platinum.
117. Sodium.	105. Oxide of chromium.
118. Antimony.	
121. Minium.	

WARM.	COLD.
124. Morphine.	106. Graphite.
125. Benzamide.	111. Oxide of silver.
126. Veratrine.	112. Common salt.
127. Pure indigo.	114. Molybdic acid.
129. Titanium.	115. Iodide of potassium.
132. Naphthaline.	119. Persulphate of iron.
133. Mineral wax.	120. Nitric acid.
137. Nickel.	122. Peroxide of manganese.
138. Copper.	123. Sebacic acid.
139. Santonine.	128. Stearic acid.
140. Iridium.	130. Massicot.
141. Tin.	131. Oxamide.
142. Cobalt.	134. Cinchonine.
148. Amygdaline.	135. Melane.
149. Mellone.	136. Hippuric acid.
150. Quinine.	143. Fumaric acid.
151. Piperine.	144. Malic acid.
156. Benzoyle.	145. Benzoic acid.
157. Urea.	146. Lactic acid.
158. Platinum.	147. Cinnamine acid.
159. Silver.	152. Superoxide of lead.
163. Eupione.	153. Gallic acid.
167. Bar iron.	154. Tannic acid.
171. Paraffine.	155. Succinic acid.
	160. Well water.
	161. Mannite.
	162. Charcoal.
	164. Starch.
	165. Gum.
	166. Sugar.
	168. Tartaric acid.
	169. Sugar of milk.
	170. Citric acid.
	172. Distilled water.

235. On examining this arrangement, we find on the side of the warm bodies almost all the metals, with potassium at the

head of the list, excepting only tellurium and arsenic, the most negative of all, and the most like to non-metallic bodies. We find here further most organic substances, and the organic bases; the highly hydrogenised compounds of carbon, and of the acids, only chromic and oleic acids. On the opposite, or cold side, we find all such bodies as sulphur, bromine, iodine, selenium, all chlorides, metallic oxides, the cyanides, and almost all the acids. As far as can be judged, there are, on one side electro-positive, and on the other electro-negative substances. It is surely surprising, and highly worthy of notice, that a girl, totally ignorant of these things, should be able to arrange by an obscure feeling alone, without seeing them, all material bodies, with the utmost delicacy and certainty, according to one of their most deep-seated and concealed properties, namely, their electro-chemical character.

236. As we were compelled to conclude, in our researches on the magnet, on crystals, on the human hand, &c., *that all substances causing a warm sensation are positive*, so we are now compelled to conclude *that all positive substances cause a warm sensation*. The same, in the opposite sense, is true of negative bodies; and we thus arrive, by a new path, at an electro-chemical arrangement of bodies, which, from our point of view, we may call the odylo-chemical arrangement.

237. With regard to the mode of making these experiments, I may here remark, that I placed all the solid bodies in the left hand of the patient, those which were in powder being laid on a small piece of very fine silk paper, which by itself had no sensible effect; and the liquids in the phials in which they were. I did not omit to control the results as accurately as possible, by repeating the same in different ways. At one time, I introduced all the bodies into one end of a wide glass tube, while Mlle. Reichel held the other end. As I changed one substance for another, so the sensation changed from warm to cold, &c. immediately.—Another time, I selected a glass rod, which felt neither hot nor cold, and one end of this, she, holding the other, pushed into powders and liquids, and laid on or beside solids. She could distinguish at all times, and accurately, the warm and cold substances; and I can

particularly recommend this method of observation, as easy to manage, applicable to all cases, and very distinct for the sensiti e observer. With two such rods of equal thickness, any two substances may be accurately compared. The trial with substances, contained in phials and thus laid on the hand, is only practicable with such bodies as act very powerfully, as sulphuric acid, potash, or caffeine. With more feebly acting substances it is inadmissible, because the glass varies much in its effect in different phials, and, according to its composition, is sometimes cold, sometimes warm, sometimes indifferent, and thus complicates the results. This often goes so far, that feeble substances, which are, by themselves, cool, may be felt even warm through the glass, and *vice versa.* Salts and other bodies in the form of crystals, must be powdered, even if only coarsely, before being tried. For no pure or unmixed result can be expected from a congeries of crystals, each of which is polar; some of the more powerful crystals act by their poles, and thus disturb the result. For example, the results from nitre, and bichromate of potash, were long contradictory, till I reduced them to powder, when they yielded constantly a cold sensation. But the powder must not be used till some hours after it has been powdered, because the mortar and pestle have communicated to it their own state, which only gradually disappears. The percussion and friction also change its natural state, the force derived from mechanical action and perhaps also from electricity being communicated to the powder. Finally, the substance must not, before trial, be left near others, especially powerful ones, which would charge it with their own influence, nor must they be exposed to sunshine or moonlight, nor held long in the hand. When several are compared, their temperature must be the same, &c. &c. Any of these rules, if neglected, would render the results more or less fallacious, as is easily seen from the preceding pages.

238. I need hardly remark, that I do not give the above series as a normal one, but merely as an example, and as a help towards completing the proof. For, that it might serve as a normal series, it would have been necessary to institute a rigid and minute enquiry into the chemical purity of the

substances; a labour which neither time nor circumstances have permitted me to undertake, up to this time. I wished only to discover and state the law; its accurate application must be left to future researches. Many other preliminary investigations are also necessary, the infinite importance and extent of which I very strongly feel; but above all things, this investigation on positive and negative bodies, which I have only been able so far to carry out with Mlle. REICHEL, must be extended to other sensitives, in different states of health and disease. We shall thus certainly find the key to the differences of their sensations; as also the effects of greater or less sensitiveness, as compared with that of Mlle. REICHEL, who must have been in a state of remarkable equilibrium, I might almost say purity, of the diseased condition, since her sensations coincided so exactly with the general chemical characters of the substances, as previously determined by physics and chemistry in quite distinct ways.

239. We shall now make some applications of the law above ascertained, namely, that substances which cause a warm or a cool sensation to the patient, and therefore are positive or negative in relation to odyle, correspond to those which are electro-positive or electro-negative. In the first place, the sun was found strikingly cool, but the moon strongly warm, by the patients; the fixed stars arranged themselves on the same side as the sun, the planets with the moon. I know not whether astronomers have yet ascertained any thing positive on this subject; so far as I am aware, nothing has been published in any way related to it, except the observations of M. KREIL* on the nature of the moon, deduced from the disturbances of the declination of the needle. It is therefore surely very interesting that we are able, by means of the human nerve and its sensations, to recognise the sun and fixed stars as being all electro-negative, while the moon and planets are found on the electro-positive side, and that heavenly bodies shining with their own light and those which shine by borrowed light are thus oppositely polar. We may possibly in this be enabled,

* Astron. Meteorolog. Jahrbuch. 1842.

if not to decide, yet to render it highly probable, that a comet, which only sends to us polarised light,* is really a body shining by reflected light, and not by its own. The severe critic may here say, that the sensations do not necessarily inform us of the electro-chemical state of the star, considered by itself, but that only the action of its emanations, of its luminous and calorific rays, &c. is recognised by the sensitive nerve. To this I have nothing to object; but I would observe, that all our knowledge of the heavenly bodies is obtained through the effects of all kinds of emanations on our senses, and rests on no other foundation than that which has thus reached us. We are thus in the same position with regard to the emanations of odyle, which tell us that the sun is odylo-negative, the moon odylo-positive, as we hold in reference to those better known emanations, from which we learn that the sun's rays contain heat, of which the moon's rays are almost or quite destitute, &c.

240. We have seen in § 147, when considering chemical action, that all artificial light or fire affects the sensitives with cold. This is worthy of notice, since we know from the investigations of POUILLET (Ann. de Phys. et de Chim. XXXV. 402), that the external surface of the flames of oxidised compounds possesses much free *positive* electricity. Not only is cold diffused by the free flames of bodies, whether positive or negative, as potassium, stearine, oil, alcohol, sulphur; but it is also perceived when the burning body is shut up, whether in positive or negative substances. Thus, when Mlle. REICHEL approached a stove, heated by means of a fire within it, she felt it indeed warm, when very near it, because its actual heat overpowered, in its effects on her, the peculiar emanations above alluded to, especially when the stove was of iron. But only a few paces further off, the stove caused a vivid sensation of cold, and that stronger as the fire burned more vigorously. In winter, when she suffered from frost, and tried to warm herself at earthenware stoves, it was only on approaching them that she felt thoroughly chilled; her fingers, already stiff,

* Qu.? positively polarised? W. G.

became rigid; and she was compelled to retire, and seek to warm herself by walking up and down the room, and rubbing her hands. In such a case, it was almost indifferent whether the stove were of iron or of earthenware. In order to explain this singular effect, we must not overlook the fact, that the phenomenon is a highly complicated one, made up of the emanations from heat, light, chemical action, the electricity set free by them, the substance of the fuel, and lastly, that of the stove itself; all of which act on sensitives, but, as we have seen, in various ways. But the resultant of all, in every observed case of combustion, is a decided impression of cold at the distance of several paces, which drove Mlle. MAIX out of illuminated churches (§ 131); and which had such an effect on Mlle. REICHEL, that, when she remained some time near a wood fire, she felt, in rapid succession, oppression of the head, stupefaction, and lastly, pain of stomach so severe, that, if she did not quickly remove, she fainted. *Fire, therefore, invariably acts on sensitives as a negative body.*

241. The next question is, what is the apparent temperature of the flames produced by odyle, and which, invisible to the ordinary eye, are perceived by the sensitive organ at the polar extremities, for example, of wires in contact with the different sources of that influence. In order to examine this, I introduced into a number of substances, first, a glass rod, by itself cool, secondly, an iron rod, by itself warm, and caused Mlle. REICHEL, in the dark, to try the flames proceeding from the other ends of both, by holding her hand at the distance of about two inches from the flames. Both rods yielded exactly the same series, which I subjoin. There flowed

COLD FLAMES	WARM FLAMES
From Bichromate of potash.	From Gold.
Sugar.	Platinum.
Sugar of milk.	Potash.
Citric acid.	Narcotine.
Oxalic acid.	Minium.
Bleaching powder.	Protoxide of lead.
Sulphur.	Cast iron.

COLD FLAMES	WARM FLAMES
From Bromine.	From Paraffine.
Graphite.	Mercury.
Charcoal.	Tin.
Arsenic.	Cadmium.
Peroxide of manganese.	Zinc.
Alcohol.	Iridium.
Persulphate of iron.	Creosote.
	Iron filings.

The apparent temperature of the flames agrees, therefore, perfectly with that of the substances producing them, when placed in the patients, or tried in a long wide tube, or with a long thin glass rod (§ 234); and we conclude here, that all negative bodies give cold, all positive bodies warm flames. The temperature of the flames yields, therefore, an expression for the relation of bodies in general to odyle.

242. Mlle. REICHEL further described as cold all emanations from electrified bodies, especially from such as were positively electrified. The conductor, the glass, and the wood of the machine, were all by themselves warm; but when excited, so as to yield sparks of only 0.2 of an inch long, she felt them all cold to a distance of ten or fifteen paces. The cold sensation was strikingly increased when I caused the plate to rotate rapidly; it was not felt instantly, but a few seconds after the bodies became electrified. The fur of the fox, warm by itself, was felt very cold after I had struck the electrophorus plate with it. The result was the same when, instead of electrifying a large surface, I allowed the electricity to be discharged directly from the conductor into the air, through a point.

But when I presented to her negatively-electrified bodies, she felt them warm. An electrophorus of resin, warm by itself, became much warmer when struck with the fur, and became, indeed, warmer with every stroke, till the heat reached a certain point, at which it became stationary.

From these observations we should be justified in concluding, that positively electrified bodies caused a cold, negatively electrified bodies a warm sensation. But as this contradicts the already established facts, we must explain it by supposing that it is the result of induction. The air, surrounding the positively electrified bodies, and in which the patient stood, became, by induction, negative, and, as the nearest body, acted on her, its effect being opposed to that of the electric state of the body employed.

243. Many experiments were made with the voltaic pile, and this would be the place, in which the temperature of the flames produced by it might be stated. But as the phenomena are too complex to admit of a mere passing notice, I must reserve this subject for a future treatise.

244. The flames caused at the ends of long wires by the flames of candles and wax lights, as well as those produced by sunshine, were all cold. Mlle. REICHEL perceived very distinctly the cold of the flame from a copper plate, illuminated by eight wax lights, at the further end of a wire, sixty feet in length. Sunshine, falling on an iron plate, and conducted to her by a long wire, gave, at the end of the wire, a flame, the cold of which was felt at a great distance. Moonlight, tried in the same way, gave a warm flame; and these results were constant at all times and on every repetition.

245. The following experiment may serve to illustrate the character of the odyle proceeding from heat; it is analogous to those described in §§ 122 and 123. I filled an earthenware vessel with cold water, into which I introduced one end of a ten foot wire, the other end of which was held by Mlle. MAIX, and covered up the water. When she was quite accustomed to its effect, I poured out the cold water and replaced it by boiling water. The wire instantly became apparently warm, and the heat increased for a few seconds, after which it was stationary. I now threw some ice into the hot water. Instantly the apparent heat of the wire began to diminish, and it sank constantly till it disappeared, and cold was then felt by the patient. The disagreeable, nauseous warmth of the wire was gone, and in its place appeared a coolness which continually increased, became highly grateful to the patient, and was felt gradually to extend through the hand, the arm, and at last the whole person. To judge from this, I must

conclude, that heating produces positive and cooling negative emanations of odyle in bodies.

246. Friction on a copper plate with a board of wood, produced in a twenty foot copper wire warm, or + O.

247. The effect of chemical polarisation was generally determined by the predominating ingredient of the compound, and in neutral bodies, by their special character and place in the odylo-chemical series. A number of experiments bearing on this point have been already described in the fifth treatise, (§§ 137, 139–142). I shall here mention some other examples.—I placed in a glass iron filings and water. A glass rod introduced into the mixture felt to the patient warm at the other end. I now added acetic acid, and instantly the wire felt cold. The acetic acid, as well as almost all other vegetable acids, is edylo-negative. But the rod soon again felt warm, the acetic acid had been neutralised by the iron, and a large excess of iron remained. I now added citric acid; and the same results appeared. Especially after I had stirred the mixture, the rod became again warm. I tried successively several other organic acids; they all produced the same effects in the same order. I next took strong solution of caustic potash, which, being positive, made the rod warm, and added a little sulphuric acid. Cold was felt for a moment, then strong heat, the alkali remaining in excess. When the acid was added in such quantity as to neutralise the base, there was warmth for a short time during the combination, but this was followed by permanent cold. Sulphate of potash, like all sulphates, is negative. Effloresced carbonate of soda, thrown into water, caused vivid cold; the absorption of water of crystallisation in the place of that lost by efflorescence was, therefore, a process which acted negatively. When the mixture was stirred, the cold increased for a short time, and then became more moderate. Carbonate of soda is itself negative. The addition of very dilute sulphuric acid had no perceptible effect on a thermometer standing in the liquid, but yet the rod became strongly warm during the effervescence, and when that was over, became again cold. Sulphate of soda is negative; but positive odyle must have been given out from the expul-

sion and gasification of the carbonic acid. The patient often said, that during decomposition she felt the effect in a pulsating or undulating form; and she thought that the undulations corresponded to those of the effervescence. We shall return to this point, when considering other similar phenomena; for in these matters, as in all, there is no effect without a cause.

All chemical action, therefore, moves in a varied alternation of + and — odyle, dependent on the place of the combining bodies in the odylic series; so that we can always predict the results, if we know their relative value in that series, and their relative quantity.

248. We now come to the consideration of organised beings; and first, of plants. I brought to Mlle. Maix some flower-pots, containing *Calla œthiopica*, *Pelargonium moschatum*, and *Aloe depressa*. I rolled one end of a long copper wire into several windings, so as to form a sort of cage, and gave her the other end till she was accustomed to it. I then laid the further end over the plants, so that they were surrounded by the wire. The effect was unexpectedly vivid, the wire becoming instantly hot to her hand, and affecting even the whole arm. Along with this, the point of the wire diffused a cool aura. The *Calla* was most powerful, the *Aloe* least powerful, so that it appeared as if the strength of the influence kept pace with the rapidity of the plant's growth. The rapid growing *Calla* produced a sensation greatly more vivid than the sluggish *Aloe*, notwithstanding the great size of the latter; while the *Pelargonium moschatum* stood, in every respect, between the others. It may not be out of place to remark here, that the *Calla* belongs to the family of Aroideæ, in which, as is well known, the greatest amount of heat is developed, and, therefore, the most intense manifestations of vital activity occur.

249. Towards the end of September, I went with Mlle. Reichel into the fields. We tried all the flowering plants that we met with. Entire trees produced, as their total effect, cold; and single plants in pots did the same. When examined in detail, she found the stems warm, the flowers cool; as, for example, in *Gentiana ciliata*, *Inula salicina*, *Euphrasia officinalis*, *Odon-*

tites lutea, *Orobanche cruenta*, *Linum flavum*, *Hordeum distichum*, *Coronilla varia*, *Rosa Bengalensis*, *Pelargonium roseum*, *Iberis*, *Impatiens*, *Alehemilla*, *Campanula*, *Daucus*, &c. So also, trees were cold above, and warm near the soil; for example, *Pinus picia*, *Abies nigricans*, *Fraxinus excelsior*, *Hippophaë rhamnoides*, *Laurus nobilis*, *Punica granatum*, *Quercus austriaca*, *Betula alba*, *Morus morettiana*, *Salisburia biloba*, *Hedera quinquefolia*, *Cassia corymbosa*, *Juglans regia*, &c. Among the aggregatæ she found several cool in the rays, warm on the disc, as in *Picris hieracioides*, *Centaurea paniculata*, *Aster sinensis*, *Aurellus*, *Dahlia purpurea*, *Senecia elegans*, *Coreopsis bicolor*, *Asterocephalus ochroleucus*, *Scabiosa columbaria*, *S. atropurpurea*, &c. In some plants, the stem was cool, the flowering head warm; as in *Plantago lanceolata*, *Salvia verticillata.* She felt a mixed sensation of warm and cold from the panicles of *Clematis vitalba*, the seed capsules of *Papaver somniferum*, probably, in this last case, the result of the simultaneous action of the alkaloids and the oil of the seeds. It appeared, *that different parts of different plants were unlike in their relation to odyle.*

250. In order to pursue this further, I took out of the soil a large turnip, and gave it to Mlle. REICHEL for examination. She found the fibrils of the root positive, the bulb negative below, positive above. The whole head, and especially the collum, where the eyes and leaves come forth, was very warm; the leaves warm below, slightly warm at the top, and in the middle, where they were most developed, very cold. A plant of *Heracleum spondilium*, nearly six feet high, was found to have the root warm, the stem to immediately below the head warm, the involucrum externally still warmer, but the umbel itself cold. A ripe cucumber and a melon were cool above at the remains of the flower, but cold below at the insertion of the fruit stalk.

251. Since it thus appeared, that there was no general polarisation of plants, as according to *caudex ascendens* or *descendens*, for example, or as in crystals, but that positive and negative influences were found in various parts of the plants, while yet the same odylic state was found in like-

named organs, and within these in like-named parts, I turned to the study of the organs specially. I first tried with Mlle. MAIX a young *Aloe depressa.* The point of the principal axis acted most powerfully, but among the individual organs, the large lower leaves were most powerful, diffusing more of cool aura than the smaller upper ones. The middle rib of the leaves, with its bundles of vessels, acted more strongly than the points of the leaves, and the under surface more strongly than the upper.—The same plant, as well as an *Agave Americana,* tried by Mlle. REICHEL, gave similar results: The stem colder at the point than below; each leaf near the soil, and on its under surface cooler; and on the upper surface and the middle rib, stronger than at the edges or in the parenchyma. The plants were of equal size, and each had ten or twelve leaves. A leaf of *Ulmus campestris,* of *Laurus nobilis, Punica granatum,* while attached to the plant and uninjured, were all warmer below than above, and at the root of the leaf stalk warmer than at the point of the leaf. Leaves of *Castania vesca,* broken off from the tree, were compared in three stages of their life; as green leaf; as yellowed by autumn; and when of the wintry brown, all of which may be had in October. The green and yellow leaves were, on the whole, cool to the patient; the green colder than the yellow; but the brown leaf was not at all cool, and seemed about as indifferent as a piece of paper, with a faint tendency to a tepid effect.

252. The total results of these experiments in plants, which are only to be regarded as preliminary, may be stated as follows: The radical fibres are warm and positive, the ends of the upper leaves are cold and negative. The point of the stem is developed into buds and leaves, and stands, therefore, on the negative side. We may therefore say, with some justice, that, in general, positive odyle prevails in the descending, negative odyle in the ascending stem. But this must be taken with considerable limitation. For within these principal conditions, there exist numberless details, an infinitely subdivided polarity, in which + odyle and — odyle alternate a thousand times. Yet we perceive this rule, *that where nature has little to do, where her formative energy diminishes,*

negativity prevails ; while positivity predominates where she is active and exerts propulsion. Thus the bundles of vessels in the ribs of leaves, the under surface of the leaves, and the part nearest the insertion, are successively more positive; while the more parenchymatous mass, the upper surface, and the point, are each more negative than the preceding. Now physiology teaches us, that leaves grow chiefly not at the point, but near their insertion; that the point is completed very soon after the opening of the bud, while the leaf long continues to grow from the side of the stalk, and of course in its under surface.* The vegetative propulsion, therefore, soon ceases towards the point, but remains active behind. Here also it seems to be positively of the imponderables, light, heat, odyle, &c., with which nature allies herself in order to produce her organisms; and when she abandons the field to negativity, life disappears.

253. We must now turn our eyes to animal life. How vast the part is which odyle here plays, we see best in the profound and puzzling phenomena of somnambulism. But we do not here discuss that subject, confining ourselves to certain effects produced in sensitives by the healthy vitality of other living beings. When I placed on a copper plate, connected with the hand of Mlle. MAIX by a long wire, a living animal, even of very small size, as a gold beetle (*cetonia aurata*), a butterfly (*bombyx mori*), or any similar creature, I was surprised to see that, after a few seconds, she could discover this without seeing the plate or the animals, by the change of apparent temperature in the wire. If I used a larger animal, such as a cat, the sensation was very vivid. The effect of my own hand, when laid on the plate, overpowered all these, as has been before stated. I have tried these experiments numberless times, and with a hundred variations; they invariably gave the constant result, that every living creature instantly produced an effect, not only on contact, but through a long wire, or a number of other conductors, which to Mlle. MAIX was warm, diffusing a cool aura, like all odylically active bodies in inorganic nature.

* ENDLICHER and UNGER, Grundzüge der Botanik, § 330. SCHLEIDEN, Grundzüge der wissenschaftlichen Botanik, 2ter. Theil, Seite 167.

When I removed the animals, the effect soon ceased, and the wire returned to its usual condition. Similar experiments with beetles, butterflies, and cats, were tried on Mlle. REICHEL. They confirmed the above in every particular.

254. When I raised my hands towards Mlle. REICHEL, she felt, even at a distance, my left hand diffusing warmth, my right coolness on her, like a distant magnet. Mlle. ATZMANNSDORFER felt the same thing more strongly. When I approached Mlle. REICHEL, so that my right side was next to her, she felt me, as soon as I entered the room, cool; but if my left side was next her, she felt me warm. Not only the hands, but the whole sides of human beings, are respectively positive and negative. I made her try me from head to foot, by holding her hand near me. She found, next to the hands, the head most powerful in its action on her, positive on the left, negative on the right side. The toes were also powerful. In regard to front and back, the forehead was altogether cooler, the hindhead especially warm till towards the neck. In the arms and hands, she, as well as Mlles. MAIX and NOWOTNY, observed the following arrangement: The points of the fingers were strongest; then the roots of the fingers; then the inner surface of the wrist, where the hand joins the arm; lastly, the inner part of the arm, where the elbow-joint is attached. In the fingers, other points of concentration were found; everywhere at the lower end of a joint, or the inside. Nature, therefore, here obviously follows the law: From the shoulder to the points of the fingers, *the point of greatest influence always lies on the inner side of the inferior or distal end of a part.* There are therefore six such chief points, increasing in power downwards: the inferior extremity of the humerus, of the forearm, of the hand, and of the three finger-joints, always on the inner side. On the outer surface, there are no peculiarly powerful points.

255. One point of very remarkable power is the mouth with the tongue. It is very cool, and therefore negative. The sensitive feel every thing most distinctly with the lips, especially the odylic influence of the bodies thus examined. And conversely, the mouth of healthy persons is a point, by means of

which all objects may be charged with odyle still more strongly than by the hands. When I took in my mouth one end of a glass tube, a wire, a silver spoon, a bar of wood, &c., and caused the patients to feel the other end, they all found them very strongly charged with odyle. When I held a glass of water to my mouth as if to drink, and gave it soon after to a sensitive patient, she took it for magnetised water. When I passed my mouth, closed so as not to allow breath to pass, along, but not in contact with, the German silver rod, so as to be about a minute close to it, the patients, Mlles. MAIX, REICHEL, ATZMANNSDORFER, and STURMANN, found it exactly so charged as if it had been acted on by the sun's rays, or with contact by a magnet, by the point of a crystal, or by my hand.

Here we obtain a not altogether uninteresting explanation of *the true nature and significance of a kiss.* The lips are one of the foci of odyle; and the flames, which our poets describe as belonging to them, do in fact play there.* The next of these treatises will clearly prove this.

It may be asked, how can this be consistent with the fact, that the mouth is negative? But in fact, the two statements harmonise very well; for the kiss gives nothing, but rather seeks, strives after an equilibrium which it does not attain. It is not a negative; but physically, as well as psychically, its state is one of negativity.

256. As odyle is, in the human body, unequally distributed in regard to *space*, I concluded that it would very probably be found also unequally distributed with reference to *time.* I had many reasons for suspecting, that, in the various bodily and mental states through which we pass every twenty-four hours, the odyle would vary in its distribution and in its intensity. Should this suspicion be verified, I hoped that we might

* Of course the author here speaks of flames visible only to the sensitive. The reader will remember the flames from the points of the fingers, described in § 92. These, and other similar phenomena, such as the one alluded to in the text, will be fully discussed in Part III. But in the meantime we may remind the reader of the very frequent observations made by persons in the mesmeric state, who are always sensitive, and by others also in some cases, of flames from the points of the fingers.—W. G.

obtain highly interesting hints at least, if not explanations, concerning sleep, digestion, hunger, the sensations of heat and cold, the mental affections in their physical effects, and hence also concerning many other questions bordering on the psychical. And even if in this way, in the first instance only comparatively small results were to be obtained, yet it would certainly open up to us a new and highly promising path towards obtaining a knowledge of things on which it is always so difficult to throw any light. With this view I caused Mlle. REICHEL to make hourly observations on me, and I represented the results graphically in curves, of which the abscissæ represented the times, and the ordinates the relative strengths or intensities of odyle. I further extended the observations to my daughter HERMINE and to Mlle. REICHEL herself. She had periods, belonging to her illness, of three weeks, during which she never slept; and I made use of these periods, to continue the observations without interruption during the night. They were made as follows: Every hour the patient took hold of my right hand, examined its relative power, and estimated this, so that I could mark on the table, prepared for the purpose, the point as nearly as possible corresponding to the power observed by her. In different experiments, this was continued for twelve, eighteen, and as far as twenty-six hours. During this time, my usual mode of life was this: I awoke from 6 to 7 A.M., then read in bed till 9 or 10 A.M.; rose, and from 10 to 11 A.M. breakfasted on cold weak tea; dined at 3 P.M., ate a very small portion of sweetmeat (jam) at 10 P.M., and retired to bed from 11 P.M. to midnight. I drank neither wine, beer, spirits, coffee, nor tea (save the cold weak tea at breakfast); and I did not smoke tobacco. I took no exercise beyond a moderate walk, which did not extend further than through the park of Schloss Reisenberg, and passed the time, for the most part, quietly at my study-table. I was in good health, tranquil in mind, and at the age of fifty-six. So much with regard to the circumstances which might have an influence on the experiments. For a quarter of an hour previous to each occasion on which the patient tried my hand, I carefully avoided touching with it metallic objects, even the locks of doors, which I had others to open for me. After dinner, during which I had touched silver utensils, I always allowed some time to pass before I gave my hand to be tried. I also avoided exposure to the sun's rays, and the vicinity of a fire.

257. Since the measures obtained rested on a sensation, that is, on the estimation of a sensation, which could not be checked or controlled by any fixed scale, they could lay claim only to a very moderate degree of accuracy. To approach as nearly as possible to the truth, I repeated one experiment five or six times; that, namely, in which my hand was examined hourly by the patient from hour to hour, and the results marked. The resulting curve is given in Fig. 1. Plate II. where the different trials are compared. The agreement between the different curves is surprising, and proves that the sensations of the observer Mlle. REICHEL have, as I have repeatedly pointed out, a very high degree of distinctness.

As soon as I had thus satisfied myself that in this way observations could be made, which should yield concordant results, I extended the experiments in various directions. I continued them by night, had them tried on females, and, among others, on Mlle. REICHEL herself, &c. I then caused her to examine different parts of the same person, as well as the opposite organs in the same individual.

258. Let me illustrate this a little in detail. In Fig. 1. Plate II., are the various observations of my right hand, tried by the right hand of Mlle. REICHEL. My right hand, as formerly mentioned, is negative, and was found so during the whole time; for this quality, in general, does not change; but the quantity of force varies, and is constantly either rising or falling. This I call the amount of the force. It will be seen, that from 6 A.M., when the observations usually began, a gradual increase took place, till the period between 10 A.M. and noon. From that period a fall or diminution began, lasting till about 3 P.M. The rise then returned, and continued till from 7 to 9 P.M.; after which a continuous fall was observed till long after midnight.

This table, with its often repeated experiments, proves, that from the period of waking, although I remained in bed for hours reading, the odyle in my right hand increased till past the hour of breakfast and towards noon. The rising day, therefore, increased the amount of force in the hand. The fall which now took place lasted exactly till the hour of dinner; and it is obvious, that it is caused by awakening hunger; for as soon as that feeling was stilled by the meal, nay, even as soon as the first food, warm soup, was taken, the fall was checked, and the rise instantly began, continuing steadily till it reached the maximum, which occurred in the evening about the time the day-light faded. Similar trials with Mlle. MAIX and M. SCHUH, as observers, yielded the same results. They both found my hands more powerful after dinner than before it. Towards nine or 10 A.M. there is a slight tendency to fall. This depends, no doubt, on the increasing hunger before breakfast; and is a collateral fact, confirming that previously described as occurring before dinner.

259. To satisfy myself of the correctness of the view I had taken of these phenomena, I repeated the experiment on an individual whose mode of life differed from mine. This was Mlle. REICHEL, who, when residing in my house, dined at 1 P.M., instead of 3 P.M. She was able to examine her right hand very well with her left, and she undertook the observation. The curve obtained was now quite different, as may be seen in Fig. 2. Plate II. Up to mid-day the rise was in general the same as before; but the fall which then began lasted, not till after 3 P.M., but only till 1 P.M., the hour of her meal, then instantly ceased, and was followed by a new increase of force, which reached its maximum when the day began to sink into evening. In this case also a slight tendency to fall before breakfast was observed, which, after the meal, yielded to the general rise of the forenoon.

260. It appeared, therefore, that hunger diminishes the amount of odyle in the right hand, while taking food increases it. Here we clearly come to the effects of chemical action, as explained in the fifth treatise. It is chemical action to which the food is subjected; digestion, that is, decompositions and combinations, begin, and movements of odyle are produced.

Chymodyle is set free, if we choose thus to express it. It is a matter of absolute indifference how much or how little share we attribute to the vital force in these changes. They are still chemical and molecular changes, and odyle is produced by them, which then spreads over the body, and charges the different organs.

261. Having ascertained the course of matter during the day, it remained to determine that of the night. What is our state, in reference to odyle, when the god of day, with his abundant supply of odyle, is absent, and we are subjected to the great and remarkable change of condition occurring in sleep? To observe this, the sensitive observer must wake while the healthy subject of his observations sleeps; and, as the observations were to be made from hour to hour, the difficulty was considerable. But I succeeded in persuading Mlle. REICHEL, by explaining to her the scientific value of such an investigation, and the merit she would have in making it, to come, as she could not sleep, every hour, during several nights, to my bedside, while I slept, to examine the state of my hand, and to note the result. There was no other way of managing it, for, in order to obtain unmixed results, it was necessary that I should sleep exactly as I was accustomed to do, and that in my own bed. Fig. 3. Plate II. shows the result of different varied observations on me and on others. From the morning the odyle rises in the right hand, excepting slight disturbances due to hunger, through the whole day, till a period from 6 to 9 A. M., according to the season, the hour of dinner, &c. It now begins to fall decidedly, and does so till 2 or 3 A. M., when it becomes nearly stationary, until the dawn of day, which at the time of these experiments was from 5 to 6 A. M. Then, as the morning twilight dissipates the darkness, the force rapidly rises, and fresh life flows into the organised world, odyle and vital force again increase during the whole day, as long as the sun shines.

262. I am here also privileged to find a confirmation of a law already discovered. The sun, a mighty source of odyle, sends it to us along with light and heat, and saturates every thing on which it shines with this force, till the maximum is

reached in the evening. *Instantly, as soon as the sun sinks below the horizon, the odylic tension on the human frame diminishes,* and with the commencement of this fall there occur, in persons leading a natural life, weariness, dulness, and sleep. *As soon as the great source of odyle is closed, the fountain of conscious, waking life dries up.* Not alone by heat and light does the sun call living beings into existence; that luminary employs yet another influence, namely, odyle, which, like heat, penetrates all bodies, and the fluctuations of which we now begin to compare with the parallel states of sleeping and waking, and are learning also now to measure.

That it is of little importance whether the sun's rays fall on us directly, or whether we are exposed only to the general light of day, or are in the shade, follows from the laws already ascertained as to the conductibility and transference of odyle. Wherever we may be, this influence from the sun will reach us in due quantity.

263. What is the state of the left hand, which is opposite in polarity to the right? Does it increase and diminish in positive force, as the right hand does in negative? Does it follow the same or another course? This could be only ascertained by observing both hands simultaneously. Fig. 5. PlateII. shows the result. The upper line is that of the positive left hand, the lower that of the negative right hand. The former exhibits in the morning a more rapid rise, and reaches in the evening a higher point than the latter. Moreover, before dinner it does not fall so low. It reaches the first maximum in the forenoon, a little later; the second greater maximum in the evening, rather earlier. This appears to indicate a greater degree of energy in the developement of its odyle.

The positive left hand follows, therefore, not a course precisely identical with, but one very analogous to, that of the negative right hand, as regards intensity of force.

264. The brain is so symmetrical in structure that, since the symmetrical hands exhibit so great a difference, I was naturally led to draw conclusions in regard to the more deep-seated machinery in the human body, of which the hands are only outstretched levers. Is the brain, which many would, and not

always successfully, explore by means of its bony case, not also pervaded by and gifted with the delicate power of odyle, and may it not render itself sensible by its action on sensitive persons? Mlle. REICHEL found my head cool on the right side, like the right hand, but considerably stronger, and warm on the left side. This was the case also with all the persons, male and female, who were examined. I may specify M. KOTSCHY, who was carefully examined by the patient, and whose head, sides, and hands, were found qualitatively to agree with mine. I thought that the head deserved a minute investigation even more than the hands, and I therefore caused the patient to examine my head during two periods of twenty-four consecutive hours, once on the 18th, and again on the 23d of October 1844. Fig. 6. Plate II. shews the result. The dotted line is that of the second observation, which was only continued till 10 P. M.

265. These results are remarkable. They show that the difference between the opposite sides of the head is still greater than between the two hands. The left side gains force in the morning far more slowly than the right. Up to 3 o'clock P. M. the left side has hardly risen, while the right at 1 P. M. has already reached its first maximum, which is little inferior to the evening one. The lowering effect of hunger is seen in both before dinner, but not so strongly as in the hands. While the right side remains nearly stationary from one to 9 P. M., the left now rises uninterruptedly from 1 to 11 P. M. The right begins to fall already at 8 P. M.; it crosses the left, and sinks far below it; whereas the left only begins to fall at 1 A. M., or five hours later. The morning rise begins in both at the same time.

The conclusions are, that the course of the brain is, in general, analogous to that of the hands, rising in the morning, sinking a little at noon, attaining its maximum in the evening, its minimum towards 4 A. M.; all these things agree tolerably well, and indicate the probable daily course of our whole frame in a person living as I do. But the brain differs from the hands in being far less affected by hunger or eating. *The organs of intellect and of the mental affections generally appear to be less concerned with the ordinary business of nutrition, than the hands which have to provide it.* Nature has providently secured, that if food should fail, the power of the mind, which devises the means of obtaining it, should not immediately be affected. The difference between the two sides of the brain, indicates that the right hemisphere inclines earlier to sleep than the left, as it also, on the other hand, rises earlier in the morning to its greatest force than the left does. It exhibits, therefore, greater excitability, but not greater power, than the opposite side.

266. There is a greater anatomical difference between forehead and hindhead than between the two sides of the brain, and I was desirous also to investigate this. The observation was twice made, each time for twenty-four hours, on the 19th and 20th October, and the result is given in Fig. 7. Plate II. The differences here appeared in the form of stronger polar opposition. The forehead, in general, was cold, the hindhead very warm, and not only in human beings, but also in animals. The patient observed it in the cat of the house, and when taken to my stables, also in horses and cows, the depression in the back of the neck in the latter being especially hot to her. The human forehead rose quickly in the morning with the grey light of dawn, was little affected by the two periods of hunger, and reached its maximum after sunset. During all this time the hindhead remained nearly stationary, so that at 6 P. M. it stood where it had been at 6 A. M. But now it began to rise, almost exactly at the time that the forehead began to fall. From this point the lines cross each other, that of the hindhead rising till 3 A. M.; while that of the forehead falls till about the same time, at which one reaches its maximum, the other its minimum. From this time they again pursue opposite courses; and while the hindhead, which at 3 A. M. is very high, rapidly falls, the forehead, which at 4 A. M. is very low, begins to rise with equal rapidity.

267. This play of forces is the image of our sleeping and waking. The forehead represents the waking state, the hindhead that of sleep. The former becomes active from the first peep of dawn, and increases till sunset. It then loses the

supply of odyle derived from the sun, and sinks continually till the new day breaks, when its force again begins to rise. On the other hand, the hindhead rests nearly motionless during the prevalence of day-light, but as the sun sinks below the horizon, its hour for labour strikes. The Morpheus then rises, and with rapid steps mounts up, till the first traces of early dawn remind him that the forehead is on its way to relieve his watch. The hindhead, at the close of night, falls as continuously and rapidly to its lowest level, as was the case with the forehead at the close of day. They are, therefore, not only oppositely polar, the forehead being cool and negative, the hindhead warm and positive; but they are also diametrically opposite in the exercise of their functions, as are day and night, waking and sleeping.

268. It also appears from the above comparison, that sleeping and waking are not opposed to each other, in the way in which action and rest are opposed, at least not in relation to odyle; but that *only the seat of activity is changed.* The force does not cease to exist, nor does it diminish in amount, but simply shifts from the anterior part of the brain to the posterior, gaining in one direction what it has lost in another. Sleep, therefore, appears, not as a failure of the vital energies, but as a dislocation of them. In the same degree as vitality was active during the day in the forehead, it predominates during the night in the hindhead. Sleep is therefore an alternation in the functions and powers of our organs, but in no way the introduction to their inactivity; poets may employ, as a metaphor, the comparison of sleep with death, but physiologists, studying the phenomena of organic life, cannot do so. Vitality is just as active during sleep as in the waking state; its direction only is changed. The phenomenon of sleep is governed by the posterior part of the brain, probably by the cerebellum, while the forehead ceases from its mental labour; and when the forehead again, under the influence of the solar rays, resumes its activity, the hindhead relinquishes its claims on the vital energies.

269. Fig. 4. Plate II. exhibits a small, but not unimportant confirmation of the above. On one occasion I became sleepy

soon after dinner, and leaning my head on the back of my chair, and fell asleep for ten minutes. During this sleep, as well as shortly before it, and shortly after it, Mlle. REICHEL had examined the state of my hand. The figure gives the result of these observations, made between 4 and 5 P. M. The force of my hand, instead of steadily rising, as on all other days at this period, made an anomalous leap downwards, after which it resumed and continued the normal rise. The short nap in which I had indulged, had been sufficient to cause a very sensible inversion of the distribution of odyle in my person. As long as I slept, the force in my hand rapidly sank; the ordinate representing it became shorter, and only lengthened again when I had awakened, and all the vital functions had once more entered on their former course.

270. The pit of the stomach, or hypochondrium, is a spot in the human body, which in somnambulists plays a very peculiar part. Anatomists have often sought there for a peculiar organ, and were surprised to find nothing remarkable, which could serve as the means of producing the extraordinary facts very often observed in persons in what is called the state of clairvoyance. Even the corpora of PACINI, which have recently been supposed to be concerned in these phenomena, are not distributed over the body in such an order as to correspond to the relative degrees of sensitiveness in different parts. But it is not at all necessary that any special organ should exist where the phenomena of odyle are particularly concentrated. In points, where nature acts by means of imponderable agents, which penetrate all things, she requires no special palpable apparatus. These phenomena are the combined resultants of numberless and complex components, which constitute the nerves, nervous tissues, and ganglia. The point of greatest concentration may then occur in the otherwise most insignificant spot on the abdomen. I was very desirous to know the course followed by the hypochondrium in relation to odyle; and I succeeded in obtaining a twenty-four hours observation of that part of my own frame. Fig. 8. Plate II. shows the result. The line differs but little, in its curves, from those of the hands. The difference is seen in some retardation or other disturbance

of the time. The effect of hunger first appears at 2 P. M., but extends long past the dinner hour to past 5 P. M. The higher culmination occurs also, not at sunset, but at 10 P. M. —The course of the hypochondrium has, therefore, nothing peculiar, but it teaches us, that the order of effects, from sunrise, from hunger, from taking food, and from sunset, extends to the abdomen, in a similar manner as the hands and sides of the head. But phenomena of another kind, produced by mental affections, and reflected in the physical phenomena of vitality, are very distinctly and strongly marked in the hypochondrium. Of these I can only speak hereafter, and must here content myself with alluding to them.

271. In order to see whether the developement of odyle followed the same course in females as in males, I had a twenty-four hours observation made of the right hand of my daughter Hermine. Fig. 9. Plate II. shows that the line obtained differs so little from that of my own right hand, that the differences may be regarded as unavoidable errors of observation. In this respect, therefore, there is no difference between the sexes.

272. Fig. 10. Plate II. is a confirmation of this. It is the line of Mlle. REICHEL'S right hand, observed during twenty-four hours by means of her left. I thought it important to obtain the course followed by the odylic phenomena in a highly sensitive patient during her illness. The line shows that it is in no respect essentially different from that of persons in health, and only varies in the times of the phenomena, because she dined two hours earlier than I did with my family.

273. But a collateral observation occurred here, which is worthy of notice. There is seen a new and sudden fall at 5 A.M. when the day had already dawned. At the very moment when this began, she had been suddenly seized with pain of the stomach, lasting till near 7 A. M. This morbid state not only checked the natural rise of the force, but rapidly gave it a decided downward impulse. As soon as the pain ceased, the normal rise went on.

274. It is also worthy of special mention, that although Mlle. REICHEL, during the whole of this night, had no sound

sleep, but only slight and interrupted slumbers every seven or eight minutes, yet, as the figure shows, the whole nightly period from sunset to sunrise exhibits the same contrast of the nightly fall with the daily rise, as was found in healthy persons, buried in profound sleep.

275. Several physiological truths may be deduced from this last series of researches. I shall here only point out a rule applicable to dietetics. If the day, with its sunlight, and the rising odyle in the forehead and hands, is destined for the voluntary and intellectual functions of animals, while the night, with its falling odyle, which now retires to the hindhead, is designed for the involuntary, unconscious vegetative functions; then must every thing in what we do or omit that corresponds to this natural order be favourable, and on the other hand every thing that opposes it injurious, to our well-being. Now, as we have seen, the taking of food is favourable to the rise of odyle in forehead and hands; the effect of the chemical changes of digestion is combined with that of the solar rays, and both promote the developement of odyle, and consequently the vital energies of the day. It is therefore advantageous for us to eat by day, as nature points out. But the effects of digestion continue for several hours. It might therefore happen that we took food at such an hour, that digestion should take place after sunset, when, therefore, the odyle in forehead and hand is falling. The rise from digestion would now, in its effects, run directly counter to the fall from the setting of the sun, which would lead to a state of opposition or contest in our economy. We have here, consequently, new theoretical reasons for, or explanations of, the old practical experience, which dictates that we should not eat in the evening, and especially not for some hours before retiring to rest; and that, if we do so notwithstanding, the consequences are imperfect sleep with restless dreams, that is, half consciousness, or in other words, half activity of the forehead. We might indeed suppose that the odyle developed by digestion should now be carried to the hindhead, and thus even promote sleep. But this is contradicted, not only by universal experience, but also by the special observations above recorded. For it appears from the

experiments and figures, that in Mlle. REICHEL, who dined at 1 P. M., the sinking of odyle in the hand and forehead began exactly at sunset, in that season; while in my daughter and myself, who dined after 3 P. M. the same fall did not begin till from 8 to 10 P. M., that is, as much later as our dinner hour was later than hers. In Mlle. REICHEL the developement of odyle caused by digestion, had already ceased when evening came on, while in us it continued some hours longer, and partly held in check, partly even overpowered the fall due to sunset. He who would go late to bed, does well to dine proportionally late.

276. But he who does this acts in opposition to the natural order, and certainly injuriously to his own well-being. We know that the sleep near midnight is the deepest, healthiest, and most invigorating, and cannot be perfectly replaced by transferring it to other periods. The above researches enable us better to perceive the rationale of this. Nature has established for the animal, and, as it appears, also for the vegetable world, this order, namely, that with darkness, with the fall of odyle, sleep comes; and conversely, that sleep ceases, when light and odyle return to the forehead in renewed energy. If we go late to bed, we must sleep till late in the forenoon. But now we have the influence of the sun and of the odyle on our frames opposed to us, and the morning sleep must be, for this reason, imperfect and unrefreshing, just as it becomes so when we go to bed after a full meal. We are, in both cases, under the influence of a wrong direction of the odylic influences. Those who go late to bed and rise late, are, on this account, generally duller and less cheerful in the morning than those who follow a more natural method. From all this it follows, that he who would ensure the continuance of good health, as far as relates to odyle, which affects our health deeply, should rise at latest with the sun, dine or take the chief meal of the day at 11 A.M. or noon, or at latest not after 1 P.M.; should afterwards eat little or nothing more; and retire to rest soon after sunset. All animals do this; and savage man, in his condition, which is in many respects a natural one, as well as all persons in poor or limited circum-

stances in the country, do so too. Only the so-called civilized inhabitants of our cities attempt to improve on nature in this respect. They sup heartily at 10 or 11 P.M., retire to rest at 2 or 3 A.M., and then send for the doctor, because they have gout, dyspepsia, or spleen.

In the whole course of the present treatise, we have seen everywhere and without exception, the observed phenomena in a state in which odyle is recognised by its producing sensations of tepid heat or of coolness. In the circumstances under which we thus observe and perceive it, we can trace certain variations very distinctly. In the first place, it is found in bodies *which have been charged with it from without*, when it proceeds from other substances, by the contact or approach of which, odyle has been communicated to or excited in the former; as in the effects of the radiation from heavenly bodies, and those of friction, chemical action, and heat. The change is transient and of short duration. Secondly, *it is durably inherent in amorphous matter, as such*, and is not very strongly marked in this case. Thirdly, it is found associated with Magnetism proper in amorphous bodies, which, such as magnetic steel, *have been charged with both forces together*, and not transient, but adhering to the body as long as it continues magnetic. (See § 23, note.) Fourthly, it appears in certain bodies, as in crystals, plants, and animals, *concentrated in certain points*, and is then permanent.

In the two former cases, odyle appears to be uniformly distributed over bodies, *like the simple electric charge;* in the two latter, it is polarised *like the voltaic pile*, and that not only in the direction of one, but, especially in organised structures, in that of many axes crossing each other.

Odyle possesses, therefore, a marked dualism, which has an unmistakeable analogy with that of electricity.

Retrospect.

a. The terms *tepid* and *cool*, or *warm* and *cold*, used by sensitives, are not to be taken strictly, but only comparatively.

They express sensations resembling those which warmth and coolness generally produce.

b. One pole of magnets or crystals almost always gives them a warm, the other a cool sensation.

c. The warmer pole is generally positive, and may be designated by + O. The cooler is negative, and, consequently, marked — O.

d. The amorphous inorganic world forms a great arrangement, at one end of which the most strongly odylo-positive body, (at present potassium,) at the other, the most powerfully odylo-negative, (at present oxygen,) are found; and which ranges from the greatest apparent warmth to the greatest apparent cold, and thus forms a series including all bodies.

e. Positively electrified bodies diffuse odylic coolness; negatively electrified, odylic warmth, or, at least, diminished coolness.

f. Heating a body excites + O; cooling — O.

g. Friction gives + O.

h. All kinds of fire diffuse — O.

i. Chemical action, as such, always produces directly — O.

k. All the odylic flames proceeding from odylo-positive bodies are warm; those from odylo-negative bodies cool. The apparent or odylic temperature of their flames, therefore, gives a measure of their position in the series.

l. In plants, the *caudex descendens in a general sense* was found positive, the *caudex ascendens in general* negative. In the details, each organ was found polar.

m. In man, the whole left side is positive, the whole right side negative. This polar opposition is most strongly marked in the hands and in the points of the fingers.

n. Men and women are in this respect alike.

o. In both men and women, the odylic intensity is different at different times, and in the different states caused by hunger, taking food, sleep, illness, &c., sometimes rising, at other times falling.

p. The forehead and hindhead alternate in odylic activity; the former being active by day, the latter by night.

q. Sensitive persons, placed in the sphere of action of bodies diffusing odyle, only feel comfortable when to the polar parts of their own frame the oppositely named poles of the bodies are opposed or brought near. If like named poles be brought near each other, unpleasant sensations, soon rising to illness, are the result.

r. There are found in bodies states of permanent and transient odylic charge, analogous to those of electric charges. In the former case they exhibit decided dualism, in the latter not.

CONCLUSION.

If I collect and compare all the experiments and observations described in the preceding seven treatises, with the deductions made from them, the following propositions, physical and physiological, present themselves.

I. The time-honoured observation, that the magnet has a sensible action on the human organism, is neither a lie, nor an imposture, nor a superstition, as many philosophers now-a-days erroneously suppose and declare it to be, but a well-founded fact, a physico-physiological law of nature, which loudly calls on our attention.

II. It is a tolerably easy thing, and everywhere practicable, to convince ourselves of the accuracy of this statement; for everywhere people may be found, whose sleep is more or less disturbed by the moon, or who suffer from nervous disorders. Almost all of these perceive very distinctly the peculiar action of a magnet, when a pass is made with it from the head downwards. Even more numerous are the healthy and active persons who feel the magnet very vividly; many others feel it less distinctly; many hardly perceive it; and finally, the majority do not perceive it at all. All those who perceive this effect, and who seem to amount to a fourth or even a third of the people in this part of Europe, are here included under the general term "Sensitives." § 60.

III. The perceptions of this action group themselves about the senses of touch and of sight; of touch, in the form of sensations of apparent (§ 217) coolness and warmth,

(§ 225); of sight, in the form of luminous emanations, visible after remaining long in the dark, and flowing from the poles and sides of magnets. §§ 8, 9, 15.

IV. The power of exerting this action not only belongs to steel magnets as produced by art, or to the loadstone, but nature presents it in an infinite variety of cases. We have first the earth itself, the magnetism of which acts, more or less strongly, on sensitives. § 60 *et seq.*

V. There is next, the moon, which acts by virtue of the same force on the earth, and, of course, on sensitives. § 118.

VI. We have, further, all crystals, natural and artificial, which act in the line of their axes. §§ 31, 33, 35, 50, 55.

VII. Also heat; § 121.

VIII. Friction; § 127.

IX. Electricity; § 159.

X. Light; § 131.

XI. The solar and stellar rays; §§ 97, 208.

XII. Chemical action especially; §§ 137, 142.

XIII. Organic vital activity; both

a. That of plants, § 248 *et seq.;* and

b. That of animals, especially of man; § 79.

XIV. Finally, the whole material universe. §§ 174, 213.

XV. The cause of these phenomena is a peculiar force, existing in nature, and embracing the universe, (§§ 213, 214), distinct from all known forces, and here called odyle. § 215.

XVI. It is essentially different from what we have hitherto called magnetism, (§ 42), for it does not attract iron, (§ 37), nor the magnet, (§§ 24, 38). Bodies possessing it do not assume any particular direction from the action of the earth's magnetism, (§ 42), they do not affect the magnetic needle, (§ 38). When suspended they are not affected by the proximity of an electric current, (§ 39); and they induce no current in metallic wires. § 40.

XVII. Although distinct from what has hitherto been called magnetism, this force appears everywhere where magnetism appears. § 43.

XVIII. But, conversely, magnetism by no means appears where odyle is found. This force has, therefore, an existence independent of magnetism; while magnetism is invariably found combined with odyle. §§ 43, 44.

XIX. The odylic force possesses polarity. It appears with constantly different properties at the opposite poles of magnets. At the northward pole, (§§ 225, 36, note), it generally causes, on the downward pass, a sensation of coolness, (§ 236), and in the dark a blue and bluish gray light; at the southward pole, on the contrary, a sensation of warmth, and red, reddish yellow and reddish grey light. The former sensation is accompanied by decidedly pleasurable feelings, the latter with discomfort and anxious distress. After magnets, crystals, (§§ 32, 50, 55, 220, 221), and living organised beings, (§§ 84 to 89, 253), exhibit distinct odylic polarity.

XX. In crystals, the odylic poles are found to coincide with the poles of the crystallographic axes, (§ 32). In polyaxal crystals, there are also several axes of unequal force.

XXI. In plants, the *caudex ascendens* is in general oppositely polar to the *caudex descendens;* but there are also innumerable subordinate polarities in all the individual organs. § 248 *et seq.*

XXII. In animals, at least in man, the whole left side is in odylic opposition to the right, (§ 226). The force appears concentrated in poles in the extremities, the hands and fingers, (§ 254), in both feet, (§ 23), stronger in the hands than in the feet. Within these general polarities there are innumerable lesser subordinate special polarities of the individual organs, both in themselves and as opposed to other symmetrical organs, (§ 254). Men and women are not qualitatively different. § 227.

XXIII. In the terrestrial sphere, the north pole has been considered as magneto-positive, the south as negative; and, consequently, the northward end or pole of the suspended needle has been considered negative, the opposite pole positive. On the same principle, I have called the odylic pole which goes along with the northward or negative magnetic pole, negative, that is, odylo-negative, $= - O$; the opposite pole, odylo-positive, $= + O$.

(§ 231). In crystals the cooler pole is, therefore, negative, the warmer positive, (§ 231). In plants, the root was generally odylo-positive, the stem and its points negative, (§ 252). In man, the left side, hand, and fingers were warm, disagreeable, and gave out red light; they were, therefore, odylo-positive. The right side, hand, and fingers were cool and pleasant, gave out blue light, and were, therefore, negative, (§§ 226, 231). It is the same, probably, in all animals. § 253.

XXIV. Of the solar rays, the red and those beyond it are odylo-positive; the blue and those beyond it negative. The latter includes the so-called chemical rays. The spectrum is, therefore, polar in relation to odyle. § 116.

XXV. Amorphous bodies, without crystalline direction of their integrant molecules, show, individually, no polarity. But as each of them acts, within certain limits, producing either warmth or coolness, and as they differ in intensity, they form a continuous chain or series, just as they do, electro-chemically considered. Thus the elements, in relation to odyle, arrange themselves so, that at one end there is found the most odylo-positive body, potassium, at the other, the most odylo-negative, oxygen. And as this natural series coincides almost exactly with the electro-chemical, we may call it the odylo-chemical series. § 236.

XXVI. The heating of bodies, (§§ 122, 245), and friction, (§§ 129, 246), exhibit + O. Cooling, (§ 123), and fire-light, (§§ 131, 244, 266), exhibit — O. Chemical action gives a result which varies with the nature of the acting bodies, (§§ 139, 142, 147). But in the greater number of cases chemical action developed negative odyle.

XXVII. All stars which shine by reflected light, such as the moon and planets, were, in their chief effects, odylo-positive, (§§ 119, 208, 239). Those which shine with their own light, the sun and fixed stars, were odylo-negative, (§§ 100, 208, 239). Their spectrum was also in itself polar. § 116.

XXVIII. The odylic force is conducted, to distances yet unascertained, by all solid and liquid bodies. Not only metals, but glass, resin, silk, water, are perfect conductors for odyle, (§§ 47, 81, 113, 118, 121, 141, 167, 203). Bodies of less continuous structure also conduct odyle, but not quite so perfectly; such as dry wood, paper, cotton cloth, woollen cloth, and the like. There is also some, although but a small degree of, resistance to the passage of odyle from one body to another. § 47.

XXIX. The conduction of odyle is effected much slower than that of electricity, but much more rapidly than that of heat. The hand, if moved rapidly, can almost follow it in its course through a long wire.

XXX. Bodies may be charged with odyle, or odyle may be transferred from one body to another. In stricter language, a body, in which free odyle is developed, can excite in another body a similar odylic state. §§ 29, 45, 72, 82, 105, 118, 143, 198, 202.

XXXI. This charging or transference is effected by contact. But mere proximity, without contact, is sufficient to produce the charge, although in a feebler degree. § 202.

XXXII. The charging of bodies with odyle requires a certain time, and is not accomplished under several minutes.

XXXIII. Bodies while conducting odyle, or when charged with it, do not exhibit polarity; which seems to be associated with certain molecular arrangements of matter.

XXXIV. The duration of the charge in bodies, after separating from the charging substance, is but short, varying according to the nature of the body, but seldom extending beyond a few minutes for strong healthy sensitives (§§ 82, 167, 169); for the most sensitive patients occasionally lasting for some hours, as in the case of magnetised water. Bodies therefore possess some degree of coercitive power for odyle. §§ 46, 83, 112, 205.

XXXV. Bodies charged with odyle, such as wires, give out at the end furthest from the changing substance sensible emanations, warm or cool, positive or negative, according to the poles from which the charge is taken. §§ 107, 114, 119.

XXXVI. Odyle has, like heat, the property of existing in two different states; that in which it is sluggish, and is slowly communicated to, and slowly passes through bodies, and that in which it is radiated to a distance, (§§ 193, 254). In this latter form, it is instantly felt by healthy sensitives, without any sensible lapse of time, at the distance of the length of a whole suite of rooms, from magnets, crystals, the human body, (§ 254), and the hands. All bodies and processes, which diffuse odyle over other bodies by slow conduction, radiate it at the same time in all directions, but with varying force; as is seen in friction, electricity, heat, chemical action, and bodies in general, (§ 201). The rays of odyle penetrate through clothes, bed-clothes, boards, walls, (§ 23, note), yet obviously with less facility than magnetism, and with a certain degree of slowness. The conduction and charging of odyle from one body to another, by mere proximity, without contact, as from the poles of magnets and crystals, from the hands, or from amorphous bodies high in the series, such as sulphur, &c. &c. seem all to depend on the radiation of odyle; and this explanation also applies to the so-called magnetising of sensitive persons.

XXXVII. Electric currents, when passed through sensitive persons, produce no perceptible odylic excitement, nor do they directly act on such persons otherwise than on all others (§ 160); but they do so act mediately, and very powerfully, when they excite the odylic state in other bodies (§ 167), which then act on the sensitives. Metals, placed within the sphere of electrical action, produce the most vivid phenomena. § 168.

XXXVIII. The light diffused by odylically excited bodies, is exceedingly feeble, and is, probably on this account, not visible to every eye. Those who are only moderately sensitive must remain a long time, perhaps two hours, in absolute darkness before their eyes are sufficiently prepared to enable them to perceive this light. During the whole of this time, the eye must not be reached by the smallest trace of any other light. But the power of perceiving the odylic light cannot depend alone on a peculiar acuteness of vision, because all those who are capable of seeing it, are, without exception, possessed also of that peculiar sensitiveness which enables them to recognise odylic impressions by the sense of feeling, and to distinguish between the odylic sensations of warmth and coolness, as well as between the pleasurable and offensive feelings they experience; and these sensations are constant. Now, since these different powers of perception are, in certain persons, namely, in the sensitive, always present together, we must regard them as associated; and they seem to depend on a peculiar disposition of the whole nervous system, the nature of which is unknown, and not on any peculiar state of individual organs of the senses.

XXXIX. The odylic light of amorphous bodies is a kind of feeble external and internal glow, extending apparently through the whole mass, somewhat similar to phosphorescence, and possibly depending on a cause common to it and to that phenomenon. This glow is surrounded by a delicate luminous veil, in the form of a fine downy flame, (§ 207). In different bodies this light has different colours, blue, red, yellow, green, purple, but chiefly white and grey. Elementary bodies, especially metals, shine the most vividly (§ 206); compounds, such as oxides, sulphurets, iodides, carbohydrogens, silicates, salts of all kinds, different varieties of glass, even the walls of a room; in short, all things give out light. § 206.

XL. Where the light is polar, as in a magnet, (§§ 3, 6), and crystals, (§ 55), it forms a kind of flaming current, proceeding from the poles, and flowing almost in a straight line in the direction of the magnetic or crystalline axes. As it extends further from the pole, it widens a little, and its intensity diminishes. It exhibits all the colours of the rainbow, (§§ 9, 13), but red predominates in the flame from the positive, blue in that from the negative pole. Besides this, magnets, crystals, and the hand, like amorphous bodies, possess also, diffused apparently through their mass, the luminous glow, which we may call the odylic glow, and this again is on all sides enveloped in a delicate, vaporous, luminous veil. § 8.

XLI. Human beings are thus luminous over nearly the whole surface, but especially on the hands, (§ 92), the palm of the hand, the points of the fingers, (§ 93), the eyes, certain parts of the head, the pit of the stomach, the toes, &c. Flaming emanations stream forth from all the points of the fingers, of relatively great intensity, and in the line of the length of the fingers.

XLII. Electricity, nay even the mere electrical atmosphere, produces, and also intensifies in a high degree, the odylic luminous phenomena, (§ 167), but this effect is not instantaneous, occurring after a short interval, not exceeding a few minutes. § 169.

XLIII. An electro-magnet exhibits the same luminous appearances as an ordinary magnet, (§ 12), and in the same degree in which it is susceptible of increased magnetic power, it is also susceptible of increased intensity in the odylic light which it yields.

XLIV. The solar and lunar rays charge with odyle all bodies on which they fall; and if wires, connected with these bodies, extend to a dark chamber, odylic flames appear at their extremity. §§ 114, 119.

XLV. Heat, (§ 125), friction, (§ 129), fire-light, (§ 134, 147, 240), produce at the ends of wires, in a dark place, flames like that of a candle.

XLVI. Every chemical action, even mere solution in water, or the recombination of water of crystallisation in effloresced salts, produce the same result at the end of a wire in a still higher degree, (§ 146). But chemical processes also, for themselves, give out odylic flames, and exhibit the odylic glow. § 145.

XLVII. The positive pole yields the smaller but brighter flame, the negative pole a larger but less luminous one. The former is more bright, because it is red and yellow; the latter less luminous, because it is blue and grey.

XLVIII. The odylic flame radiates light which illuminates near objects. This light may be concentrated by a lens into the focus, (§ 18). The flaming and nebulous emanations from bodies and their poles must therefore be carefully distinguished from odylic light in the strict and proper sense of the term.

XLIX. All these flames may be moved by currents of air, as by blowing on them, when they bend, yield, and divide themselves, (§ 20); when they meet with solid bodies, they bend round these, following and flowing along the surface, just as ordinary flame does. The odylic flame has, therefore, an obviously material (ponderable ?) character.

L. It may be made to flow in any direction, upwards, downwards, or laterally, as the body yielding it is held. It is therefore, up to a certain point, independent of terrestrial magnetism. §§ 20, 53.

LI. The odylo-luminous emanations appear chiefly at edges, corners, and points, (§ 3); and, like electricity, seem to find there an easier egress, coinciding with the resistance to the passage of odyle observed in conduction. At such places, therefore, the sensations of warmth, coolness, &c. and the luminous appearances, are especially distinct. § 114.

LII. The flames of the opposite poles of a magnet, &c., show no tendency to unite. There is no perceptible attraction between them, and in this they differ essentially from the emanations of magnetism proper. §§ 3, 9.

LIII. All odylo-positive bodies send forth warm flames, all odylo-negative cold flames (§ 223). The flames, therefore, in regard to apparent temperature, have the character of the poles from which they proceed, and the flame therefore indicates the character of the body, or pole of a body, from which it flows. § 241.

LIV. In many morbid states, especially in cataleptic fits, a peculiar kind of attraction is observed, exerted by the odylic poles of magnets and crystals, or by the hand, on the hand of the diseased sensitive (§ 23). It resembles that of the magnet for iron, but is not mutual (§§ 24, 54), that is, the sensitive hand exerts no attraction on the body by which it is itself attracted (§§ 23, 91). Even bodies charged, by conduction or otherwise, with odyles, produced, to some extent, this surprising effect. § 28.

LV. In the animal economy, night, sleep, and hunger, depress or diminish the odylic influence; taking food, day-light, and the active waking state, increase and intensify it (§§ 260, 262). In sleep, the seat of odylic activity is transferred to other parts of the nervous system (§ 268). In the twenty-four hours of day and night, a periodic fluctuation, a decrease and increase of odylic power, occurs in the human body. § 265.

LVI. Some applications of the laws regulating the odylic phenomena have been made; as the partial explanation of the facts connected with what is called "magnetised" water (§§ 27, 28, 73, 105, 112); of the light attending sudden crystallisation (§ 55); of the lights seen above graves (§ 158); of the mysterious occurrence in PFEFFEL's garden at Colmar (§ 156); of the so-called "magnetic baquet" (§ 135, 151); of certain effects of digestion (§ 152); and of respiration (§ 153); of many singular antipathies (§ 175); of the necessity of placing sensitive patients in the plane of the magnetic meridian (§§ 69, 71); of the attraction of the cataleptic hand by magnets, crystals, other hands, &c. (§ 23); of the odylic state of the human body (§ 79 et seq.); of its daily and hourly fluctuations (§ 256); and finally, of some of the properties and the probable causes of the aurora borealis. § 21.

PART II.

MAGNETIC OR ODYLIC LIGHT.

INTRODUCTION.

277. It has been objected to me, that the five girls, whom I have chiefly employed as my re-agents for the new force, are not sufficient, where the object is, to establish, with a certainty, important truths in natural science. I proceeded to act on this principle, and exerted myself to extend my investigations over a greater number of persons, in various states as to health, and in various conditions of life. In this way more than two years have been employed, and to that extent the following treatise, containing a detailed investigation into the properties of the odylic light, *as exhibited by magnets*, has necessarily been delayed. But it now appears with a train of NEARLY SIXTY SENSITIVE PERSONS AS WITNESSES, male and female, mothers and maidens, children and aged persons, low and high, poor and rich, feeble and vigorous, diseased and healthy, women during menstruation and during pregnancy; and thus we have representatives of so great a variety of physical conditions, that little more can be desired. The most essential addition, however, to the previous observations, obtained by this extension of my researches, will probably be found in this, that *perfectly healthy*, *strong persons* take their places in the list of sensitives; that persons who never were ill, and have spent their lives in continual hard bodily labour, perceive the peculiar sensations and the luminous phenomena which I have described in these researches, exactly in the same

way as females living in the seclusion of the sick room ; that neither youth, nor age, nor sex, nor position in society, make any difference in this respect ; and that sensitiveness, as the term is here employed, is not so much a morbid state, as rather a peculiarity belonging to many individuals, which only appears more or less strongly marked in different circumstances, and sometimes in a degree hardly perceptible. The list and details which here follow will show, that already more than half of the persons who, for the sake of science, have obligingly devoted themselves to these investigations, are in perfect health. I have been truly surprised to find that the proportion of sensitive persons is large, beyond all expectation, among ordinary people, and that many, who are far from suspecting in themselves the existence of any such power, perceive very plainly the odylic sensations, and, after remaining for a sufficient time in darkness, can see very distinctly the luminous phenomena. Indeed, this is so true, that I hardly any longer require patients for my researches ; and not only find healthy persons to answer the purpose, but shall probably soon exclude from my investigations, generally all who are in a diseased state, and especially those affected with somnambulism. The objections, therefore, first, that my observations are not furnished with a sufficient amount of testimony, and secondly, that the necessary confidence cannot be placed in the statements of diseased observers—these very common objections will meet, in what follows, complete contradiction, and for the future must be abandoned.*

* While it is in this, as in all similar cases, very desirable to have as much testimony, and of as good quality as possible, and while it is highly satisfactory to find that the author has succeeded in obtaining it, yet I cannot refrain from here repeating, that, in my opinion, such objections possess far less cogency than is usually ascribed to them. They are generally brought forward by those who will not, or cannot, investigate for themselves. One case, observed as our author, a practised observer in many departments of science, would observe it, may yield results so clear, that nothing material can be added, by subsequent cases, to the force of the evidence. The facts so admirably brought out in the first Part would not the less be facts, if only one sensitive person existed in Europe, although their practical applications, in that case, might be less important. And with regard to observations made on diseased subjects, and depending on their statements, in whole or in part, everything depends on the sagacity

I have already (§ 6) explained why I have not, in the first part of this work, produced a greater number of proofs and confirmations of what I have observed. I was afraid of wearying the reader with too many materials and details. But as some persons regard this as a defect in my work, I no longer hesitate to avail myself of the whole mass of evidence at my disposal, and thus to give to my propositions the broadest foundation. The reader will see the same phenomena constantly recurring under the most varied circumstances, and will be better able than before to repeat them everywhere, and with facility. I now give, for the reader's convenience, in the first place, the

List of Persons

Who have assisted me by their observations, and supported me with their sympathy during my researches, and who possess the power, in different degrees, of observing the peculiar phenomena, the study of which is the object of this work.

Madame Cæcilia Bauer, wife of an hotel-keeper in Vienna, Braunhirschengrund, No. 161.

Mademoiselle Leopoldine Reichel, daughter of an official in the palace of Schönbrunn.*

Mlle. Maria Atzmannsdorfer, daughter of a military medical man.

Mlle. Angelica Sturmann, daughter of an hotel-keeper.

Mlle. Francisca Wiegand, daughter of a hat-manufacturer in Vienna, Windmühl, No. 60.

Friedrich Weidlich, an invalided sailor.

Mlle. Josephine Winter, daughter of a painter at Gratz.

Mlle. Maria Nowotny, daughter of a subaltern official.

Mlle. Clementina Girtler, daughter of a merchant.

of the experimenter, and his experience in scientific researches; qualities which, as Part I. shews, Baron von Reichenbach possesses in a very high and rare degree. A large proportion of the most undoubted facts in medicine depend, as every physician knows, entirely on the statements of the patients. But on this point I refer to what I have already stated in the editor's preface.—W. G.

* No doubt the same who, in Part I., is described as Barbara Reichel. She has probably two Christian names.—W. G.

Mme. FRANCISCA KIENESBERGER, wife of a steward.

Mme. JOHANNA LEDERER, widow of an official.

Mlle. MARIA MAIX, daughter of an official, Vienna, Kohlmarkt, No. 260.

Mlle. JOSEPHA ZINKEL, daughter of a house proprietor at Nussdorf, near Vienna.

Baron AUGUST VON OBERLAENDER, at Schebetan, in Moravia.

M. NIKOLAUS RABE, official in the Magazine of the Imperial Mines.

Mlle. AMALIE KRUGER, daughter of an hotel keeper, Vienna, Leopolstadt, No. 27.

Mlle. WILHELMINA GLASER, daughter of the proprietor of an hotel at Bochtitz, in Moravia.

ALOYS BAIER, professionist in Vienna.

Mme. JOHANNA ANSCHUETZ, wife of M. GUSTAV ANSCHUETZ.

Dr. NIED, practising physician in Vienna, Erdberg, No. 396.

M. SEBASTIAN ZINKEL, house proprietor at Nussdorf, near Vienna.

Mlle. JOHANNA KYNAST, daughter of a baker at Waidhofeu, in Austria Proper.

Mlle. LEOPOLDINE ATZMANNSDORFER, called Mlle. DORFER, daughter of a military medical man.

Mme. VON PEICHICH-ZIMANYI, widow of a Hungarian nobleman.

JOHANN KLAIBER, carpenter in my service.

Baroness MARIA VON AUGUSTIN, wife of Major Baron VON AUGUSTIN.

Mlle. WILHELMINE VON WEIGELSBERG, Vienna, Wieden, No. 141.

Mlle. SOPHIE PAUER, daughter of the Consistorial Councillor M. PAUER, in Vienna.

Professor Dr. STEPHAN ENDLICHER, director of the Imperial Botanic Garden, Member of the Academy of Vienna.

M. FRANZ FERNOLENDT, chemical manufacturer in Vienna.

ANKA HETMANEK, a country girl, working on my property of Schloss Reisenberg.

Mlle. ERNESTINE ANSCHUETZ, and her brother,

M. GUSTAV ANSCHUETZ, painter in Vienna, Wieden, Ferdinandsgasse, No. 268.

M. STEPHAN KOLLAR, junior.

FRIEDRICH BOLLMANN, a blind carpenter.

Mme. JOSEPHINE FENZL, wife of the Imperial Custos, Dr. Med. FENZL.

Mme. VON VARADY, wife of the Court Councillor M. VON VARADY.

M. JOHANN STUDER, farmer, from Zürich.

Baroness PAULINE VON NATORP, wife of Baron VON NATORP, in Vienna.

Chevalier HUBERT VON RAINER, barrister, from Klagenfurth.

M. ERNST PAUER, Consistorial Councillor and Superintendent of the Evangelical Congregation in Vienna.

M. WILHELM HOCHSTETTER, from Esslingen, gardener at Schönbrunn.

Baroness ISABELLA VON TESSEDIK, widow of a Hungarian nobleman, who was formerly Imperial Court Secretary.

M. DEMETER TIRKA, wholesale merchant in Vienna, a Greek.

Mlle. Baroness ELISE VON SECKENDORF, at Sondershausen, in Saxony.

M. CONSTANTIN DELHEZ, philologist in Vienna, from Belgium.

M. THEODOR KOTSCHY, the well known traveller in Africa, Persia, &c.

MAXIMILIAN KRUEGER, a boy in the Orphan House, Vienna.

HERMINE FENZL, the little daughter of Dr. FENZL.

M. KARL SCHUH, natural philosopher, from Berlin.

Dr. FRIEDRICH, physician, from Munich.

Dr. RAGSKY, Imperial Professor of Chemistry in Vienna.

M. MATTHIAS MAUCH, veterinary surgeon, from Würtemberg.

Professor RÖSSNER, Councillor of the Academy of Fine Arts, Vienna.

M. EDUARD HUETTER, bookseller in Vienna.

M. FRANZ KRATOCHWILA, official in a Government office.

M. FRANZ KOLLAR, Curator in the Imperial Cabinet of Natural History.

Mlle. SUSANNA NATHER, daughter of an officer, from Basle, living in Vienna.

Professor Dr. HUSS, Body Physician to the King of Sweden, at Stockholm.

My daughter HERMINE.

Med. Dr. DIESING, Curator in the Imperial Cabinet of Natural History.

Almost all these persons live in Vienna, and may at any time be questioned, and their statements obtained from themselves. The order in which the names are here given, corresponds pretty nearly to that of the amount of their sensitiveness, beginning with the most sensitive.

---

278. I have now to prefix to the following treatise a few words on another subject.

In the whole course of the seven preceding treatises, I have uniformly (§ 34, 225, &c.) designated the north pole of the earth as magneto-positive, the south pole as magneto-negative, and I have judged of all other polarities, such as those of steel bars, of crystals, of living organisms, &c., as was naturally indicated by their relation to the terrestrial poles (of course, considering that pole of the needle to be negative which is attracted by, and points to, the positive terrestrial pole, and *vice versa;* and in crystals, &c., giving the name of negative to that pole, the action of which on sensitives coincides with the negative pole of the needle, and *vice versa.*—W. G.) But as natural philosophers are not agreed as to which pole of the earth, and consequently of the needle, is to be regarded as positive or negative, and our scientific books and manuals either pass this matter over in silence (BIOT, POUILLET, as edited by MUELLER, BAUMGARTNER, and others), or else directly contradict each other (EISENLOHR'S Physik, 3d edition, p. 461, EYDAM, Electricity and Magnetism, p. 162), one author calling + M what the other calls — M ; and as, consequently, there is an uncertainty as to the precise value of these terms, it seems to me important to explain my reasons for using them in the sense I have done.

In the electro-chemical system, introduced by Baron BERZELIUS into natural science, and by means of which he gave to chemistry the form which it yet retains, we proceed, as is well known, from the polarity of the voltaic pile of zinc and copper, and regard the electricity of the zinc pole as positive, because its properties agree with those of the electricity produced by rubbing zinc amalgam on glass, which has long been called positive. We are further agreed in designating as negative all those bodies which especially appear at the positive pole, when their compounds are decomposed by voltaic electricity,* and *vice versa.* Proceeding on this clue, I have endeavoured to investigate the odylic phenomena, which appear, on the one hand, in electro-positive, on the other, in electro-negative bodies. The great and well-marked differences between them, in regard to odyle, as I have ascertained and described these in previous treatises, I have further met with in the poles of crystals, magnets, living beings, &c. There has presented itself, with certainty, a parallel between certain properties of the one pole of such polar bodies and the same properties in one of the series of non-polar bodies, positive or negative, as just mentioned, and *vice versa*, the same parallelism between the properties of the other pole and those of the other series of non-polar bodies. Now, such poles as have properties running parallel with those of the electro-negative bodies, I have assumed to possess the same polar value, and have designated them odylo-negative, and *vice versa.* That pole in our magnetic needles which points to the south, and which I have uniformly called the *southward* pole, is that which, considered in reference to odyle, produces the same effects as

* It is necessary here to use a qualification, such as *especially*, because, with the exception of the extremes, oxygen (fluorine ?) and potassium, all the elements in the series *may* appear at either pole. Thus, for example, when sulphuric acid is decomposed, oxygen appears at the positive, sulphur at the negative pole ; but when sulphuret of potassium is decomposed, sulphur appears at the positive, potassium at the negative pole. This last occurs with all metallic sulphurets ; hence sulphur, though positive in regard to oxygen, chlorine, &c., is considered, in general, a negative element.—W. G.

electro-positive bodies. But the magnetism of the southward pole of the needle is that of the north pole of the earth (since similar poles repel, dissimilar poles attract each other.—W. G.) Assuming therefore, generally, that, in the electro-chemical series, the non-metallic bodies (oxygen, chlorine, sulphur, &c.) are really electro-negative, and that most of the metals are electro-positive, I have felt myself compelled to adopt the view of those who consider the north pole of the earth as magneto-positive, to designate it by + M, and to affix to the analogous or corresponding odylic pole, the sign + O. In exactly the same way, I have been compelled to regard the south pole of the earth as magneto-negative, and to designate it by — M, and the corresponding odylic pole by — O.

Although I have, in judging of the nature of the magnetic poles, adhered to the prevailing electro-chemical theory, and from this deduced the above conclusions, I am well aware that doubts may still be entertained as to the polar value of natural bodies; that difficulties exist, or may be raised, as to the negativity of acids and the positivity of alkalies, when we see both of them, on contact, assuming the very opposite polarities, &c. I have also, in the course of my own researches, frequently felt such difficulties; but, in the meantime, I have assumed the correctness of the prevailing theory; and according to this, the north pole of the earth must be called positive, the south negative, as long as potassium is regarded as electro-positive, and oxygen electro-negative. Should this, as is not impossible, be at any time altered, then the signs applied to the odylic poles in this work must be changed, that is, inverted throughout.

Lastly, in this introduction, I must once more refer to *the term* odyle.

It will be seen that the idea expressed by it, as fixed in § 215, very probably includes that which, a year later than I, Dr. Faraday introduced to the scientific world as a new force, under the name of diamagnetism. Doubtless, the British philosopher was not acquainted with my researches, although they appeared in an English dress in London, otherwise he would probably not have passed them over in silence, or ignored their existence. I have condensed, in the term odyle, the ultimate cause of all the phenomena described by me; in so far, namely, as they are not reconcileable with our previous knowledge of the essence of magnetism, and the other imponderables, and, in particular, are transferable from the magnet to what are called unmagnetic bodies, such as metals, glass, silk, water, salts, in short, all bodies. Diamagnetism was indeed recognised, and made known, between 1820 and 1830, by SEEBECK, MUNKE, BUCHNER, and BECQUEREL, which was also not known to Dr. FARADAY. In my researches I did not meet with the fact, that unmagnetic bodies place themselves, when suspended, across the magnetic current, and there remains, between my observations and those of Dr. FARADAY, a chasm, for the present not filled up. Yet it seems to me not impossible that we may be, as I may say, drawing the same vehicle, but by different ropes. If I do not deceive myself, Dr. FARADAY has laid hold of *one* of the numerous odylic threads, a singularly promising one, and with the force of his fertile genius, he will promote the discovery of truth in this department. This can only redound to the advantage of science. Whether magnetism, diamagnetism, and odyle, may one day be reduced to a common origin, or whether they will continue to be separated by essential differences—these are questions, the solution of which appears to me to be distant. But at all events, these influences include entirely new properties both of dead and living matter, and are, on account of their universality, and their all-embracing diffusion through the universe, of the highest physical importance.

In so far, therefore, as the interest in, and comprehension of, my subject are promoted by clearness and distinctness in the ideas developed, that is, by fixing their extent in reference to the external world, and by determining the limits of their different parts, internally considered, I consider this the proper opportunity for giving a condensed sketch of the differences which seem to me to prevail between the allied imponderables hitherto admitted, and that which I have found it necessary to designate by the term odyle. On the one hand, those things which I regard as its peculiarities will be thus brought more prominently into view than hitherto; and on the other hand, we shall thus be better able to judge whether we may hope to bring into known categories the new phenomena to be here enumerated, and thus, perhaps, to render the new term superfluous, or whether the necessity for introducing and retaining it be irresistible.

These differences, in so far as, in part, they have been made known in the preceding treatises, and, in part, will be expounded in the next and following treatises, are the following:

A.—Between Odyle and Heat.

a. Odylic emanations excite, in all sensitives, feelings of coolness and warmth, sometimes even of icy cold and burning heat; that is, they excite sensations *apparently* the same as those named. But when made to play on the thermometer, that instrument is not in the slightest degree affected; even the thermoscope of Nobili remains motionless. Neither the cold-giving nor the heat-giving pole of crystals at all affects these instruments.

b. There are many cases, in which heat and odyle produce effects diametrically opposite. A right hand feels cool in its action on sensitives: but always affects a delicate thermoscope with warmth.—The solar rays excite a sensation of cold in sensitives; but warm the thermometer.—Moonlight is felt vividly warm by the sensitive; but produces no distinct effect on the thermoscope.—Glowing charcoal, and the flames of all burning bodies, radiate cold with singular vividness to the sensitive nerve; but they radiate heat to the thermometer.—In chemical processes, cold emanations are produced, perceptible by sensitives: whereas the thermoscope often or generally indicates the disengagement of heat.

c. The conductibility of odyle through metals surpasses by far the limits of the conductibility of heat. A copper wire, of considerable length, as 60 or 70 feet or upwards, when charged with odyle at one end, produces, at the other, odylic changes of sensation; while, if heated, the heat will only be conducted along the wire to the extent of a few inches. The same is true of a rod of wood, or of glass, of a silk ribbon, of a strip of linen or cotton cloth many yards long; of bodies, therefore, which are totally incapable of conducting heat in the same way.

d. Odyle passes very easily through bodies. In the course of a few seconds, sensitives perceive the action of a crystal, a human being, or a magnet, through thick walls, without being aware of their presence there. The most intense heat would take hours, before it became even perceptible through such walls. No one, not sensitive, can perceive any effect, and least of all a cooling effect from the sun's rays through a wall. But sensitives within a house, can instantly distinguish a wall, the outside of which is exposed to the solar rays, from one which is in shade.

e. Concentrated odylic emanations are perceived by the sensitive at incredible distances. Magnets, crystals, human hands, or trees, are felt at a distance of 300 or 400 feet or upwards; whereas calorific radiations proceeding from bodies in the ordinary temperature are not indicated, at such distances, by any instrument, neither can sensitives at all perceive them.

f. Neither odylic heat nor odylic cold affect the density or volume of bodies. A thermometer may be fully charged with warm positive odyle or cold negative odyle; but its index does not move. Every one is familiar with the effect of heat and cold on the thermometer.

g. We know already that the colours of the spectrum differ materially in their relation to odyle; and we shall, in the course of this and the following treatise, become more familiar with their properties. But even when I caused the solar or lunar rays, or those of a fire, to fall, at an angle of 35° on a bundle of ten glass plates, and decomposed the transmitted light with a prism, so as to obtain a spectrum, all persons of considerable sensitiveness perceived very great differences of apparent temperature between the different colours; that, is, in places to which, as far as our present knowledge extends, no trace of free positive or negative heat could reach.

h. Metallic wires, which appear to the sensitive glowing hot, are, to the ordinary sense of feeling and to the thermoscope, of exactly the same temperature as surrounding bodies.

i. Of two glasses of water, one of which has remained in the shade, while the other has been exposed for a few minutes to the sun's rays, every sensitive declares that which has been exposed to the warm solar beams to be the cooler. § 105.

k. Nay more, a rod of porcelain heated directly over the fire at one end, and a rod of wood, burning at one end, are felt by the sensitive who hold in the hand the other end, to become very cold. This I shall, in a subsequent treatise, expound more minutely.

l. Heat itself is, under certain circumstances, a means of producing odylic cold.

Therefore, heat must be essentially distinct from odyle.

B.—Between Odyle and Electricity.

Odylic phenomena often occur where electrical phenomena either do not appear, or, as far as we know, do not exist. This is the case, for example, with sunshine, moonlight, the spectra of different kinds of light after transmission through glass, crystals, the human hand, and, in part, chemical processes.

a. Odyle distributes itself through the mass of any body charged with it; a hollow ball of metal, for example, glows when charged with odyle, not only on its own surface, but internally. A glass of water tastes to the sensitive alike odylised in every drop; and the water retains its new properties entirely, even when poured into another glass. (§ 107). Free electricity stratifies itself only on the surface. Odyle may be for a time communicated to all bodies, including the air within a room; whereas Dr. Faraday could not any where collect electricity in a room prepared for the purpose; the whole instantly escaping by the surface of the walls, &c.

b. When a body is charged with odyle, the free odyle adheres to it, so that it is not rapidly removed from the body, but requires some time—from a quarter of an hour to several hours—before it is dissipated by contact with other bodies. Free electricity is instantly removed from a body charged with it, by contact with another, such as the finger.

c. Odyle may be communicated to, and, to a certain degree accumulated, in bodies which are not isolated. Electricity can only be communicated to, and condensed in, isolated bodies; not at all in those which are unisolated.

d. All bodies, if only continuous in structure, are nearly equally good conductors of odyle; and only those which are not so coherent, such as woollen and cotton cloth, &c., are less perfect conductors. Electricity is only well conducted by metals; it is slowly conducted by many other substances, and by some even not at all. The conduction of odyle through the best conductors, such as metallic wires, takes place slowly; from twenty to forty seconds being required for its passage through a wire of fifty or sixty yards in length. Electricity flashes through distances a million times greater in an interval too short to be measured.

e. All bodies are permeable to odyle; there are, indeed, slight differences among bodies; but these are trifling. There are many bodies nearly destitute of permeability for electricity, and which oppose insuperable obstacles to its passage.

f. Action at a distance occurs much more powerfully and much farther off, when electricity excites or induces odyle in other bodies, than when electricity induces the electric state in them. Thus, a conductor, feebly charged with electricity, so as to give sparks of 0.2 of an inch only in length, excites a vivid current of odyle in a wire at six and a half feet (§ 169); whereas a charged conductor can, at that distance, produce no such induced *electrical* state in another conductor.

g. The excitement of odyle by means of electricity, is not instantaneous; but always requires a sensible time, frequently thirty seconds and upwards. This is true, both of the effects on the sense of touch, and of the luminous phenomena. An electrophorus, excited by the strokes of a piece of fur, is electrified long before the odylic flame appears. An electrified wire, and one through which the galvanic current is passing, becomes odylically glowing only when the current has lasted for some time, or some seconds after the shock of the Leyden phial has passed through it. In the multiplicator of SCHWEIGGER, the odylic light does not become visible till from ten to fifteen seconds after the deflection has taken place. All the manifestations and effects of electricity are instantaneous.

h. The duration of the odylic phenomena is again beyond comparison greater than that of the phenomena due to the passage of electricity. When a wire, which has been caused to exhibit the odylic glow by electricity, is taken out of the current, it continues to glow for half a minute, for a minute, nay, when the charge of a powerful Leyden phial has been passed through it, for two minutes, and the glow then slowly fades. In the multiplicator the needle has long returned to its normal position in the meridian, when the coil continues to give out odylic light. Certain flame-like odylic appearances on conductors, metallic plates, and wires, when they are electrified, do not at once come forth when these bodies have received the maximum of electric charge, but only after the supply of electricity has been continued for some time. But if the supply be stopped, the odylic appearances are only slowly and gradually dissipated from unisolated conductors; while the power of producing effects on the sensitive nerve of touch lasts, in many cases, as in water and in human beings, for hours.

i. But there are also cases, in which odylic light disappears sooner than the excited electricity. An electrophorus, when excited by fur, loses its odylic light in a few minutes, ten, for example, while the electric excitement of the resinous cake may last for days, and even weeks. It follows, that odyle is excited by every electrical action; but that, when excited, it pursues its own independent course.

k. Many odylic flames exhibit a constant tendency upwards, rising vertically. Electricity exhibits no such tendency, either when in motion or at rest.

l. Odylo-luminous phenomena, of great extent, appearing over electrified and unisolated metallic plates, do not adhere to the surface of the metal, but flow over it as the aurora borealis does over the earth. Electric currents always adhere closely to the metal, as far as it lies in their course. The experiments which prove this will be given in one of the next following treatises.

m. Odylic emanations do not appear exclusively at points where such are present, but often arise from the sides even of jagged bodies, and this sometimes even in large crystals. In similar cases electricity always prefers a point to any other path of egress. In the voltaic pile all the elements give out odyle, sensible to the touch and light. Of the electric current, when the circuit is closed, we only observe internal activity, and entire limitation of the current to itself.

n. Even when excited by electricity, odylic currents exhibit, in a remarkable degree, independence of it. We may take hold of with the hands, or allow to be in contact with the floor, &c., metallic plates and wires in which both forces are active, without affecting the odylic luminous emanations, while the electric emanations are at once carried away, and pursue other paths.

o. Odylic flames, from any bodies whatever, positive or negative, exhibit no tendency, when brought near each other, to unite or neutralise each other. When they cross, each carries the other with it; when they are directly opposed, they mutually repel each other, (see further on, § 401). Opposite electricities instantly neutralise each other with a powerful mutual attraction.

p. I have not been able, in studying the odylic phenomena, to observe hitherto, with any degree of certainty, induction or influence, which produce such marked electrical phenomena.

q. An electrical specimen of shorl, like every crystal, shews at its pole a lively action on sensitives, but when warmed, no change takes place; it becomes no stronger, and the electricity thus excited is not sensibly perceived.

r. The violent action of odyle on the irritable nerve of sensitives, forms a most remarkable contrast with the absence of any peculiar effect of electricity on the same individuals. The action of electricity on them is so trifling, that even in the most sensitive it does not exceed that observed in persons devoid of sensitiveness. Currents of galvanic and friction electricity, or the shocks of KLEIST'S phials, are borne by sensitives just as by others. Nay, the approach of a thunder storm, or stroking a cat with the hand, or the usual experiments on the isolating stool, were found by many of them to be agreeable.

But all this shows how great a gulf separates odyle from electricity.

C.—Between Odyle and Magnetism.

Odyle is produced and vividly manifested in a multitude of cases, in which magnetism, properly so called, is nowhere observed, or is yet unknown to exist; as, in part, during chemical changes, in the vital changes, in crystals, in the case of friction, in the spectra of solar and lunar rays and of candle light, in polarised light, and in the amorphous material world collectively.

a. In the vast majority of cases, odyle is developed alone without magnetism. Magnetism never appears alone, but is always associated with odyle.

b. Where magnetism is supposed, but not generally admitted, to exist, as in the sun's rays, in the moon's rays, it is at all events so feeble, that the fact of its presence is still in a high degree doubtful. But in these very cases odyle appears with a power and variety of effect, which is truly astonishing, and even appears capable, in certain cases, of shaking life to its very foundations.

c. The obscuring of the sun or moon by mist or clouds, instantly and very decidedly diminishes their effects on sensitives. Magnetism is arrested by nothing, and least of all by vapours.

d. All solid and liquid bodies may be charged with odyle in the very same manner, metals, steel, salts, glass of various kinds, silk, resin, water; all, with trifling differences of degree, may be charged with odyle. Magnetism can only be communicated to a very few bodies; and we do not know that diamagnetism can be at all communicated.

e. When bodies are charged with odyle, they act on sensitives exactly like the magnet. But they possess not the slightest trace of magnetism proper, and do not attract even the smallest particle of iron filings.

f. The coercitive or retaining power for odyle is found in steel only to act for a short time, at most for somewhat more than an hour, and therefore not longer than is the case in water or iron. But magnetism is retained by steel for many years, and has never been observed at all in water or iron. Magnetism therefore remains in steel, while odyle cannot do so, but is soon dissipated.

g. Odyle is conducted, to the distance of many yards, by substances such as resin, glass, wood, silk strings, cotton ribbons, &c. Nothing of the kind is known of magnetism.

h. Odyle is conducted by a long iron wire, and is perceived by sensitives at the further end, (§§ 47, 118). An iron wire, fifty feet long and 0.08 of an inch thick, stretched in the magnetic parallels, that is, at right angles to the meridian, and connected at one end with the northward pole of a nine-bar horse-shoe magnet, did not exhibit any trace of magnetic action at the other end.

i. The sphere of radiation for odyle, in the case of bodies, such as the hands, crystals, or electrified substances, extends as far as in the case of bar magnets of the same size, in many cases further. I tried and compared them, through the air, to the distance of 160 feet and upwards. But we know nothing of an equally extended sphere of action for magnetism, strictly so called, in the case of such magnets.

k. Odyle emanations appear to be susceptible of a certain kind of refraction; at least this is certainly true when they accompany the rays of light. I have already proved, and I shall, in a future treatise, publish many new and certainly very remarkable observations on this subject, that the prism, while it disperses the colours of the spectrum, also causes an analogous separation of odylic rays, which must therefore be refracted and dispersed by glass. And since odyle so constantly accompanies light, that in every colour of the spectrum there is, if I may so express it, a different colour of odyle, it is clear that odylic rays are refracted simultaneously with those of light, and that, whatever their nature may be, these rays are refracted by glass, like those of light. But this property is totally absent in magnetism, which is neither retarded nor deflected from its course, by any substance, as HALDAT has lately convincingly proved by means of his magnetometer or magnetoscope. He expressly declares, that magnetic emanations are neither refracted nor reflected. (L'Institut, 27th May 1846, p. 647.)

l. Odyle is distributed through the whole mass of bodies, as in the case of water. Odylised water may be poured from one glass into another, and is then found, to the last drop, uniformly and equally odylised. Metals, in the odylic glow, appear to sensitives translucent, glowing through and through hollow balls, when charged with odyle, exhibit strong manifestations of odylic activity in their interior. Magnetism, according to the researches of Barlow, appears to be limited entirely to the surface of bodies.

m. It appears, and will subsequently be more particularly explained, that odyle diffuses around itself alternating spherical zones of opposite polarities, like electricity. Nothing similar has ever been observed in magnetism.

n. Odyle has no attraction for, and no power of supporting, iron filings, even the smallest. But the most remarkable action of the magnet is precisely that of so attracting and supporting iron. And in reference to odylic power, crystals and hands of the same size not only equal, but often surpass the magnet. This is especially observed in the hand.

o. Terrestrial magnetism does not affect the direction of odylised bodies; but it causes magnetic bodies to place themselves in the meridian, &c.

p. The flames of odylic poles, in the inorganic world, show no appreciable attraction for each other; but magnetic poles, and their lines of force, exhibit mutually the very strongest attraction. The flames from the poles of a horse-shoe do not in any degree tend to approach each other; nay, when unlike poles are directly opposed to each other, their flames, when approximated, not only do not attract, but mutually repel each other, when we try to force them together. All this is in direct contradiction with all we know of magnetism.

q. When both limbs of a horse-shoe magnet are held horizontally, the odylic flames flow out from them horizontally for a space, but then both flames turn upwards in a curve. Such a tendency upwards is nowhere observed in magnetism.

r. A certain amount of odylic flame flows and continues to flow, (see farther on, § 458), from the magnetic poles of a horse-shoe magnet, after they have been rendered magnetically indifferent, externally, by attaching the armature. The magnetic efflux is arrested; the odylic efflux continues, although more feebly than before the armature was attached.

s. Even when two powerful and opposite poles of magnets are joined, (§§ 401, 402, 404, 405), when they unite and neutralize each other, yet, notwithstanding, currents of odylic flame continue, although less vividly, to flow from both of the united poles.

t. Magnets, placed in the electric atmosphere of the conductor, can be made to invert the position of the odylic poles, while that of the magnetic poles remains unchanged. Electricity therefore exerts a power over the former, which it nowhere exerts over the latter. §§ 373, 436.

u. Where odyle and magnetism are both excited, they never appear simultaneously. When a galvanic current acts on a multiplicator or a rotation instrument, the effect on the needle is instantaneous. But the odylic sensations only appear after several seconds, and the more slowly, the longer the conducting wire. The same is observed when the phenomena cease, the magnetic effects instantly disappearing when the current is interrupted, while the odylic phenomena continue for some time longer.

v. When a crystal, a finger, or the end of a rod, the other end of which is immersed in a mixture where chemical action is taking place, are introduced into the coil, no induced current is excited, even when these bodies are far larger, and have far greater odylic power, than a magnetic bar or needle. The latter may be ten times smaller, and may possess only a hundredth part of the odylic power, but yet it will instantly cause a current in the coil. § 40.

w. If we hold a bar magnet so, that the pole corresponding, odylically, to the hand projects from the hand, the odylic flame of the magnet, as well as its odylic power, increase.

But it does not, on that account, support one grain of iron more than before. *Mutatis mutandis* the poles of crystals may be made in the same way to act on magnets. The magnetic bar has gained in odylic force, by communication from the hand or the crystalline pole; but its magnetism, properly so called, is not in the least augmented. See further on, §§ 442, 444.

x. This may be pushed so far as to invert the odylic polarity of magnets, while their magnetic polarity is not affected. If a weak bar magnet be held with the southward pole in the left hand, the projecting northward pole not only loses its blue negative flame, but gives out a red positive flame, while it continues unchangeably magneto-negative. See further on, § 446.

y. We shall see, in the course of this work, remarkable instances, in which the odylic flame of a magnet has been extinguished by the approach of an organised living being, which in no way changes the magnetic power. § 448.

z. The moon also furnishes an excellent proof of the difference between magnetism and odyle. I have already shown, (§ 119), that the moon acts odylo-positively on all sensitives. On the other hand, M. KREIL, the astronomer at Prague, so distinguished for his contributions to science, tells us, in the first volume of his Astronomical and Meteorological Annual, p. 104, that the moon exerts an attraction on the southward pole of the needle, and must consequently possess, on the side turned towards the earth, the magnetism of the *south* terrestrial pole. This looks like a contradiction; but on closer inspection the contradiction disappears. The magnetic action of the moon on the needle is so feeble, that it can only be detected by observations of the utmost delicacy, such as a philosopher like M. KREIL can make; but is very far from being sufficiently powerful to affect, *as a magnetic action*, healthy sensitives, or those of middling sensitiveness. But all sensitives find the moon odylo-positive, and that very vividly. It is the *luminous rays* of the moon which cause its powerful action. This is best proved by the fact, that the action of the moon on sensitives may be in great part destroyed, by excluding the direct lunar rays, or those reflected from good window glass. Magnetic action, which penetrates all things, could not be excluded by excluding the rays. Thus, in spite of the fact, that the moon is proved to be negative in regard to true magnetism, she yet, by virtue of her light, sends down to us abundant supplies of positive odyle.

aa. Of diamagnetism we know only repulsions; which may finally, according to the observations of HALDAT, admit of being referred to ordinary magnetic phenomena.

bb. But the difference between odyle and magnetism is rendered most strikingly visible by the following experiment: An iron bar eighteen or twenty inches long, is fixed in a wooden support, which grasps its shorter axis. If it be now placed so, that it lies horizontally in the plane of the magnetic meridian, every sensitive person finds its northward end cool, its southward end slightly warm. If now the northward end be depressed till the bar lies in the plane of the magnetic inclination, that is, if the bar be made to revolve till it makes at its northward end an angle of about 65 with the horizon, in which position its magnetic state reaches a maximum, its northward pole should also reach the maximum of coolness, the southward pole that of warmth. But *exactly the reverse is observed.* The magnetic northward pole has now become odylically warm, and the magnetic southward pole odylically cold. Odyle and magnetism, which hitherto have preserved a certain parallelism in their effects, here appear in direct opposition. The magneto-negative northward pole is odylo-positive; the magneto-positive southward pole is odylo-negative, under the given circumstances. The two forces here follow diametrically opposite courses.

cc. We shall meet, further on, with cases, where the magnetic steel bars or lamellæ of a compound magnet, possessing the same magnetism, exhibit the fact, that while the magnetic polarity of all is the same (as is the case at each end of the magnet), yet we find the lamellæ to be alternately odylo-positive and odylo-negative. This can be rendered more intense by the approach of crystals or of hands; but the approach of magnetic poles puts an end to it. §§ 340, 344.

dd. Nay, we shall meet with an experiment in which it

occurs, that in the limbs of a horse-shoe magnet, during the process of drawing it along another magnet, in a certain limb of the compound magnet, positive magnetism and negative odylo-luminous emanations occurred at the same time. § 433.

For the present, therefore, the identity of odyle with magnetism is entirely out of the question.

The differences which, according to the above comparison, exist between odyle and heat, electricity and magnetism, have not only not diminished in the course of the investigations made subsequently to the writing of § 276, but have, on the contrary, become still more prominent. We shall find this to be still more the case, in the treatises next to follow. They are, in many points, so strong and marked, that the necessity for a special name for the new force appears more irresistible than ever. *Under the term odyle, I collect and unite all the physical phenomena, occurring in the course of these researches, which cannot be brought under any of the hitherto admitted imponderables; and also the* vis occulta, *which produces them.* It remains for future investigation to determine whether, and to what extent, these phenomena will admit of being distributed among, or transferred to, the known forces above mentioned. But a deeper insight into the essence of these things than we now possess, is necessary before such a result can be expected.

Every one is nearly convinced, that electricity and magnetism are essentially one; no one doubts that the same force acts in the one case in a half free, in the other, in a half combined or latent state. We hourly look for the grand discovery, by which their identity will be proved, and both will be brought under one common idea. But we shall never be able to dispense with either of the words, electricity or magnetism. And exactly in the same way, we shall never be able to do without the word odyle, or some equivalent term, on the adoption of which men may agree. Such a term must always be required to embrace a mass of phenomena, which cannot, with propriety or accuracy, be registered, save as a peculiar group.

EIGHTH TREATISE.

LUMINOUS PHENOMENA.

THE ODYLIC LUMINOUS PHENOMENA SEEN OVER THE MAGNET.

280. I began the present series of treatises with the most striking luminous phenomena—those of the magnet; then followed those of crystals, those of certain parts of the body, such as the hand; those of the heavenly bodies, of heat, of friction, of light, of electricity, of chemical action, and finally, of all material bodies in general. But in all these cases, I have only mentioned the phenomena *generally*, because I was endeavouring to obtain characters indicating the presence and activity of odyle, and was able to establish these, by proving the existence of the luminous phenomena. But I have uniformly confined myself to this, and reserved an examination of them *in detail*, for a separate treatise. This and the following sections will contain that detailed investigation; and thus I fulfil the promise made in § 55 and § 93.

281. I have hitherto described, in the numerous cases adduced, the impressions on the visual organs, by means of which the existence of the odyle has been established only according to the statements and descriptions of highly sensitive patients. But we shall now find that these are far from being the only persons who perceive this light; but that, in many

instances, this phenomenon appears in so vivid and marked a form, that not only those of an inferior degree of sensitiveness, *but also, and this is a point of the greatest importance, that a large number of* PERFECTLY HEALTHY PERSONS *are able to see it.*

282. I shall now, therefore, proceed to give to the physical fact, namely, the existence of the light, additional stability and greater extension, by a large addition to the number of cases in which it has been observed. I shall then consider more minutely the characters of this light in the various forms in which it appears; I shall compare these, both among themselves and with other allied phenomena; deduce from this comparison some new laws concerning odyle; and from the whole I shall endeavour to ascertain some points which may enable us to assign to this force its true physical position.

HISTORICAL DEVELOPEMENT OF THE FACT OF THE ODYLIC LIGHT SEEN OVER MAGNETS, CONSIDERED GENERALLY.

283. It has been shown in the preceding treatises that the odylic light was perceived by all the highly sensitive persons, without exception, examined by me; and that its existence was placed beyond all reasonable doubt by a series of witnesses agreeing together in every essential point. It has further been mentioned, that the different degrees of vividness presented by the phenomenon to different persons, depended on differences in the nature of their disease; and lastly, that in the same observer, different stages of his morbid condition caused different degrees of acuteness of the sense concerned, and thereby different degrees of vividness, or a different amount in the light perceived by the eye. This must be distinctly understood. Mlle. NOWOTNY, when far advanced in convalescence, saw no longer any light even on the most powerful magnet (§ 3). Some days previously when her cure was not so far advanced, she saw only transient flashes of light at the moment when I removed the armature. Still earlier, she had seen a glowing thread of light along the edges of the magnet, and a week before this she had seen a beautiful

brilliant flame on both poles of the open magnet, the radiating emanations of light from which were half an inch long. Mlle. STURMANN (§ 4) saw the flames of the same magnet about four inches long, or three times the length. Mlle. REICHEL drew them for me as of the same length as the limbs of the magnet, that is, about a foot long. Mlle. MAIX (§ 6) saw them, in her usual state, a handbreadth in height; but when she was attacked by spasms, the same magnet appeared to her bathed in fire, and the flames, in some parts, several spans in length. Mlle. REICHEL (§ 7) saw magnets even in moderate darkness, not only sending out at the poles flames as long as the limb of the horse-shoe, but entirely clothed in fine delicate lights, and this even when the armature was attached. But Mlle. ATZMANNSDORFER (§ 13) told me, that the flame of my large nine-bar magnet, in complete darkness, to her appeared to reach the height of five or six feet, so that she was often surrounded by it as if it were burning her. This magnet, as well as smaller ones, she saw enveloped on every side with a fine downy flame. Each pole gave out at its four corners flames, which were blue with red, yellow, and green. Every separate bar of the compound magnet had lateral flames of its own (§ 9). Bar magnets always had a larger flame at the northward than at the southward pole.

284. All these observations were made in 1844. I have collected, during 1845, 1846, and 1847, a much larger number, having, during these years, devoted myself uninterruptedly to the study of this attractive subject. I shall now give such of them as simply serve further to prove and establish the fact of the existence of light playing over the magnet. But I must here expressly remark, that this is chiefly intended for such readers as do not consider the observations already given to be sufficient for convincing them, and who think it necessary to have a very large number of witnesses testifying to the physical phenomenon of the luminous emanations from magnets. All those who are satisfied with the proofs already adduced, and with the accuracy of my method of investigation, may, if they please, without much loss, pass over the following pages, and go at once to § 334, where they will find the con-

clusions deduced from all the cases, without the concrete details of each.

I follow the order of the power of seeing the light, beginning with those cases in which it was feeblest, and proceeding to those in which the power was most strongly developed; I proceed from the healthy sensitives progressively to the delicate, the sickly, and lastly to the highly sensitive patients who are permanently in bad health.

A. Healthy sensitives.

285. Among these there were a few possessing in a very marked degree the power of perceiving the impressions on the sense of feeling made by the odylic influence, but who could not see the magnetic light; others who saw the light, though feebly; and some, who saw it very easily, and with such distinctness and precision, *that they surpassed in this respect not only many patients, but also some actual somnambulists.*

286. Dr. Friedrich from Munich, a young and healthy physician, of whom I had heard that he was subject to sleep-walking, was so obliging as to permit me to test his power, and to go with me into a dark chamber which I had prepared expressly for these experiments. I found him decidedly, though feebly sensitive. He saw light from other objects, but not from magnets.

287. M. Eduard Huetter, proprietor of the bookselling establishment of the heirs of Anton Doll in Vienna, a blooming, young, and perfectly healthy man, was in many ways sensitive to odylic impressions, but had no power of seeing magnetic light, with the exception of a feeble and uncertain gleam in the dark from a small but strong horse-shoe magnet.

288. Maximilian Krueger, a boy of twelve, in the orphan house of Vienna, very sensitive to the other odylic impressions, was not able to perceive any light whatever from magnets in the dark.

289. M. Carl Rössner, Imperial Councillor and Professor of Architecture in the Academy of the Fine Arts in Vienna, forty-two years of age, singularly sensitive for all the other

sensations produced by odylic action, was unable, during three hours which he kindly devoted to the enquiry, in my dark chamber, to perceive a trace of the luminous appearances.

I frequently met with the same thing, especially in men, and in a remarkable degree in Dr. DIESING, Curator in the Imperial Cabinet of Natural History in Vienna. He was in the highest degree sensitive to the odylic feelings, but utterly insensitive, even after four hours spent in complete darkness, to all the luminous phenomena.

Here, therefore, are five decidedly sensitive persons who could hardly, or not at all, perceive the light which flows from the magnet. From this we must conclude that odylic sensitiveness is not always combined with the power of seeing the luminous phenomena. But these cases are certainly the exceptions.

290. I have already often spoken of M. KARL SCHUH, when treating of the perception of the feelings excited by odyle. In reference to the luminous appearances, he stands at the bottom of the scale. He does not see the magnetic flame, but in total darkness he can clearly recognise the form of the magnets, and therefore perceives, feebly, the luminous emanations, not in the form of flame, from the mass of metal. This is what I have called the odylic glow, and it forms the first degree in the perception of these luminous phenomena.

291. The Chevalier HUBERT VON RAINER, from Klagenfurt, twenty-four years of age, barrister, a man overflowing with health and vigour, has never been ill, and knows neither headache nor pain of stomach. His sensitiveness, as concerns the odylic feelings, is very marked; but he was only able with certainty to recognise the magnetic light at the moment when I removed the armature, when he saw a flash of light.

292. Dr. RAGSKY, Professor of Chemistry in the medical and surgical Josephsakademie in Vienna, a very healthy, unusually tall and powerful man, thirty-two years of age, possessed every form of sensitive excitability. But in the dark he did not see bar magnets, while he saw horse-shoe magnets, if single and powerful, shining towards the poles, so that their form was easily distinguished. He saw also an intermitting

blue light appearing at the northward pole, but no light of this kind at the southward pole.

293. *a.* Dr. HUSS, Professor of Clinical Medicine in Stockholm, and Body Physician to the King of Sweden, saw, in the dark chamber, only the more brilliant luminous phenomena, but not those of ordinary magnets. But when I produced before him a powerful electro-magnet, by the current of a SMEE'S battery of zinc and silver, having a surface of 400 square inches, he saw a pale light appear on the negative pole, which emitted abundant vapour or smoke, ascending in the form of clouds, and of course feebly luminous.

293. *b.* HERMINE FENZL, eight years of age, daughter of Dr. EDUARD FENZL, Superintendant of the Botanical Department of the Imperial Museum of Natural History, in Vienna, a child of elegant slender build and active intellect, healthy and cheerful, showed in a high degree, sensitiveness for all the feelings produced by odyle; but she was not so sensitive for the luminous phenomena. Of the magnets, she saw three large and one small ones, of horse-shoe forms, shining with a grey light, which was brighter in the small one. She said the northward poles were brighter than the southward. Of bars, she only saw some of 0.8 of an inch long shining; others, eight inches long, she could not see. She saw neither flame nor vapour; and could perceive no bright light on any large magnets.

294. Baroness ISABELLA VON TESSEDIK, the young widow of the late M. FRANZ VON TESSEDIK from Pesth, a mother, of calm temperament, and a lady of distinguished intellectual accomplishments, saw a magnetic bar, twenty-eight inches long, and all horse-shoe magnets, feebly luminous in the metal itself, so that the forms could be distinguished in the dark; that is, they exhibited the odylic glow. On the poles of a single bar, of a three-bar, and of a nine-bar horse-shoe, she saw luminous vapour from four to eight inches high. The same appearance was visible to her on an electro-magnet of the same shape, except that, in this, and some other horse-shoe magnets, the luminous vapour appeared only on the northward pole. She saw the magnetic light brighter, when the arma-

ture was removed; when this was done, she always saw a flash of light.

295. M. CONSTANTIN DELHEZ, forty years of age, a French Philologist, residing in Vienna, healthy, but highly sensitive, after being an hour and a half in the dark chamber, saw bar magnets glowing with a vaporous light. One-bar, three-bar, and five-bar horse-shoe magnets exhibited, when open, above their poles, lights flowing out to the height of from 0.2 to two inches. The flame of the nine-bar horse-shoe rose to upwards of 18 inches, those of an electro-magnet to about 40 inches, and both of these caused a round luminous space to appear in the ceiling, 40 inches in diameter. Both the odylic glow on the steel, and the odylic flames flowing out over the poles, had always, in his case, something of a vaporous or misty character; which is stated also by all observers of the inferior degrees of sensitiveness.

296. Our respected Consistorial Councillor, and Superintendant of the evangelical community in Vienna, M. ERNST PAUER, to whose parish I and my family belong; of a tall and stately but finely made figure, fifty-four years of age, whose sensitiveness is particularly well marked in reference to the feelings excited by odyle, saw all magnets glowing through their whole length. In some of the smaller ones, he saw a flash when the armature was detached; and in small bar magnets, as well as in some horse-shoes, he saw odylic light flowing from the poles, in some cases only from the northward pole: In a five-bar horse-shoe this light or flame was stronger at the negative, weaker at the positive pole. In an electro-magnet, he saw both poles covered with a pale light, two inches high and two inches wide.

297. Baroness PAULINE VON NATORP, in Vienna, mother of two children, young and very intelligent, of fair complexion, saw the flash when small horse-shoe magnets were disarmed, and a continuous light on the northward limb. In larger magnets, she saw the flame, sometimes only of one, at other times of both poles, like a luminous vapour nearly four inches high. In large magnets, of inferior magnetic intensity, she saw over the poles only a feebly luminous thin vapour or smoke.

298. M. DEMETER TIRKA, a Greek, wholesale merchant in Vienna, forty years of age, distinguished by his love of the plastic fine arts, of powerful make, and, with the exception of occasional headache, perfectly healthy during his whole life, saw a flash of light above the northward pole of a small and of a large horse-shoe magnet, when I detached the armature. A three-bar horse-shoe exhibited a continuous light over the same pole, but no light on the southward pole. A heavy nine-bar horse-shoe did not exhibit to him anywhere a distinct flame, but all objects close to it were distinctly illuminated.

299. Madame SYLVIA VON VARADY, wife of the Imperial Court Councillor, M. VON VARADY, in Vienna, a young and blooming lady, lively, healthy, of Italian descent, saw, in the dark, all magnetic bars and horse-shoes, which appeared in a dull light. A single-bar horse-shoe gave out at the poles a vaporous light two inches high; the nine-bar horse-shoe sent out a similar light five feet high. Strongly magnetic bars of eight inches in length exhibited at the poles lights 0.8 of an inch high.

300. M. THEODOR KOTSCHY, evangelical clergyman, also a botanist, the well-known traveller in Africa and Persia, already mentioned (§§ 80, 191, 222, 232), a very powerful and vigorous, and perfectly healthy man, saw, on single horse-shoes, when the armature was detached, flames starting up on the northward pole, which soon after disappeared. On three-bar horse-shoes, he saw thin vaporous lights, constantly playing over the poles, and 1.2 to 1.6 inches high. When I attached the armature, and then suddenly removed it, they rose higher, in the first moment, but immediately fell to their first size. A nine-bar horse-shoe he did not at first see; but very soon he perceived, when the magnet was open, a very thin, widely extended light playing over the poles, sixteen inches wide and forty inches high. He did not describe it as a flame, but as a general very delicate luminous appearance, visible in the surrounding darkness. He saw a small pocket horse-shoe magnet more distinctly covered with flame-like lights than the nine-bar horse-shoe. Its magnetism was

more intense. In a three-bar horse-shoe, he saw only the northward pole covered with light.

301. Mlle. ERNESTINE ANSCHUETZ, daughter of our distinguished Court comedian, M. ANSCHUETZ, a lady of tranquil and retired turn of mind, at this time in perfect health, but having, during her earlier years, suffered from occasional spasms and headaches, saw the metal of all the magnets feebly luminous. The poles of two magnetic needles, each four inches long, had on their surface the odylic glow, stronger towards the poles, and exhibited at one end a bluish flame, at the other a yellowish flame. She saw the northward pole of a pocket horse-shoe magnet covered with a delicate flame, 1.2 inches long, but the southward pole did not appear luminous. A five-bar horse-shoe appeared with flames of a finger-length on both poles, that on the negative pole bluish, the opposite one yellowish. A nine-bar magnet exhibited also on both poles delicate luminous emanations or flames, from twelve to sixteen inches high, the one shorter and yellowish, the other longer, and more bluish or greyish.

302. Madame JOSEPHINE FENZL, wife of our distinguished botanist, Dr. FENZL, known throughout Europe for his literary labours, the young mother of several children, healthy and blooming, saw, in the dark chamber, a magnetic bar, two feet long, with the odylic glow, and lights flowing from both poles. She saw lights, in different degrees, flowing from the poles of horse-shoes of one, three, five, and nine bars. Some months later, she saw a nine-bar horse-shoe flaming to the height of nineteen or twenty inches, and above this, a luminous smoke ascended, visible half-way to the ceiling. At the same time she saw, in bars, small and large, the odylic glow, and, on the poles, flames; those on the northward pole longer, less luminous, and bluish; those on the southward pole shorter, brighter, and yellowish red.

303. M. FRANZ FERNOLENDT, from Transylvania, proprietor of a manufactory of chemical products in Vienna, Kumpfgasse, No. 285, fifty-four years of age, saw the magnets in the dark less by their odylic glow, than by their polar flames. He only saw in a few a vivid glow, and among bar-magnets only

the smaller ones, as well as a pocket horse-shoe, showed this. But he saw nebulous lights over all, which played over the poles like a luminous appearance without distinct form. Sometimes when I detached the armature, he saw a flash. Over horse-shoes of one, three, and five bars, he saw the flowing light, but always either exclusively or more distinctly over the northward pole, and usually in the form of a grey luminous nebula. Over a large nine-bar horse-shoe, he saw the flowing lights eighteen or twenty inches high; but when I acted on these emanations by the proximity of the positive electric atmosphere, he saw a mass of light gradually rise, as broad as a man, and reaching to the ceiling.

304. M. WILHELM HOCHSTETTER, son of Professor HOCHSTETTER of Esslingen in Würtemburg, the distinguished botanist and director of the well known Traveller's Association (Reiseverein), twenty-one years of age, employed in the Imperial Gardens at Schönbrunn, in acquiring a knowledge of horticulture, in perfect health, of very blooming aspect, all day long in the enjoyment of nature in the open air, saw every magnet odylically glowing in the dark. He saw the light from their poles, not as flames, but as an appearance of formless light, in small magnets from 0.8 to 1.2 inch long, increasing with the size of the magnets, and in the large nine-bar magnet about twenty inches high. The lights were larger and more luminous on the northward poles of bars and horse-shoes. On the southward poles they were very small and dull, but sent up a much more abundant smoke.

305. M. NIKOLAUS RABE, about forty years of age, a superior official in the Imperial Depot for mining products at Vienna, healthy all his life, a powerful man, of lively cheerful temperament and warm feelings, saw the metallic mass of all magnets placed before him in the dark, glowing with a delicate light, especially in horse-shoes, as long as the armature was attached. Bar magnets gave out lights at both poles; one of five feet sent out at the northward pole a flame twenty inches long, at the southward pole a flame twelve inches long. The ends of these flames passed into luminous vapour, which was brighter and thicker at the positive pole than at the negative.

All horse-shoes, when open, had flames on both poles, stronger and weaker, according to the intensity of their magnetism. The flames always ended in a luminous smoke-like vapour, which slowly ascended.

306. JOHANN KLAIBER, carpenter, already mentioned, in reference to odylic sensations (§§ 50, 191), saw the magnets in the dark glowing through their whole mass. A long horse-shoe, too feeble to carry a weight equal to its own, exhibited a very faint light, hardly in contact with the northward pole, but as if lightly floating over it at a height of nearly four inches. Other persons have observed the same appearance, of which I shall give more details hereafter. In a three-bar horse-shoe he saw a pale bluish flame of variable size on the northward pole, sometimes under an inch, at other times four inches high, according to the subjective state of the observer. The end of the flame always passed into luminous odylic vapour, mixed with occasional scattered small bright sparks. The southward pole sometimes had no flame, at other times a feeble and intermitting light. When he first examined a nine-bar horse-shoe, being quite close to it, he perceived no light, but on retiring a step, he saw it odylically glowing, and a luminous appearance, of considerable width and as long as an arm, blazed up over both poles, assuming above the form of a blue pointed flame for the most part, passing into a wide, feebly luminous vapour, slowly rising in the air, and exhibiting many isolated and transient scintillations.

307. Madame ELEONORE VON PEICHICH-ZIMANYI, young widow of the Imperial Court Secretary, M. VON PEICHICH, of Hungarian descent, healthy, but somewhat nervous and excitable, saw the odylic glow in all magnets. Bars of eight inches long poured out vaporous lights two inches high. One-bar horse-shoes exhibited luminous emanations of the length of a hand at the northward pole; half as long at the southward pole. A nine-bar horse-shoe sent out a flame nearly four feet high from the northward pole. In all cases, she saw the northward pole flaming more than the southward; generally, twice as much. When the armature was detached, she saw a

momentary flash of increased light. All horse-shoe magnets appeared to her more luminous than bar magnets.

308. STEPHAN KOLLAR, son of M. F. KOLLAR, Curator in the Imperial Museum of Natural History, fourteen years of age, slender, lively, healthy, at night often restless and speaking in his sleep, saw bar, horse-shoe, and electric magnets all glowing, more strongly at the northward pole. They had flame-like lights, varying from 0.8 of an inch high to twenty inches and upwards, and ending in smoke-like vapour, rising in the air, like heavy clouds. His father was present during the investigation.

309. Mlle. SOPHIE PAUER, daughter of the Superintendant PAUER (§ 296), (who was so kind as to assist at the experiments,) is very young, very blooming and healthy, of a tall slender figure, and full of feeling. She had the kindness to devote herself several times, during periods of some months duration, to these researches. When I took her, after her eye was sufficiently accustomed to the darkness, to where a row of magnets lay, the horse-shoes having their armatures attached, she saw them all luminous and glowing in their whole forms, and was much delighted with the peculiar delicate beauty of the spectacle. When I removed the armatures successively, she saw the odylic flames rise over the poles, and always stronger, larger, and brighter over the northward than over the southward pole. There was always brighter and larger light at the moment of detaching the armature, but the flame soon sunk to its constant size. She saw them of 1.2, 2.4, four and eight inches high, according to the power of the bars and horse-shoe magnets. The flame of the nine-bar horse-shoe was twenty inches high, and above it rose a thin vaporous column, even reaching to the ceiling. The flames were blue on the northward, reddish yellow on the southward pole. A pocket horse-shoe magnet appeared to her to exhibit the glow most intensely when laid on the palm of her hand, where its lights were reinforced by the power of the hand.

310. Dr. ENDLICHER, Professor of Botany, Director of the Botanic Garden in Vienna, forty-three years of age, too well

known throughout Europe as a naturalist to permit me to say a word here as to his qualifications for physiological investigation, was so obliging as to pass several hours with me in the dark chamber. In addition to the lights given out, more or less abundantly, by human hands, the organs of plants, crystals, and amorphous bodies, all of which were visible to him, he saw bar magnets eight inches long in a white light or glow, and sending out elongated lights at their poles. He also saw one, three, and five-bar horse-shoes, when closed, surrounded with nebulous light; and when open, pouring forth at the poles vaporous lights from two to four inches high. But the nine-bar magnet, the poles being directed upwards, gave luminous emanations more than three feet high, and they caused a luminous space to appear on the ceiling. The same appearances, but stronger and larger, were seen by him on the poles of a powerful electro-magnet. The flames were forty inches high, of varied colours, stronger from the northward than from the southward pole, and they illuminated a large circle on the ceiling.

311. M. GUSTAV ANSCHUETZ, painter in Vienna, dwelling in the suburb Wieden, in the Ferdinandsgasse, in his own house, No. 268, was the first sensitive in whom I made the discovery, that not only patients, but persons in perfect health, were able to see the odylic light. This was an unexpected discovery, and one of great value to my researches, and to the whole position, in the department of physics, of my subject, a position frequently assailed by my opponents. From that time the chains that had hitherto bound me to the sick were removed. It put an end to the pertinacity with which sensitiveness, in my sense of the term, as a natural phenomenon, was regarded, in spite of all I could say, as either mediately or immediately dependent on somnambulism. The inadmissible, but often repeated objection, that we can place no confidence in the statements of diseased persons, as if all people, as soon as they are ill, became fit for the mad-house! was driven from the field.* M. ANSCHUETZ now took his place as a very sensitive subject: a powerful man, thirty-two years of

* I beg here once more to refer to what I have said on this point in the Editor's Preface.—W. G.

age, and formerly, as an officer, hardened by a thousand military labours and sufferings; a man who was never seriously ill, of middle height, rather fair than dark, very muscular, distinguished in all athletic exercises, of lively and excitable, but susceptible and feeling temperament, in short, a true artist's nature. I found in him all forms of well-marked sensitiveness, and he was the point of departure from which, turning from the sick, I directed my researches to the healthy, of whom I found everywhere that a large proportion were sensitive, so that I might entirely dispense with nervous patients, and might, with aid of healthy sensitives alone, solve all the physical questions connected with my subject. M. Anschuetz, after being an hour in the dark chamber, saw all the odylo-luminous phenomena as clearly and distinctly as the patients had done. I shall have to refer to his observations in every section, but here I shall only give those on the magnetic light. All bar magnets appeared to him in a delicate odylic glow of a greyish white colour, shining through the black of the general darkness. The light was especially visible along the edges, and was also brighter towards the poles than near the shorter axes of the bars. Two needles, four inches long, had luminous emanations over their poles. All horse-shoe magnets, when open, gave out lights from their poles, sometimes from one pole alone, the northward, sometimes from both; in which case, the flame of the southward pole was always feebler, smaller, and duller than the other. In some cases, he only saw small luminous spots, as it were lying on the poles, which occasionally disappeared and reappeared. Some magnets, which I left with him in his house, that he might observe them, appeared to him luminous, even in a darkness less complete than that of my dark chamber. He saw flames on one pole only of a pocket horse-shoe and a three bar horse-shoe; and these flames were not steady, but often increased vividly, and again sank or disappeared. After an intermission of half a minute or a minute, they again appeared in the darkness, and varied in size, sometimes forming a concentrated mass, sometimes a larger nebulous light. The cause of these variations is purely subjective, and will be more particularly discussed hereafter. In an imperfect darkness, the nine bar horse-shoe exhibited only a luminous cloud over the northward pole. But all this does not exhaust the uncommon interest which attracts us in the remarkable case of M. ANSCHUETZ. He is a painter, and was therefore exactly the right person, not only to tell and describe to us what he saw, but, also, to do what no other had been able to accomplish, namely, to represent to us in form and colour what he had seen, to place before us an image of that which, for want of the perceptive power, we ourselves in vain long to behold. One morning, when I went to visit him, he surprised me by exhibiting a black picture, or rather tablet, on which at first, from the angle of incidence of the light falling upon it, I saw nothing. But as he turned it, a nebulous form, delicate and aerial, appeared on the dark ground. It was the countenance of his beautiful wife, as dimly seen in the depth of night by its own odylic light. Crystals, magnets, flowers, and hands, surrounded her, and I had before me an image of a natural phenomenon, such as no eye had ever before seen. I would gladly give to my readers the same pleasure which I felt on this occasion, and I have had a copy of this remarkable picture made, which the reader will find in Plate III. Unfortunately, no kind of engraving can give more than an approximation to the true appearance of so very delicate an object. We see, however, that the drawing of M. ANSCHUETZ agrees in all respects with the statements of the sensitive patients; and that his figures, as far as they concern the same objects, and founded on ocular inspection, aided by knowledge of drawing, do not, in any essential respect, vary from those formerly given. I have often had to combat the objection, that, because the statements of patients are uncertain, my observations and conclusions are unworthy of confidence. Now, although such objections, in the case of patients who are in full and clear possession of all their senses and faculties, and regularly go about their daily occupations, just as perfectly healthy people do, are quite groundless and easily met, yet the occurrence of this case alone cuts away all pretexts for refusing credit to my observations. In like manner, the figures of the luminous phenomena, given in Plate I., have often been attacked, because they are in part founded on description, and not on ocular inspection by the artist. (The reader will remember that these figures are in part from drawings by Mlle. REICHEL [§§ 9, 10].) The wish has been expressed, that the drawings might be made by one who has seen them, and is himself an artist. Now then, I produce to the scientific world such an observer, qualified *lege artis*, in M. ANSCHUETZ, whose sensitiveness no doubt will last as long as he lives, and from whose friendly and obliging disposition every stranger may be sure to obtain the full confirmation of what I have here said.*

312. Baroness MARIA VON AUGUSTIN, wife of Major Baron VON AUGUSTIN, a lady of distinguished scientific accomplishments and in perfect health, saw, in the dark chamber after a short time, all magnets, whether bars or horse-shoes, shining in the odylic glow, at first only as luminous clouds, then with their forms and outlines distinctly defined. In bars eight inches long, she saw the luminous emanations flowing from both poles; in horse-shoes, when open, she saw over the poles of a single bar, a three-bar, and a five-bar magnet, thin nebulous lights flowing out to the height of from four to eight inches, larger and stronger at the northward than at the southward pole. From the nine-bar horse-shoe, and still better from a powerful electro-magnet, she saw flowing lights rising, as tall as a man, exhibiting colours, scintillations, and luminous vapour or smoke, which rose to the ceiling, and there illuminated a certain surface.

313. WILHELMINE GLASER, aged twenty-four, daughter of an innkeeper at Bachtitz in Moravia, at present house-maid in Vienna, short but strongly built, always healthy and vigorous, and having for six years past uninterruptedly fulfilled her laborious duties, saw, after being an hour in the dark

* It was sufficiently obvious to the careful reader of Part I. that the Author's observations had been so made as to justify him in admitting the facts. The very interesting case of M. ANSCHUETZ, in reality, only confirms the facts previously established with so much care, by means of Mlle. REICHEL and the other patients.—W. G.

chamber, the odylic light of all bodies. All magnets appeared to her in a white odylic glow, with blue flames on the northward, reddish yellow and brownish yellow flames on the southward poles. A round bar magnet, nineteen or twenty inches long, exhibited at the former pole a flame of four inches, at the latter a flame of two inches in length. The nine-bar horse-shoe, with its poles upwards and open, yielded at the northward pole a flame of twenty inches, pale yellow and blue, at the southward pole a flame of ten inches, yellowish red, rising vertically, and ending in smoke that flowed upwards to a great height. On the electro-magnet she saw flames more than three feet high.

314. M. Sebastian Zinkel, an aged man of 77, who had all his life been healthy and vigorous, formerly innkeeper, now retired proprietor of the house at Nussdorf, near Vienna, No. 87, took to his house, that he might observe them during the night, magnets and crystals belonging to me. As he sleeps little, he occupied himself with these during many hours, and furnished me with very exact accounts of what he saw. A single-bar horse-shoe, whether closed or open, appeared in the odylic glow. This was strongest near the poles when the magnet was open, and strongest at the bend when it was closed. From the poles of the open magnet he saw flaming lights proceeding, which were in constant motion ; that on the southward pole, two inches long, yellow and turbid ; that from the northward pole, a finger-length in height, luminous and blue. Both ended in a smoky nebulous mass, three or four times as high as the flame, and at that height becoming invisible. The whole magnet was clothed in a shining vapour, of the thickness of a finger, which was strongest when the magnet was open, weaker when it was closed. When the armature was thus attached, it participated in the glow of the magnet, the light being red where it adhered to the northward pole, dark grey where it adhered to the southward pole. Here, then, an old man of seventy-seven saw the odylic light over the magnet as perfectly as a youth during the developement of puberty, or a young woman in the pregnant state.

315. Dr. Nied, practising physician in Vienna (Vorstadt

Erdberg, No. 396), aged thirty-two, was to me an invaluable case, because he is a physician, and must necessarily testify to the existence of the odylo-luminous phenomena in a caste, namely, that of medical men, in which my researches, contrary to all expectation, have often met with the most unfriendly reception.* Dr. NIED is a healthy, vigorous man, and in an extensive practice; he is on foot, and hard at work, the whole day; he is of a lively, cheerful disposition. Yet he saw, particularly well, all kinds of odylic light. Small and large bar magnets, as well as simple and compound horse-shoes, all exhibited the odylic glow. He saw flame-like and smoke-like emanations from bars of eight and twenty-four inches, and also from horse-shoes of one, seven, and nine bars. The flames on the northward poles were more than one-half longer than those on the southward. Open horse-shoes also appeared clothed in a luminous downy vapour. The flame over the nine-bar horse-shoes rose to more than a yard, both polar flames having united to form one column, above which the luminous vapour or cloud rose to the roof. Even the steel hand of his watch, which, no doubt, was strongly magnetic, was, in the dark chamber, so luminous, that he could tell the hour on the watch by its light.

316. Baron VON OBERLAENDER, at Schebetan, in Moravia, about thirty-five years of age, as forest superintendant (Forstmeister), constantly on horseback, and on duty in the woods in every state of the weather, of a most vigorous constitution and iron health, never having been ill, accustomed and hardened to all variations of heat and cold, rain and storms, saw light from all magnets in the dark chamber. Needles of four inches, and of different power, appeared glowing so as to shew their whole form, and gave out, at their poles, flames of 0.8 to 1.2 inch. He saw, on the poles of a pocket horse-shoe, flames of 0.4 to 0.8 of an inch, the shorter flame on the southward pole. A three-bar horse-shoe glowed through its whole mass, and

* I am not at all sure that, in this country at least, such a result would have been contrary to the expectation of any one who has experience of the medical profession, as a body, in the reception given by them to new and startling truths. There are, of course, many exceptions, but I speak generally.—W. G.

was veiled, over its whole surface, in a thin nebulous light, stronger towards the poles. On both poles he saw flickering, restless flames, exhibiting rainbow colours—at the northward pole, blue, green, whitish, and purple, of the length of a hand; at the southward pole, yellowish red, a finger-length high, and feebler; all of such beauty, that he continued gazing with delight on the spectacle, to him quite new. A large nine-bar horse-shoe, viewed at a distance of one or two paces, appeared covered, over the poles, with a great blazing light, about as thick as a man, and reaching nearly to the ceiling. This astonished him much, and his surprise abated only when I told him that others had already seen the same appearance. This prodigious light was yellow and greyish, in motion, of a delicate ethereal nature, and it could be displaced by blowing upon it. It was so thin, and its light so feeble, that when close to it he could not perceive it, but it always appeared when he stepped back. The nine-bar horse-shoe gave out, laterally, from its poles, sparkling scintillations, which flew about like little stars. He compared them to the sparks of burning pine charcoal. This is the same observation which has been already described, and in part figured, from the statements of Mlle. Reichel, in the first treatise. (See § 9.)

But the individual who, among all healthy sensitives, was the most distinguished, and whose perceptions, in strength, vividness, and duration, surpassed those of many diseased sensitives, was, beyond a doubt,

317. Josephine Zinkel, a girl of twenty-three years of age, daughter of the above-mentioned M. Zinkel (§ 314), at Nussdorf, near Vienna. She is of the tall, powerful build of the true Austrian race, that is, in the Duchy of Austria Proper, which I have met with in no other part of Germany. She is perfectly healthy, and of a calm, retiring disposition. All steel magnets appeared to her glowing, so as to shew their whole form perfectly. Two needles, four inches long, exhibited on their northward poles blue flames 1.6 to two inches long; on their southward poles, reddish flames 0.8 to 1.2 inches in length. A bar, eight inches long, had on the northward pole a blue flame of four inches; on the southward pole a reddish

flame, of 1.6 of an inch. A bar, twenty-four inches long, showed a blue flame as long as a hand over the northward, a red flame, with smoke, over the southward pole. Another bar, five feet long, gave a flame of ten inches. Horse-shoes of middling size shewed to her, over their poles, sometimes flames, sometimes luminous nebulæ, from the length of a hand to that of an arm. Stronger ones, such as a nine-bar horse-shoe, sent out blazing and nebulous masses of light, reaching to the roof, and always more abundant, stronger, and more luminous, at the northward than at the southward poles. These experiments were repeated, with numberless variations, during a long period of daily occupation with this sensitive, hundreds of times; and were sometimes attended with less strongly marked, sometimes with more striking results, from the fluctuating sensibility of the girl, subjectively considered, which varied according to her physical and moral state at the time. I cannot enumerate here more than a small part of these results.

318. Pregnancy must be regarded as a state of health, but of a peculiar kind. I have been so fortunate as to find, in the domain of sensitiveness, some representatives even for this state.

Madame Cæcilia Bauer, wife of the innkeeper Bauer, in the suburb Braunhirschengrund, in Vienna, twenty-six years of age, in the sixth month of pregnancy, of a tall, stout make, active and resolute disposition, is, and has been all her life, a thoroughly healthy and very vigorous person. But she is sensitive in such a degree as I have rarely met with; such, indeed, that she surpasses, in this form of excitability, even many somnambulists, and leaves far behind her all the other healthy sensitives, without exception. I had hardly introduced her into the dark chamber, darkened it, and called her attention to her hands, when she answered me with descriptions of luminous phenomena which she saw before her, at first feeble, but soon after so strong, that it was a satisfaction to make the experiments with her. She saw in all bar-magnets the odylic glow throughout, and blue and red flames flickering over their northward and southward poles, generally of half the length of the bar. Weak horse-shoes had blue and yellow; strong ones, blue and red flames over their opposite poles, which almost always passed into a beautiful play of iridescent colours, of which she spoke with lively pleasure. She saw, in the large nine-bar horse-shoe, which at that time was rather weak, flames pouring out to the height of nearly three and a half feet, above which was luminous smoke, reaching to the roof. I have never heard any sensitive, diseased or healthy, speak with so much certainty and precision of the odylic lights before them, as did this pregnant woman.

I must also here again mention Madame JOSEPHINE FENZL, already spoken of in § 302. The experiments of 1846 were made on her in her ordinary state; but those of 1847 belong to a period during which she was pregnant, and thus enabled me to compare these states in the same individual.

319. All these perfectly healthy persons, thirty-five in number, were entirely ignorant of their very remarkable and interesting powers of perception, and were, without exception, much surprised to discover, under my direction, in themselves such powers, of which they previously had no suspicion. The way in which I obtain indications of the existence of such persons, which I then follow up until I lay hold of them, is now simply this:—I enquire among my acquaintances whether they know of any one who frequently suffers from periodical headaches, particularly migraine; or who now and then complains of oppression of stomach; or who frequently, without any known cause, has disturbed or restless sleep; or who speaks often during sleep; rises up in bed, or even gets out of bed, during the night; or who is, in general, disagreeably affected by the moon's light; or who easily becomes faint or sick in churches or theatres; or who is very sensitive to strong odours, and to unpleasant sounds, such as shaving and sawing. I then seek out all such persons, if otherwise healthy, make a pass with a finger over the inner surface of their hands, and I hardly ever fail to find them sensitive. If they now come into my dark chamber, and remain there for one or two hours, they soon begin to be astonished at themselves, and at the perception of a multitude of luminous phenomena, of which they had

not previously the remotest conception. The number of persons who possess this degree of sensibility is, indeed, almost incredible; and I am certainly within the mark when I say, that at least one-third of people in general are more or less sensitive. For wherever I turn, I find healthy sensitives; and this not in dozens, for I could, if it were necessary, collect hundreds of them in a few days. With whatever amount of doubt or incredulity these assertions of mine may be received, the immediate future will and must prove their accuracy. Sensitiveness is not, as I myself believed only a year ago, a rare thing, but a very generally diffused property, which people will soon find everywhere, according to what I have stated, and will thereby open up a new and not unimportant field of observation in the study of the conditions of which man is susceptible.

But we must follow the course we have laid down; and we now come to

B. Delicate or Sickly Sensitives.

320. Under this head, I include all those who can go about their occupations like persons in health, yet are, from time to time, affected with discomfort and illness, so as to be compelled to keep their room or their bed.

321. Mlle. Susanna Nather, thirty-seven years of age, is the daughter of an officer from Basle. I found her ill in a convent at Vienna, with all the symptoms of highly-marked sensitiveness. When she so far recovered as to leave the convent, she paid me a visit, at my request, for several weeks, at Schloss Reisenberg, where I made daily trials with her, to which I shall often have to refer. It was remarkable that, with very great sensitiveness for all that affected the sense of touch, she could never see the magnetic light. We find, therefore, highly sensitive persons, who occasionally suffer from severe nervous ailments, yet whose organs of vision are not able to perceive the odylic light, while we have perfectly healthy persons in great numbers, who, with the utmost facility and distinctness, see these phenomena.

322. Mlle. Josephine Winter, living in Vienna, in the

Vorstadt St. Ulric, No. 60, step-daughter of the painter M. SCHMAL, in Gratz, nineteen years of age, tall, of full habit, vigorous, blooming, full of mirth, at present in perfect health, had, two years ago, a severe nervous illness, during which she for some time suffered from spasms and somnambulism. She retains, since that illness, a degree of sensitive excitability which is very easily affected. She saw all magnets placed before her in the dark, glowing with a whitish light. Two needles, of four inches, gave out at both poles flames of an inch and more, blue and larger at the northward, red at the southward pole. A long single horse-shoe appeared to her with a flame of six inches on the negative or northward pole, and one of four inches on the opposite or positive pole, both of which ended in thin luminous vapour. She saw a three-bar horse-shoe in a white glow, with a delicate nebulous light covering it, and blazing at both poles 8 or 12 inches high, vividly blue at the one, and yellowish red at the other pole. These flames flickered backwards and forwards when she blew upon them.

323. Madame JOHANNA ANSCHUETZ, née STEINER, wife of the above named M. GUSTAV ANSCHUETZ, twenty-eight years of age, mother of two children, a lady of delicate and refined habits, naturally retiring, of acute and intense feelings, had suffered, during nearly her whole life, from many acute diseases, which always produced spasms, and were frequently accompanied by fits of somnambulism. For some years past she has been indeed healthy, but slight agitations of any men tal kind are sufficient again to give rise to the spasms and the sleep-walking. She was so obliging as to allow me to try her power of seeing luminous phenomena, and saw most of them not only in my dark chamber, but also in her own house by night. In two needles of four inches, she saw flames from 0.4 to 0.8 inch long on both poles. A pocket horse-shoe sent out lights at both poles. A single-bar horse-shoe twenty inches long was seen by her glowing throughout, and giving forth at both poles, a moving flame like luminous vapours. She saw a three-bar horse-shoe also glowing throughout, with a brighter fringe round all its edges, especially at the poles.

Another time, during menstruation, she saw over the same magnet at both poles, thin luminous appearances flaming up to the height of eight inches. A five-bar horse-shoe appeared covered with luminous vapours, a handbreadth in height. A seven-bar horse-shoe only exhibited light on its northward pole. She saw a nine-bar horse-shoe shining most brightly at the edges, while thin lights sixteen inches high played over the poles, which, as she assured me, in order to be distinctly seen, had to be viewed from a certain distance, became dim as she approached, and disappeared to her eye when she was close to the magnet. She described the flame as so thin and etherial that it could not be compared to an ordinary fire. It was a pale aerial light, so bodiless, that it could not be closely examined, without, so to speak, dissolving into nothing, "and, like the baseless fabric of a vision," leaving "not a wreck behind." They only became visible at a certain distance by contrast with the surrounding darkness.

As not only Mme. ANSCHUETZ, but her husband, saw the odylic light of magnets, they sometimes amused themselves, in the dark, one of them hiding magnets, which, however, the other soon saw and discovered.

324. Mlle. LEOPOLDINE ATZMANNSDORFER, the younger sister of the so often mentioned Mlle. MARIE ATZMANNSDORFER, and whom I shall, for the sake of distinction from her more sensitive sister, MARIE, hereafter call Mlle. DORFER, is nineteen years of age, of small stature, lively, of healthy aspect, without any external marks of disease. But she suffers often from headache, spasms, and fits of somnambulism, in which she walks about, and speaks to others so naturally, that no one who did not observe her eyes to be shut, would suspect any thing unusual. After being an hour in the dark chamber, she saw a pocket horse-shoe, covered with pale thin lights of 0.4 inch, a large single horse-shoe with lights of six inches, a three-bar horse-shoe with lights of eight inches, a seven-bar one with lights of 1.6 inches at one time and of ten inches at another, all of which disappeared when I attached the armature, and shot up again when I removed it. They were in fluctuating motion, scintillating, always longer and bluish at

the northward, smaller, duller, and yellowish red at the southward pole. They were also brighter and more distinct at the edges and corners of the poles.

325. Mlle. WILHELMINA VON WEIGELSBERG, about twenty-three years of age, living in Vienna, Vorstadt Wieden, Fleischmannsgasse, No. 451, with her aunt, looks well, but suffers much from spasms, and has variable health. She saw two four inch needles shining through their mass with a pale whitish glow, brightest towards the poles. Both poles had flames, bluish on the northward, yellowish red on the southward poles. She saw also horse-shoe magnets glowing. A pocket horse-shoe which I had given to her, showed in her own house by night thin smoke-like flames, 0.8 to 1.2 inches long, on the poles, but shorter on one pole than on the other. They moved, and appeared sometimes brighter, sometimes more nebulous, sometimes smaller, occasionally disappearing on one pole. In the dark chamber she saw, on a five-bar horse-shoe, only a short feebly luminous spot on the southward pole; on the northward, a restless nebulous flame 1.6 inches high. She saw the large nine-bar horse-shoe shooting out flame-like light, twenty inches high, which illuminated the nearest objects.

326. A very singular case, belonging to this section, is that of a *blind sensitive*, the master carpenter, JOHANN FRIEDRICH BOLLMANN, in Vienna, Vorstadt Wieden, Ferdinandsgasse, No. 268, renting the house in which he lives from M. GUSTAV ANSCHUETZ, to whose obliging attention I am indebted for my acquaintance with this remarkable man. He is fifty-six years of age, a native of Kiel in Holstein, and was thirty years ago servant in the laboratory of our highly esteemed natural philosopher PFAFF in that place. He has long had disease of the lungs, became affected with cataract, and was operated on, without success, by Professor FRIEDRICH JAEGER of Vienna. For years he has been blind; that is, quite blind to all shape or form of things, but not entirely insensible to the impression of light generally. The poor man has no longer any crystalline lens, but the retina is healthy. Luminous rays, falling on his mutilated eye-ball, can therefore no longer be concen-

trated into a regular image, but penetrating, in their diffused form, through the turbid humours, reach the healthy retina, and through it produce impressions on the internal visual apparatus, thus giving rise to the mental perception of light. It follows, as a necessary consequence, that he can perceive dimly diffused light and colour, but no form. If any one wears a bright blue or bright yellow dress, or if a lady has on a green or red shawl, he sees the colours, if they are strongly illuminated; but he would experience exactly the same impression from a green branch or a red door. Now it happens that this blind man is sensitive. He was brought to me at Schloss Reisenberg, where I prevailed on him to remain all night, and took him next morning into my dark chamber. After having been there for an hour, the blind man saw a number of luminous phenomena, which I, with good eye-sight, could not see; and when we moved about among the odylo-luminous objects, *it happened, probably for the first time since men have existed, that the blind led him who possessed his sight;* for Master BOLLMANN led me, and thus we exchanged places. I was deprived of the day-light necessary to me; but to him the odylic light had dawned, which to me was invisible. I shall have to refer hereafter to the details; and here I shall only mention, that he saw a small pocket horse-shoe as a luminous spot on the table where it lay; that he did not at first perceive, when close to it, a long single-bar horse-shoe, but when I removed it to the distance of one pace from him, and detached the armature, he saw a sudden flash of light, which disappeared after a few seconds. This he saw only on one pole; and when I made him place my hand on it, I found it, even in the dark, by means of the sign upon it, to be the northward pole. A three-bar horse-shoe appeared to him steadily luminous. He could not perceive the feeble odylic glow of the mass of metal; but when the armature was removed, he saw a permanent light, this time also on the northward pole only. He described it as a round spot, of an inch or an inch and a half in diameter. A nine-bar horse-shoe exhibited a large luminous cloud, which diffused light over the surrounding objects over a circle of more than six feet in diameter. Odylic light therefore penetrated through the vitreous humour, and was received on the nervous expansion of the retina like ordinary light; but he had no sense of the forms of the magnets, of flames, vapour, or sparks, but only that of diffused light. This case is certainly very rare, and it is one which bears strongly on the really luminous nature of the odylic emanations.

327. One of the sensitive females had told me that, in her childhood, she had been affected with green sickness, and had been then very fond of certain kinds of food, especially of those which were raw. As I already knew that very sensitive persons much liked raw food, it occurred to me that possibly chlorosis might be accompanied by well-marked sensitiveness. In order to try this, I looked out for chlorotic patients, and soon learned that, among my own labourers on the farm at Reisenberg, there was a girl who had been permanently chlorotic for the last three years. I immediately proceeded to try her. ANKA HETMANEK, twenty-one years of age, short, but strongly built, plump, much valued as a good and industrious girl, of quiet disposition, a very dexterous silk-spinner, has only once menstruated, is free from headaches; but, along with her chlorosis, suffers much pain and oppression of stomach, which attack her at all times. She proved to be a thoroughly sensitive subject. She experienced all the peculiar sensations in a high degree. All magnets appeared to her to glow with a whitish light, even when closed. When they were open, she saw flames over the poles, stronger and blue over the northward pole, but with more smoke over the southward. The flame from the electro-magnet exhibited rainbow colours, &c. &c.

328. Madame FRANCISCA KIENESBERGER, thirty-nine years of age, wife of an hotel-keeper, mother of two grown-up sons, living in Vienna, Vorstadt Schaumburgergrund, Mittelgasse, No. 97, looking well and stout, very lively, excitable, and giving way to her feelings. She suffers much from headaches and oppression of stomach, and is occasionally affected with spasmodic fits. She is very highly sensitive. At my request she, from time to time, devoted some weeks to visits at my

house. All magnets appeared to her to glow strongly. Two four-inch needles appeared covered at both poles by flames of an inch and more. A bar magnet five feet long shewed on its northward pole a flame of eight inches. All horse-shoe magnets appeared luminous, even when closed; when they were enveloped in a thin luminous down, just as described by Mlle. REICHEL. When I removed the armatures, the poles exhibited flames. A pocket horse-shoe, four inches long, had flames of 0.8 of an inch. A single bar horse-shoe appeared in a whitish glow, with flames of a finger-length on the poles, ending in luminous vapour. A seven-bar horse-shoe exhibited on the poles tongues of flame of the size of a walnut, waving backwards and forwards. At another time, during menstruation, she saw these flames covered to the height of eight to twelve inches with a flame-like vapour. She saw, in her ordinary state, a nine-bar horse-shoe in a white glow, and with polar flames a foot high, ending in a luminous vapour, which rose to a much greater height; but during menstruation the flames appeared more than five feet high, ending above in a bright vapour, reaching to the ceiling. When I detached the armature of this large magnet before her in the dark, she uttered exclamations of wonder and delight at the splendour of the flames, and the showers of sparks and fire, of rainbow colours, which suddenly rose from it.

C. DISEASED SENSITIVES.

329. Mlle. AMALIE KRUEGER, thirty-seven years of age, the daughter of a head-waiter in an hotel, living in Vienna, Vorstadt Leopoldstadt, grosse Ankergasse, No. 27, a young woman of gentle disposition and pious turn of mind, well educated, particularly in languages, and thus better able than many others to describe her observations distinctly, is of full habit, and looks well. From her early youth she has suffered from many nervous disorders, has frequently walked in her sleep, although this affection went and came; and has often had spasms, which are easily excited in her. At such times she saw, on magnets, vivid flames from four to eight inches high.

She sometimes, at my request, visited me for several days at a time, and most obligingly devoted herself to these researches. I shall hereafter have frequent opportunities of reporting her observations. In the dark chamber she always saw the magnetic light most distinctly flash up when I removed or replaced the armature. The nebulous lights which she saw above open magnets were, in proportion to her high sensitiveness, not large, generally only an inch or two long, and visible only on one pole, which was uniformly the northward. This was the case with a pocket horse-shoe, a large single-bar horse-shoe, and similar magnets of three and seven bars.

330. FRIEDRICH WEIDLICH, thirty-two years of age, formerly in the English navy, now invalided, living in Vienna, is affected with severe and incurable hypertrophy of the heart, accompanied with spasms and periods of somnambulism. This man has, whether justly or not I do not know, acquired no very good name among the Vienna physicians. I do not enquire into such matters, but I have seen, that in the trials which I made of his sensitiveness, he has acted throughout with truth and honesty, and has given me genuine statements of his perceptions. It is absolutely impossible for a sensitive, in consequence of the mass of observations I have made, and the experience I have acquired in the matter, to address to me even a single untrue sentence, the untruth of which shall not be instantly detected; either because it is already controlled beforehand, since, with every new subject, I go through all the investigation from the beginning, even those parts of it on which no doubt any longer rests; or because, if I hear today a new statement from one sensitive, I repeat the experiment to-morrow with others. Besides, such a sensitive, even were he magister in physics, cannot perceive or even guess the meaning and object of the cross-examination to which I unceasingly subject him, purposely disconnecting the questions. It is sufficient to say, that every statement made by WEIDLICH bore the impress of truth and accuracy, and was uniformly confirmed by every kind of control I could apply to it. I do not meddle with his conduct in other respects.—He saw all magnets in a pale reddish white glow, brighter towards

the poles, and nearly null at the shorter axis. From the poles of four-inch needles he saw flames flowing, larger at the northward than at the southward pole. At a later period he stated the size of the flames on the northward pole to be two inches; and on the opposite poles 1.3 inch; the former bluish, the latter yellowish. He described a long single bar horse-shoe as having a very thin flame on each pole, that on the northward pole stronger, larger, and blue, that on the southward pole smaller, duller, and reddish yellow. A three-bar horse-shoe had on the northward pole a flame which was of the length of a hand, iridescent, but with predominating blue; on the opposite pole a somewhat more turbid, small, reddish yellow flame. These flames rose close and parallel to one another, vertically, and ended in thick smoke. In a large nine-bar horse-shoe, he saw at first, when close to it, only a flame of eight inches, with much vapour above it; but when he made a step backwards, he then first perceived the great column of light which was not before visible to him. He described it as of the height and thickness of a man, mixed, above, with clouds of smoke, which rolled up to the ceiling, and there illuminated a large space. The colour of the flames was bluish and reddish yellow; when I blew into it, the column was disturbed, but very soon recovered its place. In a later experiment, he described the size of it as rather less, but every thing in proportion; corresponding, therefore, either to an inferior subjective perception, or to an objectively inferior intensity in the magnet.

331. Mlle. CLEMENTINE GIRTLER, eighteen years of age, daughter of M. GIRTLER, Cloth-merchant in Vienna, Vorstadt Wieden, Hauptstrasse, No. 63, a gentle and affectionate young lady, suffered long from hepatic disease, and fell into intense somnambulism. During this time, the moon acted very strongly upon her, and through the kindness of her physician, Dr. HORST, Junior, who completely restored her, I had frequent opportunities of witnessing the most singular attacks. She saw luminous appearances flowing from open horse-shoe magnets, both from a small one, and from one of seven bars. I had no opportunity of trying her powers in the dark cham-

ber, so that I could not make detailed observations with her. But the above facts are sufficient in this place.

332. JOHANNA KYNAST, twenty-two years of age, daughter of a baker in Waidhofen, living in Vienna, Braunhirschengrund, Schmidtgasse, No. 127, with her brothers and sister, looking healthy and well fed, was ill of a nervous fever five years ago. Since that time, she has always been affected with nervous complaints, and has from time to time attacks of somnambulism, lasting for some weeks, disappearing for some weeks and months, and then returning, and so forth. She paid me a visit, and remained some days in my house. In the dark chamber she saw at first not much, and the little she did see, with frequent intermissions, even after having been an hour there. But all at once, and unexpectedly, she fell into the somnambulistic sleep, which lasted half an hour. I allowed her to sleep quietly till she requested me to awake her. As soon as I had done so, she now saw all the odylic lights, from persons, crystals, and other substances; and the magnets, although closed, all lay before her in a delicate whitish glow of light. Needles and bars of various sizes, as well as horseshoes, exhibited the polar flames. A long single bar horseshoe had flames on its poles 2.4 and 3.2 inches high; those with three, five, or seven bars had, in proportion to their strength, longer and brighter flames. The nine-bar horseshoe exhibited flames, more than twenty inches high on the northward, under twenty inches on the southward pole, and above these, a luminous greyish smoke flowed to the height of five feet. She described the flame of the northward pole as larger and more bluish; the other as smaller and reddish or red. They illuminated all surrounding objects.

333. FRANCISCA WEIGAND, twenty-seven years of age, sister of the hat-manufacturer WEIGAND, living in Vienna, Vorstadt Windmühl, obere Pfarrgasse, No. 60, a native of the district of Königshofen in Franconia, suffered from bronchitis, and had periodical fits of somnambulism. I found this girl in a state of remarkable sensitiveness to the odylic light, and from her very obliging disposition and good-will I might have looked for the most interesting results. But unfortunately all

my efforts were impeded by a fanatical physician, devoid of all sense of the scientific value of the case which had so unluckily fallen into his hands. The poor somnambulist was made to prophesy or tell fortunes, and her sufferings were made a source of pecuniary profit to others; so that the case soon acquired a deplorable notoriety in the whole of Vienna. This is precisely the way in which, both in France and Germany, the truly interesting phenomena of somnambulism have been covered with opprobrium, and have been rendered disgraceful in public opinion. I was myself present when our worthy Professor, Dr. LIPPICH, administered to the physician a severe and well-merited rebuke, but, as the sequel too clearly proved, without effect.—I had brought to her some small magnets of both forms. She saw them all, in the dark, very plainly glowing, less bright at the axis, brighter towards the poles. A four-inch needle had on the southward pole a flame of two inches, on the northward pole a flame of six inches long. The horse-shoe, the armature of which, when attached, also glowed, gave out, when open, flames from both poles; that from the southward pole as long as the limb of the magnet, that from the northward pole twice as long. The whole magnet was also enveloped in a delicate fiery down, as formerly described by Mlles. REICHEL, ATZMANNSDORFER, MAIX, and others of the more highly sensitive.

334. I have now *added to the six or seven formerly mentioned, nearly fifty new witnesses*, and I as well as the reader begin to tire of enumerating always the same things, and exhausting myself in repetitions. I could easily go on, to adduce the testimony of many more new sensitives; but I think I have already done more than enough.* Every *rational* doubt or hesitation, which might possibly be felt in regard to the five sensitive girls chiefly alluded to in Part I., must give way before the variety and trustworthiness of the observations made on so many persons, differing so widely in age, sex, rank,

* Even in the list, given in § 277, the reader will find the names of several other sensitives, such as ALOYS BAIER, Mme. LEDEREZ, Baroness SECKENDORF, M. MAUCH, Prof. RÖSSNER, M. KRATOCHWILA, M. KOLLAR, senior; all more or less sensitive.—W. G.

position, and occupation, and of the facts confirmed, as they are, by persons of the highest possible character; facts, moreover, which I have furnished the means of everywhere and easily testing, controlling, and confirming by repetition in other places. I well know, that notwithstanding all this, people enough will be left, to whom all that I have done will not appear enough; for there are such things as *irrational* doubts; there is an *absurd* incredulity; and lastly, there is also an *evil-minded* scepticism.* These I am unable, and do not wish to refute. I only concern myself with thoughtful, reflecting, and rationally judging people, with those who are friends to the quiet progress of science; and such persons will, I trust, be satisfied with the evidence here laid before them, in so far as it bears on the fundamental proposition, *that light emanates from magnets in the dark;* and that not all, but very many persons, both *healthy* and diseased, perceive this light with certainty and distinctness.

If we now unite all these observations and testimonies in a

* A small association of Vienna physicians has lately given us a deplorable instance of this. These gentlemen, after an examination lasting for half a year, came to the edifying result, that Mlles. Reichel, Krueger, Nather, and others, were merely impostors and liars! I sincerely pity these gentlemen (there are, altogether, not less than twenty-three of them, doctors and professors of medicine), that, in the course of twenty-two sittings, they did not know how to get nearer to the truth, and by degrees lost themselves and their sensitives in such a monstrous labyrinth of confusion, that the whole investigation resolved itself finally into mere lies and imposture. I shall name no names. When we read the protocol which, under the Ægis of the Journal of the Society of Physicians in Vienna, they have published in Nov. and Dec. 1846, we cannot escape from a feeling of pain in reflecting that men of such ability, who, if they had chosen, could have rendered essential service to science, have wasted their powers in a manner so deplorably useless, nay, obviously hurtful to the progress of knowledge. For, instead of confirming and discovering any truths, they have, by a series of experiments, very badly made, arrived at the most absurd and perverted conclusions, and done their best to involve anew in doubt and obscurity facts which might be regarded as established. I shall not fail, as often as the course of my work leads me to notice these blunders, to do them full justice in notes; not that I believe that, in the eyes of those who understand scientific researches, they require contradiction, for to such persons they refute themselves by their own absurdity; but because it is my duty to protect foreigners, and those not familiar with such subjects, against the delusions which these ill-made experiments are calculated to foster.—R.

kind of collective evidence, we obtain the following *well-established leading propositions.*

a. Every steel magnet gives out light, independently; *it is odylic light in the general sense.* It appears under different forms, and in these it exhibits different degrees of intensity, different colours, different degrees of thinness or density, and different kinds of motion.

b. Every eye cannot perceive the light. *Only a certain class of persons can do so*, the individuals of which may be either healthy or diseased. Certain morbid states increase this power to a high degree, but the same or nearly the same degree of it is also now and then found in healthy persons.

c. The odylic light is very feeble, and is so overpowered, or, as it were, killed by every other light, as to be thus rendered invisible. In order to see it distinctly, therefore, the eye must be prepared by remaining *for one or more hours in absolute darkness.* The slightest trace of light that penetrates into a room darkened for the purpose, almost always renders the observation of the light impossible, and at all events makes it uncertain.

335. In order to establish, as firmly as possible, the fundamental position, that light, (i. e. a new, hitherto unknown something, not reconcileable with our notions of magnetism, and whether it be ordinary light, or possessed of other peculiar and inherent properties), does emanate from the magnet, I was resolved to spare no labour. And I felt the more bound to confirm it by all the means at my disposal, because as yet we do not possess any visible proof of the same cogency as we can employ in other physical experiments, so as to represent it generally. But having once fixed and established it as a natural law, by the inductive method, and with the enumeration of a superabundantly great number of indisputable special cases, I do not think it necessary for the future, and it would certainly be regarded as a tiresome and useless superfluity, to corroborate all the further observations on this subject, in its measureless extent, and every detail and peculiarity of each separate division of it, by adducing, as I have hitherto done, several dozens of experiments, which only repeat and confirm

each other, and which are made on as many different persons. I shall therefore drop the very detailed and voluminous form of proof above employed for the fact of the existence of the light from magnets, and be satisfied with a smaller amount of testimony to my further observations. Yet in all things of any considerable importance, the reader will find, that I have seldom called as witnesses less than ten or twelve individuals. I would request the reader to judge of the account here to be given of my further researches into the phenomena of the odylic light in its various forms, with reference to the principles just enumerated.

FORMS OF THE EMANATIONS OF ODYLIC LIGHT IN THE MAGNET.

336. The odylic light of magnets appears, as far as my researches at present extend, in five forms, exhibiting themselves as distinct to the eyes of the sensitive. These are—

1. The odylic glow.
2. Odylic flames.
3. Threads, fibres, or feathery down.
4. Luminous vapour or smoke.
5. Scintillations.

We shall consider each of these forms of odylic light successively.

1. ODYLIC GLOW ON STEEL MAGNETS.

337. We have seen that one of the first and most general phenomena, observed by almost all sensitives in complete and long-continued darkness, is a peculiar kind of luminous state, in which steel magnets appear as if feebly glowing or red hot, and which I know not how to call otherwise than the *odylic glow.* Of the nature and appearance of this glow I have already spoken in the preceding treatises, and have produced publicly the names of more than fifty earlier and later ocular

witnesses to the fact of its existence. The next question is, whether the odylic glow remains at all times alike, or whether it is subject to variations, and what these are.

338. I showed to the boy STEPHAN KOLLAR, in the dark chamber, a simple bar magnet, one and a half feet long, lying in the magnetic meridian, with its northward pole in the normal position towards the north. He saw only one-third of the bar glowing; that, namely, which includes the northward pole. The remaining two-thirds escaped him, so that he only had glimpses of them occasionally. I showed the same bar, under the same circumstances, to the healthy sensitives Mlle. ZINKEL, Mme. BAUER, Dr. NIED, Baron OBERLAENDER, Mme. VON VARADY, M. RABE, Mme. VON PEICHICH, and many others. They saw it glowing throughout, strongest at the poles, and diminishing towards the middle. But the colour of the glow was not everywhere the same. The northern half glowed in bluish, the southern in yellow red light, which agrees with the colours of the flames on the opposite poles. Mlle. ZINKEL also described the division between the colours as unequal, the northern bluish half being rather shorter, the opposite one longer. I inverted the bar, placing the southward or positive pole towards the north, the northward or negative towards the south, so that the same poles of the bar came to be in the same position as those of the earth. The colours of the glow became now somewhat turbid, and were modified in some degree. The reddish half of the bar was now towards the north, the bluish towards the south; but the blue was duller than before, and had a tinge of red; the yellowish red was turbid, and mixed with much grey. The northern half was again the shorter of the two. In every change of position, and of intensity of colour, the half which happened to point towards the north was, therefore, always the shorter. I now took *an open horse-shoe magnet.* Professor RAGSKY, M. SCHUH, Chevalier VON RAINER, M. HUETTER, and M. DELHEZ, saw the limbs towards the poles only in a feeble dark grey light. I placed before more highly sensitive persons both poles in the meridian, first pointing both northwards, then southwards. In the former case, they saw the blue glow of the northward pole

increased and brighter, the glow of the southward pole compressed and dull red, with a tinge of greyish blue. In the latter case, the blue glow of the northward pole appeared enfeebled, dull, with a tinge of reddish grey ; but the red glow of the southward pole was larger, and more lively and bright. Analogous results were obtained in several other experiments, in which I showed to Mlle. ZINKEL a horse-shoe magnet, sometimes standing with its poles upwards, at other times lying with its poles *towards the east or west.* It appeared first, that both poles, when lying in the magnetic parallels, gave a duller light than when turned upwards. It next appeared, that when turned towards the east, the glow had more of a greyish blue tinge ; when towards the west, more of a reddish yellow tinge. Further, whenever, in these three positions, the magnet lay with the northward limb on the north side, the glow, both blue and red, was more vivid ; but uniformly, in the opposite case, when the limbs lay unconformably, the colours of the glow were duller and troubled. (I borrow the terms *conformable* and *unconformable* from geologists and miners, who call a stratification opposed to that of the general lie of the surfaces in a mine unconformable, in contradistinction to the conformable stratification which forms the general parallelism. In the same way, I call that position of a bar or horse-shoe, or any other form of magnet, conformable, in which the northward pole lies towards the north, the southward towards the south, and the opposite position unconformable). I made the experiments with Mme. KIENESBERGER, with a three-bar horse-shoe, placing the poles upwards, and turning them alternately to the north or to the south. I also repeated them with a nine-bar horse-shoe, with Mlle. ATZMANNSDORFER, besides making numerous repetitions with other sensitives. The results were always the same. I therefore omit the repetition of the description of them. These phenomena kept, in some degree, pace with the respective intensities of the magnetism of the earth, and of that of a bar magnet; concerning which M. GAUSS has taught us that, in our latitude, they are to each other inversely as the weight of one-eighth of a cubic metre of the earth's mass and that of a steel bar weighing one pound.

In other words, the magnetism of a common steel-bar magnet is, in our latitude, generally more intense than that of the earth, and outweighs so greatly, that the former cannot be, during an experiment, overpowered by the latter. But yet the earth's magnetism makes its presence known in so far that, in unconformable positions, the colours of the odylic glow of the magnet visibly lose in point of clearness and purity. We know that, in virtue of the connection between magnetism and odyle, the allied phenomena always appear together in magnets, of which we have had many examples.

339. A similar result was obtained, in a later trial, in which I set up before Mlle. ZINKEL a bar magnet, twenty inches long, *vertically*, the northward pole upwards. The upper half of the bar appeared with a dull yellowish bluish grey glow, the lower half reddish bluish grey. When a simple horse-shoe was used, both poles retained their colours, whether pointed vertically upwards or downwards, but they varied in tinge and intensity. But in all the trials, the magnetism of the bar, and its associated odylic state, overpowered, to a certain extent, that of the earth, and uniformly the more decidedly, the stronger the magnet employed.

Fig. 1.

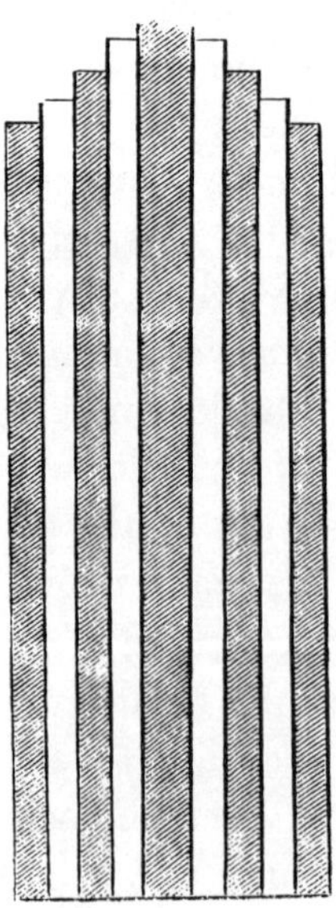

340. The results were more complicated *with a compound magnet.* I made many detailed experiments on this point with Mlle. ZINKEL. I placed before her, in the dark chamber, a large nine-bar horse-shoe of great power, vertically, with the poles upwards, and in a conformable position, the armature being removed. She saw the side of the northward pole in a blue, that of the southward pole in a red odylic glow. This was the case when she looked at the broad side of the lamellæ, and could therefore only see one at a time. But when she stood so as to look at the narrow side, and then had a lateral view of all the lamellæ at once, as in Fig. 1, the appearance was different. She now found that all the lamellæ had not the

same colour of glow. The longest middle lamina, which projected beyond the rest, and was the bearer of the united magnetism of the whole luminous bundle, appeared *on the northward side* in a blue odylic glow. The next, on each side, or the first pair (white in the figure), were not seen in a blue, but in a red glow. The second pair of lamellæ was again blue, the third pair again red, the fourth and last, or external pair, again blue. Only the middle plate, the second and the fourth pairs, were blue; the interpolated first and third pairs were red. But exactly the converse took place on the southward side. Here the middle plate was seen in a red odylic glow. The first on each side, or first pair, had a blue glow, the second pair had again a red, the third a blue, and the fourth again a red glow. The colours of the glow of each plate were therefore regularly opposite on the opposite sides, and alternated with those of the same side. Since, now, all the plates had been originally arranged with the like-named magnetic poles, and, of course, also like-named odylic poles, on the same side, and had been thus fixed together, there must have happened, during their contact, a reversal of the polar states of the first pair of plates, an exchange of the opposite polarities. But, on trying the plates with the needle, I found nowhere a corresponding change of the *magnetic* polarities, all of which were northward on the northward side, and southward on the southward side. Consequently, it was not the magnetic, but only the odylic polarities which had been reversed in the alternate plates; and this took place with the polarities, not only not *because* the magnetic polarities were reversed, but *in spite of* the magnetic polarities not having been reversed. The intensity of the glow was greatest near the poles, and gradually diminished towards the curve of the horse-shoe. The blue, in this direction, became more and more dull, passed, in the middle of the limb, into grey, and towards the middle of the curve, the glow disappeared. On the opposite side, the red passed through reddish yellow into yellow at the middle of the limb; further towards the curve, into whitish grey; then into grey, and disappeared at the middle of the curve. Of the lamellæ, or plates, the middle single one and the fourth or outer pair, glowed at both poles the brightest; the intermediate pairs less brightly. Finally, when I placed the horseshoe with both limbs in the parallels of latitude, so that the broad sides of the outer plate were turned towards the terrestrial poles, then, at each pole, the plates nearest to the opposite pole, and turned towards it, were uniformly the brightest, those furthest from the opposite pole, and turned from it, were the dullest.

341. I find the same experiment, as made with Mme. Cæcilie Bauer, thus noted in my Journal of Experiments:—"She saw the large nine-bar magnet in alternating bands of odylic glow, like Mlle. Zinkel, the middle plate on one side blue, on the other red; also the two external plates of equal intensity, blue and red, when the magnet stood conformably vertical, the poles upwards." Further on, I find,

a. "*On the collective northward limb;* the first pair, one on each side of the middle plate, appeared red, the eastward plate more turbid and dark, the westward brighter, with a tinge of violet.

"The second pair, that lying in the centre, appeared everywhere grey, the eastward plate darker, the westward lighter, yellowish grey.

"The third pair from the middle was again chiefly red, greyish red on the eastward, pale orange on the westward plate.

"The fourth, or external pair, blue on both sides, dark greyish blue on the eastward; azure blue, to pale yellow, on the westward side."

b. "*On the collective southward limb;* the middle plate red; the first pair externally, next the middle plate, she saw blue, the westward plate more greyish blue, the eastward light blue.

"The second pair grey, the plate on the east side darker, that on the west side brighter, with a tinge of yellowish red.

"The third pair again blue; the plate on the west side

more dark grey, that on the west lighter, tending from blue to yellow.

"The fourth, or outer pair, on both sides red; greyish red on the eastward, yellowish red on the westward plate.

"The bands shewed a distinct alternation of brighter and darker plates, but not of blue and red alone, as Mlle. ZINKEL had seen; but an alternation of these colours, with a general mixture of grey on the east, and of yellow on the west side; so that all these tints on the east side of the middle plate, both of the north and of the south limb, were, so to speak, slightly but sensibly pervaded by a veil or breath of grey; those on the west side were in the same way pervaded by yellow. This was a kind of transversality, and consequently a further complication. On the greyish, or east side, the former vividness of the colours became throughout turbid, depressed, dull; on the yellowish, or west side, it was enlivened, heightened, beautified."—Such were my notes of this experiment, made at the time. We shall return to the subject, when treating of the colours of odylic light more specially.

342. I repeated the chief part of this experiment with the far less sensitive Mme. JOSEPHINE FENZL. She perceived no distinct difference of colour between the lamellæ, but a distinct alternation of brighter and darker. The brighter plates appeared feebly reddish grey, the darker feebly bluish grey. Her observations, as far as they went, served to confirm those of the sensitives previously tried.

343. I was enabled to produce the same phenomenon in another way. I attached to the nine-bar horse-shoe, in the manner of an armature, four magnetic bars of equal size, exactly fitting on its poles, that is, on the ends of the projecting middle plate, Fig. 2. p. 284. But I placed them one above the other, so that their northward poles lay on the southward pole of the horse-shoe, their southward poles on its northward pole. The whole was left in this position for twenty-four hours, the poles of the horse-shoe being vertically and conformably placed. When I brought Mlle. ZINKEL to the magnet in the dark chamber, she saw the bars serving as a compound armature, alternating in red and grey odylic glow, as represented

Fig. 2.

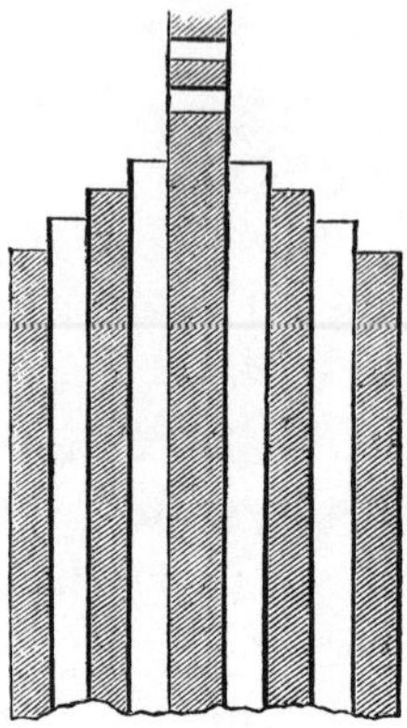

in Fig. 2. On the blue pole of the horse-shoe lay a bar, with a red glow and a red flame; on the opposite red pole the same bar shewed a grey glow and a blue flame. The second bar had, at each end, a glow of a colour opposed to that of the first; the third was of an opposite colour to the second, and the fourth to the third. But on removing and testing the bars, I found them all south polar where they had rested on the northward pole, and north polar where they had rested on the southward pole of the horse-shoe. There was therefore no magnetic alternation in the bars, as they lay on the horse-shoe poles, yet they shewed, notwithstanding, an alternation of the colours of the odylic glow.

344. The odylic *polarities*, therefore, *are reversed*, when several like poles are approximated or coupled together. They then form alternating strata, and this, in spite of the *magnetism proper*, which *is not reversed.* In this case, therefore, there were seen magneto-negative poles in red odylic glow, and magneto-positive poles yielded blue odylic glow, but alternately stratified with negative magnetic poles in the normal blue glow, and positive magnetic poles in their normal red glow.

This very striking phenomenon is one of the very obvious distinctions between magnetism and odyle, which present themselves to the enquirer.

With the view of intensifying this phenomenon, I approached *a bar magnet, twenty inches long*, near to the sides of the horse-shoe, using *unlike poles.* But the result did not correspond to my expectation. When I brought the northward pole of the bar near the southward pole of the horse-shoe, at about two inches from it, the red glow of the latter was indeed perceptibly increased; but the alternation of colours in the lamellæ disappeared, and Mlle. ZINKEL saw nothing but a series of red plates. When I made the same experiment, *mutatis mutandis*, at the opposite pole of the horse-shoe, all the plates appeared in the blue glow. The increase of the magnetism by induction from the bar, therefore, had put an end to the stratification and alternation of colours; that is, in regard to odyle, as well as to magnetism, it had given the entire preponderance to one pole, where both poles formerly alternated.

When I approached to each other, in the same way, *like poles*, the result was striking and unfavourable. All odylic glow was extinguished, and neither red nor blue colours could be seen.

I was desirous next to test the effect which *crystalline poles* might have in a similar experiment; for as the bar magnet, acting on the magnetism of the horse-shoe, interfered with and destroyed the luminous odylic phenomena, it was quite conceivable, that an unmixed odylic influence might affect in quite a different way the odylic appearances presented by the compound horse-shoe. The result proved, in fact, quite different from that obtained with the bar magnet. For when I brought from the side the positive (southward) pole of a very large rock crystal to within about four inches of the northward pole of the horse-shoe, the stratified bands not only did not lose in point of brightness and contrast of colours, but the blue of the alternate lamellæ increased greatly in distinctness and beauty, while the red turned to greyish red. On the opposite side, *mutatis mutandis*, the red lamellæ acquired a finer and brighter colour than before, while the blue ones became decidedly greyish blue.

But when I brought near to each other, like poles of the crystal and the horse-shoe, north to north, south to south, all the odylic glow belonging to the northward or southward poles respectively became dark, and even disappeared; while, on the contrary, the glow of the opposite polarity was vivified and intensified; that is, when the northward or negative pole of the crystal was approached to the same pole of the horse-shoe, the glow of all the blue bands disappeared to the observer, Mlle. ZINKEL, while that of the red bands became brighter; and if the southward poles were brought near each other, the opposite result took place.

This led me to try the effect of the odylic *poles of the animal body.* I turned first my right and then my left side *towards the compound positive or southward* limb of the horse-shoe. In neither case was the odylic glow extinguished; on the contrary, in the first case, the red lamellæ increased in intensity of light and colour, while the blue became darker; in the second, the blue became brighter, the red darker. When I did the same at the *negative limb*, my left side caused increased brightness of the red lamellæ, and dulness of the blue; my right side intensified the blue lamellæ, and rendered the red darker. Each pole of my body, therefore, increased the light and colour of the opposite pole, and depressed those of the like pole; and thus exalted generally, to the eye of Mlle. Zinkel, the contrast of colours in the stratified bands.

From these observations it follows, *that the influence of the magnet on the colours of the odylic glow in steel magnets, is quite different from that of the odyle of crystals, or of the human body. In other words, odyle, when coupled with magnetism proper, has a different effect on the phenomena of the odylic glow in steel magnets from that produced by odyle, when not associated with magnetism, as it emanates, for example, from crystals, and from the living human frame.*

345. *The closing of the magnetic circuit, by attaching the armature*, had a powerful influence on the strength and distribution of the odylic glow in steel magnets. It has been already frequently stated, that when the magnets are open, the glow is strongest towards the poles, and weakest in the middle, in the shorter or magnetic axis; but when the armature is attached, the appearances are changed. In the dark chamber, before Mlle. Zinkel, I attached the armature to a simple horse-shoe, standing vertically. The first thing she noted was, that the armature, which previously, like iron in general, showed a feeble whitish glow, now became much more luminous; and next, that the colour of its light changed, dividing into two halves, a bluish and a reddish. The end lying on the northward pole was reddish, the other end bluish. The same experiment was repeated, some months later, with a five-bar horse-shoe, its poles pointing to the east, (and the limbs

lying in a conformable position?—W. G.) The result in the armature was the same; and occurred also when the poles pointed towards the west. M. SEBASTIAN ZINKEL saw the armature red at the end lying on the negative pole, bluish grey at the other. The same was seen by Mme. BAUER, Mlle. WINTER, and others. Mlle. SOPHIE PAUER hardly saw the armatures when detached; but when placed on the magnet, they instantly became luminous, most brightly towards the ends. These observations are explained by the fact, that the armature is not a mere conductor of magnetism, but instantly becomes, by induction, itself a magnet, when attached; and that consequently its poles must be opposite to the magnetic poles to which they adhere. This proposition, derived from the theory of magnetism, was here practically confirmed by the observations of the sensitives, (who were, no doubt, quite ignorant of the theory.—W. G.)

346. The effect, on the single horse-shoe, of attaching the armature was, that the poles instantly lost much of their light, while the curve, which was previously hardly or not at all luminous, rapidly increased in brightness, so that a kind of uniformity or equilibrium in the glow appeared over the whole magnet, but with distinctly greater brightness at the curve. The bluish glow of the negative limb, and the reddish glow of the positive limb, extended now much more uniformly to the curve, and seemed almost to touch where they passed into each other. The curve itself shared in these differences of colour. Its two halves, the red and the blue, were subjected to the same influences as the like limbs, and changed as the like poles. The same experiment, with a five-bar horse-shoe, yielded the same results. The poles, which, when the magnet was open, far surpassed the curve in glow, became, when it was closed, much less bright, while the curve now surpassed them in light. M. DELHEZ saw, in the open magnet, the poles far brighter than the curve; but when it was closed, the light spread far more uniformly over the whole magnet, without, however, becoming quite uniform. These statements agree well with the theoretical views which we usually take of what happens in the magnet. The concentration of the

magnetic influence at the poles is permitted by the armature to enter into circulation; and thus the opposite polarities are enabled to come into a certain degree of equilibrium along the channel through which they flow.

347. In regard to *intensity*, the luminous appearances on the closed magnet are not constant in their distribution on the different parts, but vary in different ways, partly on the one hand, inasmuch as the magnetism of the earth, and probably other agents, such as terrestrial and atmospheric electricity, sunlight, human hands, and other substances near the magnets which diffuse odyle, more or less act on them; on the other hand, inasmuch as magnetism and odyle are naturally unequally distributed over a magnet, in consequence of their own peculiar attractions and repulsions. All these influences act, now on one side, now on the other, and dislocate more or less, the odylic poles, and the force of their luminous energy. It will never be an easy matter, in all time to come, to calculate the effects due to all these factors. But I have, in the meantime, collected some observations, which I subjoin.

348. Mme. Cæcilia Bauer saw a number of magnets lying on a table in the dark, but observed that they differed much in intensity of glow. Many appeared beautiful and bright, others dull, turbid, and deadened. I arranged them according to her account of the intensity of the light, from the dullest to the brightest. On examining them in daylight, it appeared that the intensities of light did not keep pace with the weights supported by magnets of different size, but uniformly with the magnetic intensity. Simple horse-shoes, which affected the needle at a great distance, glowed brighter than a nine-bar horse-shoe, which carried ten times more iron, but did not act on the needle at so great a distance.—*The intensity of the glow was therefore in proportion to the magnetic intensity.*

349. All sensitives possessed of tolerable power of perception for the light, saw bar magnets brighter at the poles than at the magnetic axis. This was the case with Mlles. Reichel, Weigand, Atzmannsdorfer, Glaser, Baroness von Varady, M. Rabe, Baron von Oberlaender, Baroness von Na-

TORP, M. ANSCHUETZ, and many others, who observed it most distinctly.

350. The same thing occurred with horse-shoe magnets. I find the following persons enumerated in my journals as bearing testimony to this fact: M. DELHEZ, Mme. FENZL, Mme. VON PEICHICH, Mlle. PAUER, Mlles. ANSCHUETZ, WEIGAND, GLASER, Baroness VON AUGUSTIN, Mme. BAUER, M. ZINKEL, Dr. NIED, Baron OBERLAENDER, Mme. VON VARADY, JOHANN KLAIBER, Prof. RAGSKY, and M. HOCHSTETTER. They all agree in this, that in an open horse-shoe, the two limbs are brightest towards the poles, and that the intensity of the light gradually diminishes towards the curve, where it is least. Mme. FENZL saw this in a simple horse-shoe, in one of five and one of nine bars; Mme. BAUER on a simple one; Mme. PAUER on a pocket horse-shoe and on one of three bars; KLAIBER and M. HOCHSTETTER on several horse-shoes, &c.

351. I showed, in the dark, a large electro-magnet, excited by a voltaic pair, to Mme. BAUER, Mlle. ZINKEL, Baroness VON NATORP, Mme. VON TESSEDIK, Mme. KIENESBERGER, Baroness VON AUGUSTIN, and JOHANNA KYNAST. They all saw the open poles glowing brightest, the curve also glowing, but far less intensely.

352. *The attaching of the armature changed this.* All the persons above named then saw the light at the poles diminish, and that of the curve rapidly increase, while the intensity of the glow at the same time increased over the whole magnet, including the armature, and approached in some degree to a uniform distribution of light over the whole magnetically active surface. In most cases this went so far that the curve surpassed the poles in brightness, and in particular, Mlle. PAUER, Mlle. ZINKEL, and Mme. FENZL, decidedly declared the glow at the curve to be the more intense. With the two latter, the experiments were varied and tested in many ways on horse-shoes of three, five, and nine bars.

353. The armature shared in this increased glow; it was brighter at the ends where it lay on the poles, and darker in the middle not in contact with the magnet. Mlle. PAUER and Mlle. ZINKEL saw, in some cases, the poles, in some others

the armature, glowing brightest. Mme. KIENESBERGER, Baroness VON AUGUSTIN, and Mlle. ATZMANNSDORFER, saw in every case tried with them, the poles brighter than the armature. It seemed to me, that in all cases, where the ends of the armatures were well fitted to the poles, and well polished, and thus offered many points of contact, the poles were duller and the armature brighter, and this inasmuch as the armature then conducted the magnetism more perfectly, and discharged the poles more rapidly and completely. The more rapid the conduction, the duller was the light at the poles; the slower the conduction, the greater was the accumulation of magnetism and odyle at the poles, and the brighter their light. The light in the armature is probably always the same, but in the one case is overpowered by that of the poles, in the other, it exceeds that light in intensity.

354. When I closed a horse-shoe magnet, not with its armature, but with another horse-shoe, the results were somewhat modified. Both curves indeed became rapidly brighter, but did not now surpass the poles in intensity of light; on the contrary, the light of the poles was still superior to that of the curves. Mlle. PAUER and Mlle. ZINKEL, two perfectly healthy, exact, and trustworthy observers, saw this distinctly. While closing by the armature gave the greater intensity to the curve, the closing by a horse-shoe magnet still left the poles brighter than the curves.

355. We can see the cause of this difference. When two horse-shoe magnets close each other, a double quantity of magnetism and of odyle must be conducted through both in the same time. The opposite poles also excite and set in motion, when in contact, a still greater amount of magnetism and odyle. There is therefore, on the one hand, a much larger amount of these imponderable influences brought into circulation; on the other hand, the channel through which they must pass has become less open; for soft iron is a much better conductor for them than steel, which has so great a coercitive power for them. Now the armature is of soft iron; but the horse-shoe supplying its place is of hardened steel. Again, the armature is only one-tenth of the length of the horse-shoe; consequently

the former permits a rapid passage, the latter, one ten times slower.—Thus, a much larger amount of the imponderable influences must pass through a more difficult channel, when a horse-shoe is closed by another horse-shoe instead of an armature. The consequence is, that the imponderables, impeded in their egress, accumulate at the poles, and then of course produce brighter odylic glow. This somewhat interrupted or retarded state of circulation approaches to that of the entirely interrupted circuit in the open horse-shoe, when no means of egress are afforded; and in that case, we know that the polar glow is the brightest, and far surpasses that of the curve.—All this, therefore, agrees well with theory.

356. We have already seen how powerful an influence is exerted on the *colours of the glow* by the position of the magnets, as to north and south. The *intensity of the glow* is also affected by this circumstance. All the sensitives saw in every *bar magnet* conformably placed, both poles beyond comparison glowing brighter, than in one placed inconformably, in which the glow was dull and turbid. The magnetism and the odyle of the terrestrial poles, in the former case, act along with and reinforce the odylic emanations of the bar; in the others, they act contrary to, and thus enfeeble these emanations, the force of both terrestrial poles disturbing the phenomena. This simple experiment has been repeated so many hundred times, and is so often necessarily adduced in describing others, that I shall not here waste any space in enumerating the names of observers.—There is sometimes more complication in the case of *horse-shoe magnets;* we have already considered them with reference to this point when open (§ 338); I shall here state the observations made on closed horse-shoes.

A. When a closed horse-shoe lay in the meridian, the poles turned northward, the northward or negative pole was brighter, the opposite pole duller. With the poles turned southward, the intensities were reversed, the northward pole was duller, the southward brighter.

B. When the magnet stood vertically, closed, and the limbs conformably placed, *the poles being upwards*, Mlle. ZINKEL saw the poles became brighter, the curve duller.

C. But when the closed horse-shoe was inverted, the limbs still conformable, and *the closed poles downwards*, the intensities were dislocated, the greatest intensities appearing on both sides of the middle of the curve. *That part, therefore, of a closed horse-shoe* (no matter whether curve or poles) *which was turned downwards, always glowed* LESS BRIGHTLY, *that which was turned upwards, always* MORE BRIGHTLY.

D. When a closed horse-shoe lay in the magnetic parallels of latitude, the limbs conformably placed, Mlle. ZINKEL saw, when the poles pointed eastward, the curve become brighter, the poles duller. But when I turned the poles *westward*, this was reversed. Hence, *that part of a closed horse-shoe which was turned eastward* (whether poles or curve), *always glowed less brightly, while that which was pointed westward always had the more intense glow.* Mlle. ZINKEL also saw the armature brighter when lying westward, duller when lying eastward, as might be expected from its representing the curve. Taken together, therefore, *the curves* and poles of a closed horse-shoe glowed brighter when turned upwards or towards the west, less brightly when turned downwards or towards the east (see § 536.)

357. These experiments were repeated at many different times, with simple and compound horse-shoes, and Mlle. ZINKEL always gave the same consistent account of the phenomena. On the last repetition, she added a slight distinction. When the poles lay eastward, she saw *the northward limb, which lay conformably on the north side, rather darker, the southward limb rather brighter.* But when the poles pointed westward, *the northward limb, conformable as before, appeared rather brighter, the southward darker.* The difference, she said, was not great, but distinctly perceptible. These differences in the phenomena of odylic light are uniformly delicate, and require, for their observation and testing, an observer who is very sensitive, calm, and exact; but still more, a truth-loving, unprejudiced, cautious, and patient investigator. People who value victory more than truth, who are prejudiced in favour of any view whatever, and strive to gain for it a triumph over another view, are not qualified for such investigations. They can only confuse a subject of such delicacy, and tear the

entangled threads which connect the phenomena, instead of disentangling them.*

358. When I *closed the horse-shoe magnet with another horse-shoe*, with the unlike poles, as before, in contact, the results were again modified in some degree. The poles became indeed duller, and the curve brighter than before closure. But Mlle. ZINKEL saw the limbs of both horse-shoes, when lying conformably in the magnetic parallels, still rather brighter at the poles than at the curves. In both, the *glow was most intense at one-seventh of the distance from the poles to the middle of the curve;* from which point the intensity gradually became less towards the bend. The poles, therefore, had lost less of the intensity of their glow than when closed by the armature. In

* Something like this happened in Vienna. It was the object and determination of some persons, with one blow, to make an end of the offensive subject of Magnetism. All the physicians who were most strongly hostile to the very name of magnetism, availing themselves of a convenient opportunity, assembled, and *calling themselves* a committee of the Society of Physicians, began what they intended as their work of annihilation. But the very title was an unauthorised impertinence ; for not only did the Society of Physicians appoint no committee, but that body knew nothing whatever of the matter. Nay, when at last the society became acquainted with it, after the protocol of the "committee" had been, *contrary to rule*, received and printed in the "Journal of the Physicians of Vienna," *it formally disavowed*, in its sitting of 16th November 1846, by a large majority, *all connection* with this one-sided proceeding of some of its members, adopted on their own sole authority. I myself, although an honorary member of the society, first heard of the matter long afterwards, when the "committee" had already held many sittings with Mlle. REICHEL. Indirectly, and under the cloak of some seeming compliments, my work (Part I. was then published) was the object of attack. But although I was so near, it was not thought fit to invite me to the sittings. It was to be feared that I might clear up and rectify contradictions and incongruities in their labours. But the object of the "committee" was not to acquire knowledge, or to ascertain scientific truth ; it was to gain a triumph, in the sense desirable to a coterie of practitioners of the healing art. An impartial committee, chosen by scrutiny out of the whole society, and in which all opinions should have been represented by men *acquainted with the subject*, would have been an excellent and praiseworthy thing, would have, beyond a doubt, brought to light most valuable truths, and would have established these by the weight of its authority. A one-sided party club, chiefly of young men, who, by their own confessions, were *utterly ignorant of the matter in hand*, has no authority, and deserves no authority. I shall take opportunity, from time to time, to illustrate the value of their labours.—R.

a later experiment, she compared the state of the curve in compound horse-shoes to a white jelly; that is, the white transparent glow of the mass of the steel in the dark caused it to resemble a translucent jelly.

359. When I tore the horse-shoes asunder, all the poles became brighter, and rapidly acquired their original intensity of glow; while the curves returned to their dull or dark state. This was confirmed by all the repetitions I made of the experiment, at many different and distant times.

Fig. 3.

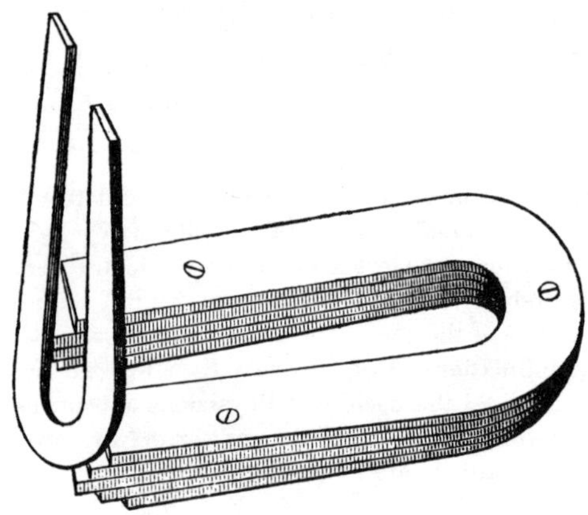

Fig. 4.

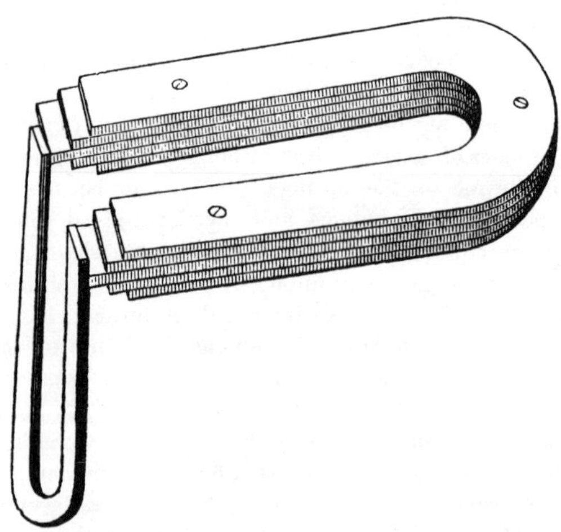

360. The phenomena just described were also confirmed, and seen more distinctly, when I closed the large horse-shoe, not with the poles, but with the curve of the other, as in Fig. 3. The result, in the large horse-shoe, which was lying in the parallel, was just the same as when it was closed by its armature; namely, brighter curve and duller poles. When I drew downwards, vertically, the curve of the smaller horse-shoe, till its poles formed the connection between those of the larger magnet, in the position represented in Fig. 4, I again obtained, in the horizontal magnet brighter poles and a duller curve.

Both results confirm those before stated, and were observed three months later by Mlle. ZINKEL.

361. They may be referred to two causes. First, to the shorter road which the imponderables have to traverse, when the circuit is completed by the curve alone, instead of the whole horse-shoe. Secondly, to the stronger mutual action, and the larger amount of magnetism and of odyle, which is set in motion by the polar contact of two magnets; in which case, the magnetism is accumulated or concentrated at the poles, and is less quickly carried off on account of its greater quantity; which also comes in the end to diminished conduction. And as odyle moves rather sluggishly through matter, we have, in this, an addition to the theoretical probability. It is, however, at present premature to enter on such speculations; we must first collect and arrange the facts.

362. It may serve to shew how delicate the distinctions may become in these matters, if I add, that the observer also noticed a difference in the intensity of the light, when the limbs of both horse-shoes were so placed as to form one line, as in Fig. 5., and when they were placed so as to form a right angle, as in Fig. 4. In the former case she saw the poles glowing rather brighter, the curve duller; in the latter case, the reverse took place. In fact it must be supposed, that when in one line, the magnets excite each other more powerfully than when at right angles, when the directions of their forces cross each other. The fact that the odylic flame proceeds in the prolongation of the line of the magnet or of its limb, shows, that the odylic impulse acts in the straight line of the bar or of the limb, as the case may be.

Fig. 5.

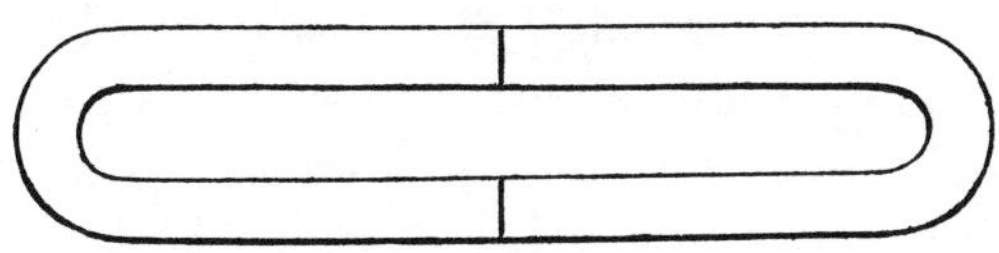

363. Let us now attend to the course of the intensity of the glow *on the other horse-shoe, that, namely, which is used as armature.*—When placed as in Fig. 3., the curve of the closing magnet, previously hardly visible to Mlle. ZINKEL in the dark, quickly acquired a brighter glow, blue on the one side, red on the other. Its poles were also seen above in brighter glow.

When it was now slowly drawn downwards along the poles of the other, its own curve became gradually still brighter, its poles duller. When at last I reached the position of Fig. 4. the curve was at its brightest, the poles at their darkest, although they still glowed pretty brightly. In this position the curve is the brightest part, the limbs the darkest, the poles of intermediate intensity; yet the curve was never so bright as when its poles were closed by an armature. All this agrees perfectly with the preceding observations.

364. The intensity of the odylic glow is further different, even at equal distances from the poles, in different parts of steel magnets. Mme. JOHANNA ANSCHUETZ, and Mlles. ATZMANNSDORFER, WINTER, WEIGAND, NOWOTNY, GLASER, and others, always saw *the edges* brighter than the flat surfaces; *the corners* again brighter than the edges; and *all points* brighter than the corners. They saw this also in the attached armatures. Edges and solid angles were always brighter than flat surfaces; nay, they often saw the edges glowing, when the flat surfaces were dark or invisible. I showed to Mlle. ATZMANNSDORFER a twenty-inch bar magnet, formed with three armatures into a parallelogram. She saw the whole glowing, the magnet brighter than the armatures, but all appeared as it were fringed on the edges with glowing threads on all sides. The healthy sensitives, Mlle. PAUER, M. RABE, M. ZINKEL, and others, often gave the same account of it. —Mme. BAUER saw a simple horse-shoe, when open, more strongly and brightly fringed with light than when closed; in which latter case, the luminous threads on the edges immediately became duller. She saw these threads or borders on both internal and external edges, but brighter on the latter. They were also brighter on the negative than on the positive limb.—Mlle. ZENKEL described this phenomenon most clearly. I showed her horse-shoes of one and three bars, both open and closed. She saw the *simple one, when open*, enclosed in a bright border of light on all its edges. The border on the outer edges was the most luminous, both at the poles and at the curve, so that, in the open simple horse-shoe, the greatest intensity of glow ran round the external edges. *When it was*

closed, it was also bordered with glow; but at the curve the inner edges were brighter than the outer ones. This greater intensity, however, diminished gradually towards the poles, where it was transferred to the outer edges, and attained its maximum about one-seventh of the length of the limb below the pole.—*In the three-bar horse-shoe, when open*, she saw the inner edges of the curve brightest, but diminishing in brightness and even losing their glow towards the poles, where again the outer edges were most luminous.—*When the three-bar horse-shoe was closed*, the outer edges were more bright than the inner, as was also the case with those of the armature. We must not forget, that in the three-bar magnet only the projecting middle bar is directly closed by the armature. Hence the slight differences between it and the simple horse-shoe in the brightness of the luminous border. She saw all armatures, when attached, brighter on the external than on the internal edges.

365. From these observations we may deduce the rule, that the fringe of light on the edges *is more intense on the external edges*, when the magnets are open, and their odylic tension is (probably) greater; and that this brightness of the external edges diminishes when the magnets are closed, their odylic tension diminished, and the dynamic activity set in circulation, and thus directed more internally, *so that the luminous borders on the inner edges acquire increased intensity.*

366. Mme. FENZL, Mme. BAUER, Mme. KIENESBERGER, M. ZINKEL, and Mlle. GLASER, pointed out, as the brightest part of the bar or of the open horse-shoe magnets, always a spot, not on the pole itself, but a little backwards from it, *about one-seventh of the distance to the axis.* This was the case when the magnets were open, or closed either with their armatures or with other magnets; or in any arrangement whatever. When closed, the curve was indeed brighter than the poles, but this spot was always brighter than the curve. But it is well known that at this point the focus of magnetic attraction lies; that focus coincides, therefore, with that of the greatest intensity of odylic glow.—Mlle. ZINKEL observed a remarkable case of change of place in this most luminous point.

When a closed horse-shoe was placed vertically, the poles and the armature downwards, these spots left the poles, and fixed themselves on both sides of the middle of the curve, and again about a seventh of the distance to the poles. When I closed one horse-shoe with another, the most luminous spots on both magnets were one-seventh within the poles, and were more distinct, because now the polar ends were more luminous than the curve. The same thing happened with soft iron bars; when magnetised by induction, Mlle. ZINKEL always saw the bright spots one-seventh of the half length of the bar within the poles.

367. *Stroking one magnet with another excites an artificial, and very bright odylic glow.* At the point where the magnet used for stroking may happen to rest with its poles on the other, it produces in it a very luminous spot, which moves with the stroking magnet as it is drawn along. The point touched is brought into a bright glow, which always fades away as the magnet is shifted, and appears where it is placed. This observation was repeated, in many varied forms, with Mlle. ZINKEL.

368. Not only magnets, but every other object which yields odyle, produced this effect. When I drew slowly, along the surface of the steel lamella of a magnet, the point of a crystal of chrome alum, arragonite, rock crystal, or heavy spar, the point of contact became much brighter, over a circle of 0.4 to 0.8 inch in diameter. Even the human hand has the same action. When I laid my right hand on the curve of a horse-shoe magnet, and passed it along towards the southward pole, there appeared a brighter spot wherever I touched the steel. The same thing happened when I drew my left hand along the northward limb. Mlle. ZINKEL saw this very often, as did also Mlle. GLASER, Dr. NIED, and Mme. VON PEICHICH-ZIMANYI, most distinctly. ANKA HETMANEK saw it on a bar magnet. When I placed a pocket horse-shoe, which only exhibited to the sensitive a feeble glow at the poles, on the palm of the hand of Mlle. PAUER, of Mlles. WEIGAND, ZINKEL, GLASER, and of Mme. BAUER, they saw the whole magnet slowly become luminous, and in a few minutes acquire

the full odylic glow. We shall hereafter meet with a similar action of magnets on other odylo-luminous bodies.

369. Heat was found not to increase the odylo-luminous phenomena. I laid on the hot stove in the dark chamber, before Mlle. PAUER and Mlle. GLASER, some small bars and horse-shoes, which, when cold, shewed to them a beautiful glow and distinct flames. When they had acquired nearly blood-heat, the girls both found the glow much duller and more turbid. Mlle. ZINKEL saw a large horse-shoe become gradually duller as it grew warmer, and at the temperature of 167° F. the glow was strikingly reduced; but it returned, as before, on cooling. We know that under such circumstances the magnetic intensity follows a similar course; so that it would appear that in this case the odylo-luminous phenomena run parallel with it.

370. When I brought a magnet into the electric atmosphere of the conductor of a machine, at the distance of forty inches, when the sparks were two inches long, the glow increased considerably in intensity, even when the magnet was not isolated, but held in the hand or lying on the table; and the effect increased as it was brought nearer the conductor. The light did not change perceptibly, whether the magnet were electrified itself or not. These things were ascertained by a series of experiments with Mlle. ZINKEL, made with short bars, with larger ones, with one of five feet, and also with smaller horse-shoes, and with the large one of nine bars.

371. The effect produced by the electric atmosphere on the odylic polarity, appears to me still more worthy of attention, in so far as it affects the colours of the glow. If, for example, I turned towards the positively charged conductor, the negative, northward, and blue glowing side of a magnet, bar or horse-shoe form, the intensity of light was increased, and the blue became brighter. But if I turned round the magnet, and placed towards the conductor, the positive, southward, and red-glowing side of the magnet, not only did the red colour of the glow fade, but it soon passed into blue. *The odylic polarity was reversed.* The + O, turned towards + E, was changed into — O; the blue — O of the opposite

end of the magnet became red glowing + O. Both Mlle. ZINKEL and Baroness VON AUGUSTIN saw this.

372. I reversed the experiment, giving to the conductor positive electricity. It now produced reversed effects on the odylic glow of the magnet; changing, when turned towards it, the negative blue-glowing pole of the magnet into a positive red-glowing pole, and the opposite positive and red pole to a negative blue one, all at forty inches distance. But when I ceased to electrify the conductor, in a few seconds the light at both poles faded, and returned to its original colours, corresponding to the magnetic polarities. I could repeat and vary backwards and forwards every minute, this reversal of the colours. The power of electricity, merely by its atmosphere, to intensify, in a conformable position, the odylic polarities even of unisolated magnets, and to reverse them for a time, when unconformably placed, was thus demonstrated.

373. Since every one, familiar with this department of physics, knows that, in these circumstances, the magnetic poles of a magnet are not reversed, it is almost unnecessary to detail the experiments which I made to satisfy myself of that truth. As the approach of another needle could not here decide the question, as it might possibly be affected by induction, I adopted another method. I hung at the further end of a bar magnet, twenty inches long, a small needle, so that its point was merely attached to the extreme point of the bar, and was therefore so easily detached that a slight shaking of the floor caused it to fall. If the attracting power should be in the slightest degree diminished, it must instantly fall, and that long before the power was reduced to null. But such a diminution, and final passage into a state of indifference, must necessarily occur, if the magnetic polarities underwent reversal. I now charged the conductor, and then discharged it; the glow of the bar changed its colours by the effect of the electric atmosphere, and then returned to its original state; all this being several times repeated. But the needle never stirred nor fell off. The magnetic polarity of the bar, therefore, suffered no appreciable change, while the odylic polarities, as was shewn by the colours of the glow, were repeatedly reversed.

These facts supply one of the most striking proofs of the distinction which exists between magnetism and odyle, since they prove that cases occur in which these forces are, in one and the same bar, and at the same time, in a state of direct opposition. We shall return to this point when considering the odylic flame.

374. Just as the armature becomes glowing by induction, so also does *an electro-magnet.* A soft iron horse-shoe, the limbs of which were thirteen and a half inches long, 1.6 inch in diameter, and round which was coiled a copper wire, 0.032 (about one-thirtieth,) of an inch thick, was connected with a battery of six pairs of zinc and silver, according to SMEE, each pair having a surface of sixty square inches. All the sensitives to whom I showed this in the dark, namely, Mme. KIENESBERGER, ANKA HETMANEK, Mlle. ZINKEL, STEPHAN KOLLAR, Mme. VON TESSEDIK, Prof. ENDLICHER, Mme. FENZL, M. DELHEZ, and others, saw the iron, as well as the armature, when attached, very soon become more or less glowing. The open poles were brightest, and the light diminished towards the curve. The colour of the glow was bluish at the northward, reddish yellow at the southward pole, exactly as in a steel magnet magnetised by stroking. The armature exhibited, when attached, the colours reversed, red at the northward pole of the magnet, blue at the southward. It had therefore been converted into an induced magnet by the induced magnet which it closed.

375. Not only did the soft iron, thus converted into an induced magnet, exhibit the odylic glow, but this was also the case with all those parts of the apparatus which temporarily acquired magnetism, as the coil, the connecting wires, and even the battery itself. I only mention this here to render the section on the odylic glow complete; but must reserve the details for a subsequent occasion.

376. Even a simple bar of soft iron, although in itself constantly exhibiting a feeble glow, as all bodies do more or less, is not unaffected in its glow by changes in its position, with reference to the magnetic meridian. Mlle. ZINKEL saw a soft iron bar, twenty inches long, increase in blue glow towards

the north, and in reddish yellow glow towards the south. When turned round, the colours changed places in the bar, so that the blue glow was always seen on the north side. When the bar was placed vertically, she saw it, contrary to all expectation, glowing with a bluish grey light at the upper end, and with a whitish red below. (This apparent anomaly will be hereafter discussed.)

377. We already know, from my previous treatises, that the magnet possesses the power of communicating to other bodies its own inherent odylic force. But later researches first taught me, *that its power of exciting the odylic glow could also be transferred to other bodies.* Mme. KIENESBERGER, Mlles. ATZMANNSDORFER, GLASER, and ZINKEL, and FRIEDRICH WEIDLICH, went through numerous experiments, in which I placed over the flaming poles of magnets iron and copper wires, rolled up into flat coils, with projecting ends, of the length of a foot or more. These wires immediately absorbed the odylic flame, *and their own odylic glow became much more intense.* I shall detail these experiments further on, when considering the odylic flame, (§ 485.) Even glass, such as large lenses, bell jars, &c. when brought near the poles of strong magnets, acquired the glow. See § 481.

378. If we now collect, in a brief form, what has been said above concerning the odylic glow, we obtain the following condensed summary :—

a. The proper light, or glow of all magnets, of whatever form, appears in the dark, when of the lowest degree of luminousness, dark grey; it gradually rises to whitish and yellowish, and generally assumes at the northward pole a bluish, at the southward pole a reddish tint.

b. These colours exhibit many varieties of shade, according as the poles of the magnets are open or closed, conformably or unconformably placed, turned eastward or westward, upwards or downwards; that is to say, according as their struggle with the magnetism of the earth was carried on under different conditions. But the phenomena uniformly follow a law, which in each position decides their nature, and the effects of which never fail to appear.

c. An electro-magnet obeys essentially the same laws.

d. A bundle of bars or of horse-shoes, placed with like magnetic poles in contact, has its odylic poles reversed, so as to alternate with each other.

e. The armature shares in these effects. As an induced magnet, it interferes both actively and passively, altering the intensity and colour of the light.

f. An unmagnetic iron bar plays to some extent the part of an induced magnet, formed by the influence of the magnetism of the earth, and exhibits the phenomena of odylic glow corresponding to that state.

g. The intensities of odylic glow, in the same magnet, keep pace with the magnetic intensities. They are of different degrees in different magnets, and in different parts of the same magnet; and this according as the magnets are open or closed, with the poles conformable or unconformable, turned towards east or west, upwards or downwards. In these different circumstances, the points of greatest intensity change their places at the magnetic poles or axes. The light is strong enough, to the eyes of the sensitive, to be reflected by ordinary mirrors.

h. Magnets, when closed by other magnets, instead of their armatures, exhibit dislocation of the points of greatest intensity, according to the same laws, but in a different manner.

i. The poles of magnets, laid on any part of other magnets, produce there partial increase of intensity in the glow, or brighter spots. Crystals and hands produce the same effect.

k. The electric atmosphere intensifies the odylic glow.

l. It has so powerful an action on the odylic light, that it can produce reversal of the odylic poles, where no change in magnetic polarity occurs.

m. Heat diminishes the odylic glow in steel magnets.

n. The odylic glow is transferable from the magnet to other non-magnetic bodies, such as copper wire, &c.

o. Although the odylic intensities run parallel with the magnetic intensities, yet their polar values do not always do so. There are frequent cases in which + M appears associated with — O, and *vice versa.* These contrasts afford strong proof that magnetism and odyle are not identical.

II. THE ODYLIC FLAMES OVER STEEL MAGNETS.

379. In the preceding experiments on the glow, the flames flowing from the magnetic poles followed everywhere the same course. This phenomenon forms the second degree in the scale of the odylo-luminous appearances, and consists in a light, which, to the more feebly sensitive, appears as a vague gleam over the poles, but which the more highly sensitive, according to the degree of their perceptive power, saw gradually passing into the aspect of a real flame, under which name they described it. There are magnets which exhibit the glow without the flame, but none which have the flame fail to exhibit the glow. The glow is always the first luminous appearance. I once had a horse-shoe, which had become so weak as not to carry its armature, but still retained perceptible traces of magnetism. I showed it to Mlle. ZINKEL in the dark chamber, at a time when she was very highly sensitive to odylic light. She saw the whole magnet glowing, but could perceive over its poles no flame, only a feebly luminous smoke. When the odylic intensity is increased, the flame is added. We are entitled to assume that the flame exists everywhere, but we can only speak of it where it becomes visible. We shall study its characters, as far as I have been able to extend the investigation into these beyond what has been formerly described. Let us go through the experiments which I have made with magnets, on this branch of the subject.

380. I must, for the sake of connection, refer to the details given in the first treatise, § 3 to 20. A year afterwards, in 1845, I went through a series of researches with Mlle. ATZMANNSDORFER, who was then living in my house. These researches partly confirmed and partly extended those formerly described, and must be added to them. She saw, on all magnets, without exception, flames varying in size or in intensity; the former according to the size, the latter according to the amount of magnetic or odylic charge of the magnets, and varying also according as the fluctuations in her state of health heightened or depressed her sensitive powers. Out of hun-

dreds of experiments I adduce only a few. Towards the end of summer, when she is generally better, she saw, on a weak bar magnet of twenty inches, flames little more than an inch in height. At the same time, she saw, on the poles of a seven-bar horse-shoe, flames of eight inches high. Afterwards, when more sensitive, the largest of my magnets, a heavy horse-shoe of nine cast-steel bars, was placed before her in the dark. Although she did not know which magnet was used, she saw again, as she had done a year before, flames of fully five feet in height burning on both poles. They were so large, that when the poles were upwards, they rose, and united into a column of fire. She could see, by the somewhat different colours of the flames given out at the two poles, that this column was formed of two, the one yellowish white, the other bluish; the former smaller than the latter. The whole dark chamber was so illuminated by them, that she could see in it the outlines of all objects. Between the limbs, which had a white odylic glow, she saw the whole space filled with threads of flame, and the outer surface of the steel enveloped all round in a fiery down, which undulated, and appeared to flow sometimes towards one pole, sometimes towards the other. At the planes of junction of the lamellæ, and at their edges and covers, where they form the poles, there were separate small flames flowing out laterally, and strongest on the outer corners, where they at last ended in sparks, which flew singly away. As all this corresponds exactly in the details with what Mlle. REICHEL had, a year before, described in a somewhat inferior degree, and on a weaker magnet, I shewed to Mlle. ATZMANNSDORFER, after her observations and the description of them had been finished, the drawings given with the first part of my researches, in Plate I. She found the whole to correspond pretty exactly with what she had herself seen, and thus confirmed the accuracy of my earlier investigations. But she said that the flames were all much lighter and more delicate than they appear in the drawings. They were, as Mlle. REICHEL had observed, in motion, yielded to every breath, and shewed a beautiful play of many colours.

381. I have to report, almost in the same words as were

used by this patient, what a healthy sensitive, Mlle. ZINKEL, described. She also saw, more distinctly under certain circumstances, from bar magnets flames of 0.4 inch to four inches long, and those of the large nine-bar magnet from forty inches to five feet high. The presence of the catamenia was one of these circumstances. She saw, too, both in bars and horse-shoes, the small lateral flames on the edges and solid angles; the play of colours; the strong illumination of surrounding objects; the sparks flying singly, all just as Mlle. REICHEL had described them, two years before.

It has already been stated, § 316, that a healthy sensitive, Baron VON OBERLAENDER, saw these flames from the nine-bar horse-shoe of the same size. The same has been stated of FRIEDRICH WEIDLICH, § 330, of Mme. KIENESBERGER, § 328, of the healthy sensitive Mlle. GLASER, § 313, of Dr. NIED, § 315, of Baroness VON AUGUSTIN, § 312, of M. FRANZ FERNOLENDT, § 303, of Mme. VON VARADY, § 299, and in a degree very little inferior of the thoroughly healthy Mlle. SOPHIA PAUER, § 309. *All of them saw flames about the height of a man, burning up over the nine-bar horse-shoe.*

382. Let us now consider some of the properties of these flames; and first, *their extent.*

We have seen that the size of the flames depends, objectively, in part on the size of the magnets, in part on their intensity, or amount of charge; but that, subjectively, the perception and recognition of them is determined by the state of sensitiveness of the observer. As among men, in reference to ordinary vision, there exist such differences, that while one man does not recognise his brother when he passes, while another, with falcon glance, counts the larks in the sky; as, further, there are persons who can see nothing even in a moderate degree of darkness or obscurity, while others find their way in the darkest night; as there are patients, suffering from intense hemeralgia, who only see in the brightest light, and become quite blind in a feeble light, and are thus directly opposed to sensitives; as there are Albinos, and persons affected with what is called Daltonism, who are insensible to certain colours, nay, some persons who are insensible to all colours, and to whom the world around appears like an engraving; so is it with the power of perceiving the odylic light. According to the degree of their perceptive power, the sensitives see the flames of one and the same magnet larger or smaller. This is not only the case in different persons, but in the same person on different days, at different hours; nay, pretty frequently the variation occurs in a surprising manner from one minute to another, so that, for example, in three successive moments, the same odylic flame may be at first imperceptible, then visible, but small, and immediately afterwards large and strong, without any change in the object, in consequence of variations in the sensitiveness of the observer to the light. I shall make known the cause of these variations in one of the subsequent treatises, when the power of seeing the odylic light will be specially treated of.

383. The lowest degree of perception of the odylic flames consists in perceiving a flash of light, which quickly disappears, at the moment of suddenly detaching the armature. This was seen by Dr. NIED, Baroness PAULINE VON NATORP, M. KOTSCHY, Mme. VON VARADY, Chevalier VON RAINER, Baroness ISABELLA VON TESSEDIK, M. DEMETER TIRKA, Mme. VON PEICHICH-ZIMANYI, Mlle. AMALIE KRUEGER, Mlle. GLASER, M. PAUER, Prof. ENDLICHER, M. DELHEZ, Baroness VON AUGUSTIN, as well as Mlle. NOWOTNY and the other early sensitives, (§ 3).—Mme. CÆCILIA BAUER gave a more detailed account of this phenomenon. At the moment of detaching the armature she saw a flashing scintillating light, almost like that produced by flint and steel, which instantly disappeared. But at that moment, the flame over the poles was not at its largest, it was on the contrary at its smallest, and at first hardly perceptible. But it immediately began to appear, small at first, then growing, and soon reaching its stationary maximum. This required, in all, about a minute. Some months afterwards, Mlle. ZINKEL gave me exactly the same account of the phenomenon.

The different degrees of light exhibited by the lights seen over the magnetic poles, from a mere glimmer of light to the distinct flame, are described with slight differences by almost every observer. At the bottom of the scale we find MM.

HUETTER, SCHUH, and Prof. RAGSKY. Next to them come Prof. HUSS of Stockholm, who only saw lights over the electro-magnet. In the middle we find, among others, Prof. ENDLICHER, who begins to see the luminous emanations more extended and more dense. Then come Dr. NIED, M. RABE, Baroness VON AUGUSTIN, Mlle. PAUER, Baron VON OBERLAENDER, ANKA HETMANEK, and M. ANSCHUETZ. Still higher stand Mme. BAUER, Mlle. ZINKEL, and above all, the sensitives affected with somnambulism, who always see the lights over magnets as decided flames.

384. Among the objective conditions *for the size of the flame, is the difference between northward and southward poles.* In most cases, the two flames are of unequal size, at least under the 48th parallel of north latitude, which is that of my residence, when the horse-shoe magnets stand upright, or when their poles are pointed to the north, or in bars when they are conformably placed in the meridian. It is probably different, and proportionally so, in other places of different latitudes. Within the tropics, the difference between the polar flames of conformably placed horse-shoes will become imperceptible; and under the magnetic equator these differences will no doubt disappear. Beyond the tropics, for example at the Cape, or in Van Diemen's Land, or Buenos Ayres, they will be reversed; the southward flame will be the predominant one, instead of the northward, as with us. When a bar magnet is placed in the meridian, the northward pole towards the north, all my sensitives, diseased and healthy, have found, in numberless trials, the flame of the northward pole larger than that of the southward pole. The difference amounted, on an average, to nearly one half; so that the northward flame was about twice as high as the southward. I say *about* twice as high, because my innumerable questions in the dark could not be answered with a measuring rod. The observers usually made use of spans, finger-lengths, hand-breadths, and thumb-breadths, to explain their meaning. Very great accuracy was here not required, since, for the present, we only give the general outline of the phenomena. But the sensitives agree in regarding the proportion of the size of the northward flame

to that of the southward as two to one. This is in an inverse ratio to the odylic glow of the bars; for, as we have seen, in that phenomenon, the blue northward side is shorter than the red southward side. (But this may be an optical delusion; because the blue glow is much less luminous than the yellowish red, and, therefore, in the lower degrees of sensitiveness, becomes invisible to the eye sooner than the other; which may happen without its being really shorter.) Were I to produce all the testimony I possess on the subject of the relative length of the flames, I should have to name almost every sensitive I have tried, which would be useless and tedious. I, therefore, confine myself to mentioning Dr. NIED, Baron VON OBERLAENDER, Baroness VON VARADY, Mme. VON PEICHICH, Prof. ENDLICHER, M. DELHEZ, Baroness VON AUGUSTIN, Mlle. PAUER, all healthy; Mme. KIENESBERGER, Mlles. ATZMANNDORFER, REICHEL, WEIGAND, WINTER, all more or less in bad health; and further, the strong and healthy Mlle. ZINKEL, and the healthy carpenter KLAIBER, with whom, in particular, I have multiplied and varied these experiments.

385. The result is analogous, but altered, when a bar magnet is placed, not conformably in the meridian, but in such a position that its southward pole points to the north, its northward pole towards the south, or unconformably. In this case, the sensitives observe a diminution in the size of the flames, both of them becoming duller, less luminous, more turbid, shorter and narrower, but also being modified in their colours. This has been observed in numerous experiments by Mme. KIENESBERGER, Mlle. ATZMANNSDORFER, Mlles. GLASER, REICHEL, WINTER, and ZINKEL, just as it may be deduced from the general theory of magnetism.

386. These sensitives also observed a horse-shoe, laid horizontally in the meridian, as stated in § 338, first with its poles towards the north, then with its poles towards the south. In the former position, the bluish northward flame appeared larger and of a brilliant blue, while the southward flame was smaller, muddy red, almost bluish red. In the latter, the

northward flame was dull, of a greyish blue with a tinge of yellow, and one-third smaller than before; while, on the contrary, the red southward flame was one-half larger, of a vivid and brilliant red.

387. When I placed a five-bar horse-shoe vertically, with its limbs and poles in the conformable position, northward towards north, &c., Mlle. ZINKEL described both flames as large, vivid, and brilliant. But when I reversed the limbs into the unconformable position of southward towards north, &c., both flames were smaller, dull, turbid, and discoloured.

388. When I inverted the horse-shoe, by resting the curve on a bar of copper, so that the poles pointed vertically downwards, while the limbs lay conformably, the northward flame was shorter, the southward longer.

When both poles were placed in the oblique direction of the magnetic inclination, the results were nearly the same.

389. All transverse positions, with the poles towards east or west, yielded varieties of intermediate results, in which, however, the size of the flames showed less variation. This took place in the following way:—

390. When both poles lay towards the west, Mlle. ZINKEL described both the flames as short; with the poles towards the east, they were rather longer. But they were always shorter than in the vertical position with the poles upwards.

391. It must further be mentioned, that the flames appeared on the northward pole when towards the

east—longer, but duller;
... ... west—thicker, but brighter.

On the southward pole when towards the

east—thicker, but brighter;
... ... west—longer, but duller.

We shall see further on, (§ 405, where the conflict of two odylic flames is treated of,) what is the meaning of these differences of longer and thicker; we shall there see proof that the former is connected with an attraction, the latter with a repulsion, of the odylic flames; and we may here anticipate the important fact, to be afterwards more fully developed, *that east, in its relations to odyle, stands in some degree on the side of north, and west, in the same way, on that of south.* See § 536.

392. In a similar way, I tested, in numerous experiments with the most sensitive subjects, the effect produced by the magnetic inclination, on the character and size of the odylic flames on steel magnets. Most of these experiments were made with Mlle. Atzmannsdorfer, Mlle. Reichel, Mme. Kienesberger, and the healthy sensitive Mlle. Zinkel. It first appeared that the results arrived at in § 11. are not unconditionally, but only conditionally exact. I have there stated, on the authority of Mlle. Reichel, that the magnetism of the earth appears to have no perceptible influence on the size of the flames. This is correct, when, as was the case in these first trials, *the changes of the poles and of inclination are rapidly made, and when some time is not given for the developement of the flame.* At that time I was not yet acquainted with a certain degree of slowness, with which nature effects odylic changes; requiring, for the completion of such changes, in all cases, a short interval, half a minute, a whole minute, and sometimes several minutes, as I shall have frequent opportunities of pointing out. When the experiments in § 11. were more slowly made, and carefully observed, it certainly appeared that the flames changed gradually, not only in their size, but also in their form, becoming smaller or larger according to the position; and that the magnetism of the earth undoubtedly exercised a slow but very decided and considerable influence on them. The suspended needle, as is well known, instantly alters its position when the earth's magnetism can act on it, and this action produces instantly its full effect. I at first took for granted that this applied equally to odyle, and my first experiments were made under this impression; but longer experience has taught me that it is not so, and that the effects of odyle only sluggishly attain their maximum. When I afterwards varied the experiments with Mlle. Reichel, so that the bar magnet was not held in my hand as before, which complicates and adulterates the polar flames, but was supported by a wooden stand, which allowed

me conveniently to turn it in any direction; and when I placed it in the direction of the magnetic inclination, allowing it to remain there for a minute, I expected to find that the blue flame of the north pole would increase in size, brightness, and purity of colour. But the result did not at all agree with my preconceived opinion. On the contrary, the observer informed me that, instead of this, the north polar flame had lost very considerably, both in strength, purity of colour, and intensity of light. The blue became dull, turbid, and grey, and at last so dark as nearly to disappear. When I reversed the bar, so as to place its north pole in the direction opposite to that of the inclination, the flame appeared, *at the distance of forty inches*, large, vivid, and bluish. The same experiment, repeated at a different time with Mlle. ATZMANNSDORFER, gave the same unexpected result. In the inclination, the north flame not only did not increase, but it nearly disappeared to her eyes; but when the bar was reversed, she again saw the blue flame nearly twice as large as before. Finally, I tested these very striking results most carefully with Mlle. ZINKEL, devoting to their thorough investigation many entire days, which were passed in the dark chamber. The result was still the same. In the inclination, where I had expected the flame of the northward pole to reach its maximum, it fell, on the contrary, to a kind of minimum; and instead of brightening into the most vivid blue, all colour faded, and there remained only a dark grey. We shall see, by and by, that these surprising visible phenomena were accompanied by corresponding effects on the sensations of all sensitives, who recognised and confirmed them in this way, even when their sensitiveness was much inferior. Now, since this appears directly opposed to the action of terrestrial magnetism, whereby the northward or negative pole of the needle, when brought nearer to the inclination, is attracted, not repelled, by the positive or north pole of the earth, we see from these, as clearly as from any preceding experiments, that the odylic flames, or the odylic phenomena generally, are indeed influenced by magnetic attraction, but are not unconditionally determined or regulated by it. *Our globe, therefore, includes other properties or influences, besides those of magnetism, which affect odyle and its characters.* I shall give, when discussing the odylic colours (§ 489, et seq.) what I have, up to this time, succeeded in ascertaining ; but here, where I only treat of the effect of the earth's magnetism on the flame of the needle when placed in the inclination, I shall content myself with stating the results above mentioned.

393. When I used, in these experiments, *horse-shoe magnets*, the result was altered. When both poles, in the inclination, pointed to the north, the blue flame of the northward pole was of its full length, but the opposite southward flame appeared depressed, turbid, bluish red, and smaller. When the poles pointed to the south, the red flame of the southward pole was now the larger, that of the northward pole was now smaller, duller, and reddish blue. When both poles stood pointed upwards, the northward flame was increased, the southward diminished. (This confirms the results in the preceding paragraphs.) This went so far as to yield a fact of importance for the practical observation of the phenomena ; namely this, that frequently the odylic flame of the unconformably placed pole was entirely suppressed ; that is, it became invisible to the observer, as if one pole alone possessed a flame. My journals are full of such instances ; and there is hardly one among my numerous sensitives with whom it has not frequently happened, under the proper circumstances, especially in those less sensible to the luminous impressions, or in imperfect darkness, or when the observer had not remained long enough in complete darkness. I mention the names of Baroness VON NATORP, Mlle. DORFER, Baroness VON TESSEDIK, Mme. FENZL, Mlle. PAUER, M. HOCHSTETTER, M. FERNOLENDT, and the chlorotic ANKA HETMANEK.

394. It was not a matter of indifference, whether a magnet stood with its poles free in the air, or lay flat on a table. In the latter case all the observers found the flames decidedly larger, sometimes nearly twice as large. Mlle. GLASER saw the flames on both poles of bars become twice as long when laid on a polished walnut table. Mlle. PAUER observed horse-shoes on an unpolished fir table. They formed on the table a

long stream flowing in undulations, at the southward pole shorter, thicker, and red, at the northward pole longer, narrow, and blue. The undulating motion was most distinctly seen in the profile of the surface of the table; when seen from above, it surrounded the poles like a saint's glory, which, close to the pole, seemed to rise in the air. Mme. BAUER and Mlle. ZINKEL, the last especially, often saw this appearance, from which it would appear as if the flames found on the table a kind of support, promoting their flow.

395. The results were exactly the same whether I took my largest magnet, or those of middling size, or the smallest of all, which could be concealed between two fingers. The phenomena were in this case smaller, indeed, but were in kind precisely the same as before. For such researches we do not therefore require prodigious masses, but for most objects small instruments are sufficient, and with such the greater part of my observations may be repeated and confirmed on a small scale. *A small magnet, charged to saturation*, seems to possess not less odylic tension than one of much greater size. The luminous phenomena appear of smaller dimensions, but the results are qualitatively the same.

396. The odylic flame, in compound magnets, presents the same appearances as I have already explained in § 340. et seq. in reference to the glow, namely, *alternate stratification of colours.* I must refer, for the sake of brevity, to the details there given. The middle plate of a nine-bar horse-shoe gave out at the negative pole a blue flame; those on each side of it gave a red flame; the next pair again a blue; the third again a red; and the last, or outer pair, a blue flame. The red flame of the alternate and included plates was somewhat compressed, especially when the magnet lay conformably. All these things appeared reversed at the southward pole, where the two included blue flames were compressed, and rather grey than blue. Seen laterally, the flame had, next to and above the steel of the poles, for a short space a stratified appearance of red and blue. But the eye could not trace this to a considerable height; the colours mingled and were soon lost in the tint of the predominating pole, so that at the negative

pole the red flame from the second and fourth pairs very soon was lost in the prevailing blue; and *vice versa*. The colours only continued to be seen in certain red and blue sparks and threads, which rose into the general flame, and which will be treated of in § 454.

397. The phenomenon here described includes that described in § 9, and represented in Fig. 10. Plate I., but which was described less minutely by Mlle. REICHEL than by the very exact Mlle. ZINKEL. She saw also the stripes of flame from each lamella; but the intervals between these were not empty, but also filled with flame, only less luminous. Mlle. REICHEL'S observations were made in an extempore dark chamber (the stair formerly mentioned, and in a much less perfect darkness than Mlle. ZINKEL), as I had in the mean time had a perfectly dark chamber constructed. Mlle. REICHEL, therefore, saw only the bright stripes, and the less luminous intervals appeared to her void of light. Mlle. ZINKEL saw the whole upper surface of the poles of the nine-bar horse-shoe covered with flame; but from the edges of the lamellæ flowed brighter lines of fire, which she could not trace to the upper part of the general flame. The banded or stratified lateral aspect of the flame was thus increased.

398. The INTENSITY OF LIGHT in the odylic flame is very various, extending from a feeble degree, in which it appears to the most highly sensitive as a luminous vapour, to flames which are bright, and often so brilliant that the sensitives could not conceive how I was unable to see them. But they never reached a point at which I was able to perceive the slightest trace of them; although, during my frequent and long-continued experiments in perfect darkness, I was certainly not deficient in attention to this point. When the flame was gently breathed upon, especially in the direction of its flow, its intensity increased, more particularly on the parts where the breath fell upon it. This was seen by Mlles. GLASER, ZINKEL, PAUER, Prof. ENDLICHER, M. HOCHSTETTER, M. DELHEZ, and Baroness VON AUGUSTIN, on bars, horse-shoes, and electro-magnets of various forms. I shall return to this in § 409.

399. *The restlessness and continual motion*, which prevail in the odylic flame, was confirmed in many new experiments by Mlle. ATZMANNSDORFER, Mlle. ZINKEL, Mme. KIENESBERGER, Mlle. WEIGAND, Mlle. DORFER, FRIEDRICH WEIDLICH, M. RABE, Baron VON OBERLAENDER, Mlle. SOPHIE PAUER, Mme. VON PEICHICH, Baroness VON NATORP, Baroness VON VARADY, Mme. VON TESSEDIK, Prof. ENDLICHER, M. DELHEZ, Baroness VON AUGUSTIN, Dr. NIED, JOHANNA KYNAST, JOHANN KLAIBER, and others. The first and last mentioned persons compared it in some degree to the positive luminous bundle of the electric machine, but said it was more delicate, fuller, the northward flame bluer, but equally moving and flickering. They found it cool, of feeble light, sometimes almost moniliform. It shared in the slightest motion of the air in which it appeared.

400. The *direction* taken by the flame was frequently studied. Instead of giving many statements, I shall confine myself to two, which include all the rest. I placed before Mlle. ATZMANNSDORFER the nine-bar horse-shoe on a chair, the poles towards the south. She saw the flames flowing out to the length of an arm, first horizontally in the prolongation of the line of the limbs for a space, then curving upwards till they formed a quadrant, with their points upwards. Mlle. ZINKEL saw the same with horse-shoes and bars. The flame, therefore, leaves the poles with a certain force, of a projectile nature, which carries it away from the poles; but it has, on the other hand, an innate tendency to rise in the air, and its material substratum must therefore be lighter than air at the earth's surface. From the composition of these forces results the rising quadrant of the flame. But this experiment only succeeds with very powerful magnets, the flames of which are long. When the flames are short, they are always seen to proceed in the line of the limb. From the various directions which we can give to magnets, towards different points of the compass, from the various degrees of inclination to the horizon in which we can place them, arise, therefore, compound increments and decrements of the flame, according as these positions are more or less conformable to the normal one.

401. I must here mention a series of experiments with several sensitives, the object of which was to ascertain the effect which different odylic flames might have on one another, when brought into conflict. This might give some further information as to the relation of the magnet to the odylic flame. We regard positive and negative magnetism as having a powerful tendency to attract each other, and by mutual interpretation to neutralize each other, or produce an equilibrium. But it would appear that the odylic flame, when we consider the rising quadrant of the last paragraph, and the phenomena described in § 392, where it appears opposed to the magnetic inclination, has no such inherent attraction. We here meet with distinctions between the two forces in their special manifestations. To Mlle. von Weigelsberg, Mlle. Winter, and Mme. Anschuetz, all in delicate health, and to the healthy sensitives, Mlle. Ernestine Anschuetz, M. Anschuetz, M. Delhez, Mme. Fenzl, Mme. von Peichich, M. Hochstetter, Dr. Nied, and Baroness von Augustin, I showed, in the dark chamber, two bar magnets, both four inches long, holding them in the magnetic parallel, horizontally, with their unlike poles turned towards each other, at a distance equal to twice the length of the bars. I attended to the indispensable precaution of holding in my right hand the southward pole of one, while I held in my left hand the northward pole of the other, the opposite poles projecting from my hands; the reasons for which will soon become obvious. All the observers saw, that on thus placing the bars opposite to each other, the flames of both became narrower and longer, as if striving to meet. When I moved them nearer, the flames returned to their former shorter but thicker form, which increased as they were brought closer. At the same time, the intensity of the light increased. When at last I united the poles, both flames nearly disappeared, but at the further ends they immediately became stronger, and nearly twice as long as before. M. Hochstetter and Dr. Nied also stated, that the thickness or breadth of the flames much exceeded that of the bars, when the poles were near, and that they were brightest when the poles were nearest. Mlle. Reichel saw better, and described more minutely, these

phenomena. I gave her, in 1844, two small bars of equal length, and desired her to hold one end of each in each hand, and gradually to bring them nearer in a straight line. Even at the distance of thrice their length, she saw that the flames of the poles turned towards each other were larger than those at the further ends; they became narrower and longer, and stretched towards each other as if striving to unite. When brought nearer, the interior flames grew at the expense of the exterior ones as the nearness increased. *When at last the flames came in contact, they did not destroy each other.* They became thicker; while by the approach of the poles giving them out, they became shorter. At the distant poles, they disappeared almost entirely, leaving only feeble and dull flames. When the poles at last touched, the flame between them disappeared in great part; but then much larger flames immediately appeared at the further ends; that is, the two bars were now united into one magnet of double size, which gave out, at its poles, flames twice as large as each bar separately had given. The flames, therefore, increased with the increased magnetism, first in the poles turned towards each other, and then, after contact, in the distant poles. Professor ENDLICHER saw the flames of both bars, held opposite to each other at some distance, become somewhat longer; and when the poles came near, the flames contracted themselves to disks of 0.4 inch in diameter, each round its pole. When the poles were brought in contact, the flames disappeared. M. PAUER observed the same experiment. He perceived both flames when the bars were two inches asunder. They increased in intensity till the poles touched, when the light instantly fell, but did not disappear. The poles, now in contact, appeared enveloped in a luminous veil, about 0.8 inch long on each. Baroness VON VARADY saw the same. Mlle. GLASER made the same observations, and added, that just before the contact, she saw the thickening and mutual repulsion of the flames increase so much, that she perceived a partial turning back of the flames over their poles. The account of these phenomena was completed by observations which I made in 1845 with Mlle. PAUER and Mlle. ATZMANNSDORFER, and in 1846 with

Mme. KIENESBERGER and FRIEDRICH WEIDLICH. I used in all these cases the same bars, while the observers sat in the dark chamber, their backs towards the north, the bars being held horizontally in the magnetic parallel. The appearances presented themselves in the same order, and of the same kind, as already described; at a distance, lengthening and narrowing of the opposing flames, as if they tended strongly to unite; but when brought nearer, they did not unite, but assumed their former thick shape, which thickness increased on nearer approximation, the length always diminishing at the same time; and when the flames should have reached each other, they showed so little tendency to unite, that they rather mutually repelled, shortened, and compressed each other. They behaved as if, so to speak, they were influenced by mutual shyness, collected themselves, each round its own pole, like the flame of a candle when gently blown upon from above; and when only a few lines apart, each flame formed a sort of flat, compressed envelope round its pole, distinctly forced back by the opposite flame. This went so far, that when the distance was diminished to about one-twelfth of an inch, both flames retired over their poles, and were in a manner inverted, so as to enclose them. The four small corner flames did this sooner than the central flames, which became first flattened and retreated, till they also became inverted. After the poles came in contact, and firmly held each other, this appearance lasted for several seconds, during which it gradually faded, the inverted flames shortening till they disappeared, or seemed to be extinguished. But they were not entirely extinguished, their light only became feebler; for attentive observers, such as M. PAUER, Mlle. SOPHIE PAUER, and Mme. KIENESBERGER, still saw them enveloping the opposite poles with a feeble light. During this, the flames at the opposite ends of the bars increased, till they attained double their former size.

402. I tried, with Mme. BAUER, bars of eight inches, or double the length. She saw beautiful red and blue flames on the positive and negative poles, the former two inches, the latter four inches long. When I caused the friendly or unlike

poles to move towards each other, in the parallels, using the hands conformably as before, the flames at first lengthened out towards each other, then on a nearer approach became shorter and broader, then became partly inverted over *their own* poles, and after contact the inversion disappeared as before, and the flames enveloped the opposite poles in a feebler light.

403. But *when like-named or hostile poles were made in the same way to move towards each other*, for example, the two negative poles, the first lengthening and attenuation did not occur; but, when the magnets were of equal force, on a nearer approach they mutually forced each other back, and became at last inverted over their poles, as when unlike poles were used. But if the magnets were unequal in power, the stronger flame sooner forced back the weaker, and caused it to accumulate round its pole, in the form of a small disk, at right angles to the long axis of the magnet. Mlle. ATZMANNSDORFER compared the effect to that of blowing gently downwards on the flame of a candle, when it spreads out, yielding to the stronger current of air in a direction opposite to that of the flame.

404. These experiments were made in all sorts of varied forms, with the healthy sensitives, Mme. BAUER and Mlle. ZINKEL, and here we obtained the clearest results. In the first experiments with the latter, at a time when her vision was not very acute, she saw no flames, but only a luminous vapour or nebula on the poles of the bars. But the result was the same as in the case of the flames seen by others. The vapours became longer, with apparent mutual attraction (when unlike poles were used) at a distance, and when nearer did not meet, but flattened out, yielded, and became inverted on the poles, and on contact slowly disappeared, while the vapours doubled their size at the further ends of the bars. On another occasion, I availed myself of a period when her powers of vision were infinitely more exalted; and I used two small bars, one stronger than the other. The appearances were those already described. I extract from my journal the following notice of them, written at the time:—" ZINKEL, Exp. No. 453.—*a*. The two positive poles being approached closely

to each other, the flame of the smaller magnet nearly disappeared, that of the larger became feebler, and was inverted. *b.* With both negative poles, both flames were seen to become inverted. *c.* With unlike poles, the negative of the larger, and the positive of the smaller, made to approach, the smaller flame is extinguished, the larger becomes feebler, and is inverted. *d.* With the negative pole of the smaller and positive pole of the larger, both flames are extinguished on contact. The stronger, therefore, always overpowered the feebler pole, and the residual light on the former became inverted. But the positive pole in the same magnet is always, in this latitude, feebler than the negative." In these experiments, the feeble residual flames, enveloping the opposite poles, escaped the eye of the observer.

404. *b.* The appearances were still more distinct in subsequent experiments with Mlle. ZINKEL, and in one which yielded the same results, with Mme. BAUER. With bars of eight inches:

a. Like poles mutually forced back their flames; and on contact the flames disappeared.

b. Unlike poles first, at a distance, stretch out their flames towards each other; but on a nearer approach, the flames are pressed back, and flattened out, till both become inverted.

c. When the contact took place, the inverted flames disappeared, and each pole was now completely enveloped in the flame from the opposite pole.

d. This envelope extended five or six times as far as the inverted flames had done. When the inversion was one twenty-fifth of an inch deep, it was followed by flames enveloping the opposite poles, of which the blue was one-fifth of an inch long, the red rather shorter.

e. The intensity of light in these enveloping flames was such, that the glow of the bars could not be seen through it, but was only perceptible beyond the point to which the envelope extended. These envelopes surrounded the ends of the bars like a dense luminous nebula, concealing the bar, as far as they covered it.

f. As long as I made these experiments with the bars lying transversely in the parallels, Mlle. ZINKEL only saw with certainty the inversion of the red flame; but when I placed the bars on the meridian, and conformably, the light became more intense, and the blue flame was also distinctly seen.

g. When the unlike poles of four-inch bars, of high magnetic intensity, were made to approach each other in the parallel, the blue flame increased more in size than the red; while the red flame increased in intensity more than the blue. In one flame, therefore, the light, in the other, the size, did not much increase.

405. I examined this point also in *horse-shoe magnets*, with the aid of several sensitives. M. HOCHSTETTER observed in the dark chamber, the effect of approaching a three-bar and five-bar horse-shoe to each other. All the four poles had flames, the negative flame blue, and six inches long. He saw the lengthening at some distance, then the flattening and forcing back of the flames, which became much broader than the limbs of the magnets. On contact, the flames disappeared to his vision, which was only moderately acute in reference to odylic light. By Mlle. PAUER, all these observations were made with greater distinctness and detail. She saw the inverted flames change, after contact, into flames enveloping the opposite poles. With Mlle. ZINKEL, I examined the matter most minutely. I placed two five-bar horse-shoes on a table, in the meridian, with unlike poles opposite each other. I shall here give the whole details of the observation in one view. *a*. At two feet distance, she saw the northward poles of both magnets in blue, the southward poles in red light, both in the glow on the limbs and in the polar flames; the flames of the poles which lay conformably were the brightest. The blue flames flowed to eight inches on the table, the red to six inches. *b*. Between the flames she saw a luminous vapour or smoke rising to the height of two inches, and then disappearing. *c*. Illumination was observed, to the distance of twelve to sixteen inches all round, on the table. This light also extended above the table, so that the poles were surrounded by a kind of glory. *d*. The limbs, as far as the curve, were covered with a luminous veil of thin flame, reddish on the

southward, bluish on the northward limb. *e.* When the poles were far apart, the middle space between the flames appeared grey; but if the vapours of the two flames came in contact, they caused a luminous transverse band on the table. *f.* When the poles were fifteen or sixteen inches asunder, the lights from the flames met on the surface of the table; the united smokes now rose in the middle, of the breadth of a hand, and to the height of a span. *g.* At twelve inches, the flames met, and began to contract. *h.* At eight inches, they acted more strongly, and rose, in consequence of their increased density, higher above the table, to nearly two and a half inches, above which, luminous smoke rose to the height of six inches. *i.* At four inches, the flames became still denser, and rose higher, especially those of the northward poles; the smoke also rose much higher. *k.* At two inches, the flames turned back, each over its own pole, on the northward side, to the depth of two inches. The smoke now appeared, *not, as before, in front of the poles, but behind them*, directed towards the curve. At 0.8 inch, the inverted flames were so large, as to reach back to the centre of the curve, and even to four inches beyond it. The smoke now rose at four inches behind the curves. *m.* On actual contact of all four poles, the inverted flames disappeared, and the enveloping flames appeared on the opposite poles. These extended nearly to the centre of the curve, and rendered the limbs hardly visible within them. Those limbs which a moment before had been clothed with their own inverted blue flames, were suddenly enveloped in the red flame of the opposite poles, and *vice versa.* All four poles, therefore, changed their colours. The centre of the curves acquired a brighter glow, and were again compared by the observer to a white luminous jelly. The smoke was no longer visible, even in profile.

I shall have to return to some of these details, others of which have been already mentioned, as under the head of odylic glow, when I come to the sections to which they belong. I have not here separated these from the rest, in order to give a full account of the phenomenon.

406. When the poles were again separated, it was observed

that the enveloping flames did not at once disappear. In the last mentioned experiment, when the poles were separated, these flames continued for a time, and did not disappear till they had been removed to more than an inch; when the smoke also began to rise, and the curve to become darker. A similar experiment with unequal horse-shoes, of three and five bars, gave, with Mlle. ZINKEL, the same result, the enveloping flames continuing when the poles had been separated to one twenty-fifth of an inch. This occurred also in bar magnets, when in the meridian. They were two feet long, and had a cross section of one square inch. When they lay in the meridian, with their unlike poles in contact, Mlle. ZINKEL could not see the enveloping flames, probably from the feeble intensity of the magnets, and of their odylic light, and the state of her sensitive vision at the time. But as soon as I separated the poles by the thickness of paper, the enveloping flames instantly became visible to her, the northward pole of one bar enveloped in its blue flame the southward pole of the other, and was itself enveloped in the red flame of the latter, to a distance of about four inches.

407. All this renders it highly probable that the odylic flame is a real projection of something, which, close to the outer surface of the magnet, is charged with odyle, becomes thereby luminous, and is then projected. If it meet with obstacles, it is turned aside or forced back. Such an obstacle may be found in an opposing current of flame. But on contact of the poles, this obstacle is removed, and the enveloping flames appear. They are the result of the residue of the magnetism and odyle heaped up at the poles, which, owing to imperfect contact, cannot be conducted away rapidly or completely enough. Now, the nearer the opposite poles remain to each other, the more easily will the currents pervade them, and be kept up; we have already frequently seen, that they have no perfect continuity. And the better the polar ends fit one another, the more numerous and close the points of contact, the more perfectly does the magnetism appear to be conducted away, and confined to its own internal circulation, and the more feeble is the intensity of the luminous phenomenon both inverted and

enveloping flames approach the point of becoming invisible. We shall see similar phenomena hereafter in the odylic light of crystals, and we shall there obtain further insight into them.

408. When the flames of bar magnets, whether positive or negative, pass close to each other crossways, or even touch each other without coming into full contact, there appeared neither attraction nor repulsion. We formerly saw something analogous with the flames of horse-shoes, which flowed from the two poles, and remained parallel, neither attracting nor destroying each other. But when two unequal flames actually met, Mlles. REICHEL and ATZMANNSDORFER both observed, that *the stronger carried the weaker with it*, but only when its projectile force or current was stronger. But according to numerous observations by both these sensitives, the stronger propulsion is always found in the flame which is nearest to its source; and that flame, which at the point where they cross, happens to be most distant from its pole, is carried along by the other. It is indifferent which is the larger; a small flame, if crossed near to its pole by that part of a larger flame which is more remote from its own pole, will carry the latter with it, and determine the direction of their united course.

409. *The motion given to the flame by blowing into it* has already been mentioned, § 20. Since then I have endeavoured to establish this phenomenon firmly in numerous experiments with a large number of persons. Mlle. STURMANN, even in the early experiments, pointed out to me that a current of air displaced the flames of magnets.—Mlle. REICHEL, in her sleepless hours, amused herself by fanning the flame hither and thither with her hand, and by giving to it all kinds of shapes with her breath.—Mlle. WINTER and Dr. NIED caused the flame to flicker by blowing on it.—M. HOCHSTELLER blew it asunder, and could strengthen or scatter it, according to the mode of blowing.—Mlle. SOPHIE PAUER did the same, and observed that it shone brighter as long as she blew.—In the presence of FRIEDRICH WEIDLICH, I blew over such parts of horse-shoes as I believed to have flames. He told me that they were instantly displaced and disturbed, but that they

soon recovered.—JOHANN KLAIBER often moved the flame by blowing on it.—Prof. ENDLICHER saw the flames of a five-bar horse-shoe flowing out to the length of four inches, becoming brighter when he blew on them. If he blew along the limb in the direction of the flame, they became both brighter and longer, but flickered restlessly.—M. DELHEZ breathed into the flames of the nine-bar horse-shoe; they became brighter and flickering.—Mme. FENZL blew into the flame of an electro-magnet, and saw the flame yield and bend.—STEPHAN KOLLAR caused the flame of an electro-magnet to divide and flicker by blowing into it.—Mlle. GLASER breathed over a bar conformably placed, in a direction parallel to that of the bar; the flame became larger and brighter. Another time she did the same with a nine-bar horse-shoe; the flame divided, became larger, and whirled round.—Mlle. ZINKEL blew from above on the flame of a nine-bar horse-shoe, and caused the flame to spread on all sides; when she ceased, it resumed its former state.—Baroness VON AUGUSTIN blew downwards along the limbs of a five-bar horse-shoe. She not only saw the flame flicker and become brighter, but made the curious observation, that when she blew by jerks, a portion of flame every time separated from the magnet, and flew off to some distance before it was extinguished, just as we often observe portions of flame to do in a common fire, playing for a moment in the air.—Mlle. ATZMANNSDORFER saw precisely the same phenomenon when she blew in jerks. On all sides, therefore, the observations were confirmed, *that the magnetic odylic flame may be affected by the breath or by a current of air, and thus mechanically set in motion.*

410. It is now the proper opportunity for explaining an observation, given in § 13. It is there said that the flame of a common magnet yielded to that of an electro-magnet. This was really the case, but not, as I there omitted expressly to state, in the stationary condition of the apparatus, but during the rotation of the electro-magnet. In consequence of the rapid rotation there arose a brisk wind, which, as in the cases just mentioned, blew outwards the flame of the permanent

horse-shoe. The fact remains the same; but its significance must now be more precisely ascertained.*

* On this, as on all former occasions, we see the accuracy of the statements made by Mlle. REICHEL, which continue to be exact, even in cases like the present, in which I did not at first rightly apprehend them. And this is the same Mlle. REICHEL whom her countrymen, the Vienna physicians, formerly mentioned, have not been ashamed publicly to brand as a liar and impostor. She is a simple, but intelligent and well-principled girl, belonging to an order of nuns; and during three months, which she spent in my house, her conduct was entirely blameless, and such that all of us felt attached and kindly disposed towards her. Nothing is easier and more convenient, as a cloak for ignorance, than to get rid of, by declaring it to be imposture, a phenomenon which, for want of knowledge, we cannot understand, or, for want of dexterity in investigation, we cannot lay hold of. But then, I must say it openly, there is nothing more unmanly and dishonourable than, abusing our superiority, recklessly and unconscientiously to deprive a poor, sick, defenceless girl, of the only treasure she has, her good name, and to brand her with disgrace. When the accusation, besides, is a falsehood, a mere groundless calumny, as I shall prove to these gentlemen out of their own account of their deplorably bad experiments, it cries to heaven for redress, and every honest heart, with a sense of truth and duty, will share in the indignation I feel at such unworthy conduct. I shall here give *one* proof of what I have said. It is said to be a falsehood, that Mlle. REICHEL ever saw magnetic light at all. This makes my statements, indirectly, to be also false, since they rested, at first, chiefly on the statements of Mlle. REICHEL. I now invite the gentlemen of the (self-styled) Committee of the Society of Physicians of Vienna, to go to M. PAUER and his daughter Mlle. SOPHIE PAUER, two persons whose truth and honour no one in Vienna will venture to attack, and to ask them what they saw when I put before them, in the dark, a dozen of magnets? Should this not be sufficient, I further request them to ask Baron AUGUST VON OBERLAENDER, M. NIKOLAUS RABE, M. GUSTAV ANSCHUETZ, and M. SEBASTIAN ZINKEL, whether or not they saw fire and flame over magnets? I challenge them to call on Baroness VON TESSEDIK, Mme. CÆCILIE BAUER, Mme. JOHANNA ANSCHUETZ, Baroness VON NATORP, Mme. KIENESBERGER, Mlle. WINTER, and the Baroness VON AUGUSTIN, and to hear what they saw blazing over magnets in the dark. They will further have the goodness to ask MM. ENDLICHER, KOTSCHY, TIRKA, Chevalier VON RAINER, FERNOLENDT, KOLLAR, SCHUH, and HOCHSTETTER, whether they saw light appear over magnets or no? And I beg them to compare, with what they shall thus hear, what they may thus learn on the same subject from Mme. JOSEPHINE FENZL, Mlle. VON WEIGELSBERG, Mlles. DORFER, GLASER, WEIGAND, ZINKEL, KYNAST, the carpenter KLAIBER, and the blind M. BOLLMANN, &c. &c.; all people who live in Vienna, and may be daily seen and spoken with. And should all this not suffice, I refer them to their own colleague, Dr. NIED, from whom they may learn the phenomena he saw in my dark chamber. They may also learn the necessary truth from Dr. HUSS, the body physician to the King of Sweden. If all these honourable persons do not unanimously say that they saw unusual luminous appearances, of considerable size, partly as luminous nebulæ, partly as

411. If we now view, collectively, the observations from § 399 down to the present place, we have many and various proofs that the odylic flame, in itself, is not magnetism, as has been repeatedly stated. It does not obey the laws of magnetic attraction and repulsion, and is so far material, that it may be moved about, like ordinary flame, by currents of air. It rather appears as *a phenomenon which accompanies magnetism*, but follows the magnetic current only in part. It resembles a projectile, which, when once thrown out, pursues its own course in the direction of the impulse originally given to it, and is affected by the obstacles and new impulses which it meets with.

412. It appeared to me interesting, *to make observations on the effects produced by stroking steel bars with magnets*, and apply to them the collected results obtained in regard to the odylic glow and the odylic flames. These lights must thus necessarily appear in many forms and degrees, and must yield many confirmations of what has been described. It was also to be hoped that we might obtain new information as to what takes place in the process of making steel magnets, and in general, on the transference of magnetism and of odyle from one body to another. With this view, I made the following researches with Mlle. ZINKEL.

413. I began with bar-magnets, and afterwards proceeded to horse-shoes.

flames, over magnets in the dark, then I will admit that Mlle. REICHEL never saw magnetic light, that she is a liar and impostor, and that I am her dupe; whereby the real but veiled object of these physicians will be attained. But if the contrary takes place, and if the testimony of all these persons establishes immoveably the fact of the existence of the magnetic light, then I beg these gentlemen not to be offended with me, if I plainly tell them, in the face of the world, either that they are wretched experimenters, incapable, even in the way of imitation, of performing successfully the simplest physical experiment; or that they must submit to have the suspicion that any dishonesty has occurred in the transaction, removed from Mlle. REICHEL, and applied to themselves. Nature is eternally the same. After millions of years, the odylic light will shine as it does this day. All attempts to suppress such a truth, when once it has been disclosed by successful experiments, are poor and insignificant.—R.

The case of the spontaneous somnambulist, Miss M'AVOY, in Liverpool, many years ago, was a perfect parallel to this, in the conduct of certain medical men.—W. G.

The first result was, that the magnet used in stroking, whenever it was held perpendicularly on the bar, caused the latter, round the point of contact, to exhibit a much brighter odylic glow.

Fig. 6.

The pole of the magnet, therefore, acted with its whole magnetic and odylic force so strongly on the steel bar, as to cause it to shine more brightly wherever it touched it. Fig. 6. will show this more clearly. This luminous spot appeared at any and every part of the bar. And as it immediately disappeared when the magnetic pole was moved from the spot, it follows, that in stroking the bar from one end or pole to the other, the bright spot also traversed the bar from one end to the other, constantly accompanying the pole of the magnet. Now since the bar, as we have seen, when placed in the meridian, exhibits on its northward half a bluish, on its southward half a reddish glow, so also the moving spot appears bluish on the northward half, reddish on the southward. The effect of drawing a magnetic pole along a steel bar, therefore, appears obviously to be a local one, attached to the immediate presence of the pole of the magnet, as far as the odylic glow is concerned. It does not, at least not visibly, spread over the whole extent of the bar. In this it apparently differs from the proper magnetic effect, which at every stroke is spread over all parts of the bar, and not only appears in all parts, but adheres to them.

I used in these experiments two bars ; the magnetic bar or stroker, was twenty-four inches long, the steel bar to be stroked eighteen inches long.—Let us now follow the results, first in the steel bar, then in the magnetic bar ; and in the first place, with reference to the glow, in the next place, in reference to the flame.

414. A. *Odylic glow of the steel bar.* I placed on a table

in the meridian, a steel bar, hardly at all magnetic. It had a feeble glow in the dark, its northern half appearing grey instead of blue, its southern half whitish yellow instead of yellowish red, the light of both being turbid and dull.

I laid the bar magnet, on the west side of the other, in the parallels, on the table, and brought its northward pole in contact with the north end of the steel bar, so as to form a right angle with it. I thus began exactly as is generally done when we wish to magnetise a steel bar by stroking with a magnet, but with this difference, that both lay in a horizontal plane, whereas the operation is commonly performed in a vertical plane. This is a point which is probably not altogether indifferent, in reference to the essential part of the change produced.

The results were as follows.

a. When the two negative poles were in contact, the steel bar appeared in a blue odylic glow over more than the northern half, while the red glow of the southern half was rather shorter.

b. When the magnet had reached a quarter of the length of the bar, the same condition continued, but the blue glow extended to three-fifths of the length from the north end, while the red extended over the remaining two-fifths.

c. When the magnet had reached the middle, the blue glow extended to three-fourths of the length; (but in a later experiment, when the steel bar had acquired some magnetic power from previous stroking, the blue only reached to the middle, where the red began.)

d. The magnet, having reached three-fourths of the length of the bar, was now in the red glow of the latter.

e. When the northward pole of the magnet touched the south end of the bar, the blue glow extended to the middle. The red glow was now much brighter, and more intense.

415. B. *Odylic flame of the steel bar;* and first, the

α. Blue northward flame, originally feeble, and two inches long,

a. When the two northward poles were in contact, disappeared entirely.

b. When the magnet had been moved only 0.4 inch from the end, a small blue flame appeared, about 0.4 inch long.

c. With the magnet at one-fourth of the bar, the blue flame was 1.6 inches long.

d. With the magnet at the middle, the flame was 2.4 inches long.

e. At three-fourths, the flame was 3.2 inches long.

f. When the magnet reached the south end of the bar, the blue flame reached its maximum of 4.8 inches.

β. *The red southward flame*, originally 1.2 inches long,

a. With the northward poles in contact, immediately increased to its maximum of 2.8 inches.

b. With the magnet at one-fourth of the bar, it was only two inches long.

c. With the magnet in the middle, it had sunk to 1.2 inches.

d. At three-fourths, the red flame was only 0.8 inches long.

e. When the magnet reached the south end, all red colour was gone.

The course followed by the southward red flame was, therefore, entirely the same inversely, as that of the northward blue flame.

Let us now attend *to the magnetic bar*.

416. C. *Odylic glow of the magnetic bar*.—Before contact, while lying alone in the magnetic parallel, it had a bluish glow on the northward half, a reddish glow on the southward, both of about equal extent.

a. When both the northward poles were in contact (as above stated, in a horizontal plane, the bars forming a right angle, with the steel bar on the meridian), but so that the corners or solid angles alone touched each other, as in Fig. 7, there appeared at the touching point of the magnet, which had originally a blue glow, a small red spot, about one-twelfth of an inch long; beyond this, the magnet had a blue glow for six inches; then a grey or nearly indifferent and feebly luminous portion of two inches, while the remainder of the magnet, 16.8 inches long, had a red glow. It was truly surprising, and

Fig. 7.

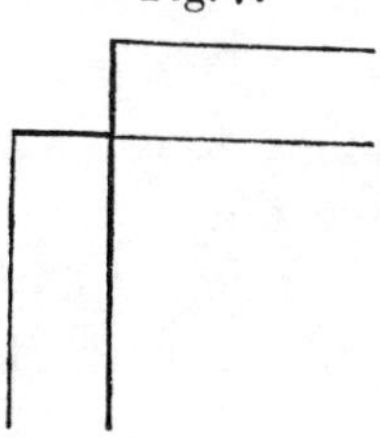

well worthy of notice, that the magnetic bar, so superior in magnetic force to the other, yet had its odylic polarity, although only for a very short distance, reversed, and its blue glow changed to red, so that zones of different colour appeared, and its northward polarity was enclosed between two southward polarities. There was, therefore, at this part of the experiment, a three-fold state of the magnet.

Fig. 8.

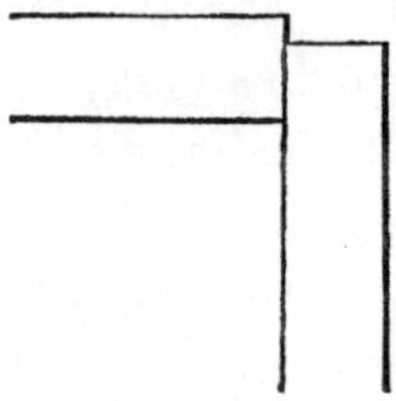

b. When the magnet had been moved only to the extent of two-thirds of its cross section, or two-thirds of an inch, along the other bar, so as to occupy the position represented in Fig. 8, the small red spot on the magnet had disappeared. At the same time, a blue luminous covering came into view on the cross section of the steel bar, which, when the magnet was advanced further, developed itself to the blue flame referred to above in B, *a*, and so forth. As soon as the red spot had vanished, the magnet only exhibited the two portions of glow, the blue on the northward, the red on its southward half.

c. When at one-fourth of the bar, the blue part of the magnet extended to two-fifths, the red to three-fifths of its own length.

d. When it had reached the middle of the bar, the blue part was somewhat larger.

e. When at three-fourths of the bar, the blue and red parts were nearly equal, the blue a little longer.

f. When the magnet had reached the south end of the bar, the blue glow on the magnet extended to three-fourths of its length, the red to one-fourth.

417. D. *Odylic flame of the magnetic bar :—*

Its blue northward flame, originally four inches long, was extinguished in contact with the bar. We have only, therefore, to consider the red southward flame. The magnet lay, in all these experiments, in the magnetic parallels, with its southward pole towards the west.

a. When the two northward poles were in contact, the red flame of the magnet was 2.8 inches long.

b. At one-fourth of the length of the steel bar, the red flame of the magnet had fallen to two inches.

c. At the middle of the bar, the red flame of the magnet had fallen to 1.2 inches.

d. At three-fourths, it had rapidly increased to 2.8 inches.

e. When the magnet had reached the southward end of the bar, the red flame of the former had increased to 3.6 inches.

When I now turned the magnet out of the parallel, and brought it into the meridian of the other bar, and joined the northward pole of the magnet to the southward pole of the bar, in a straight line, thus forming a compound bar of more than double the length of the unmagnetic bar, the northward blue flame of the latter, and the southward red flame of the former, both became twice as long as before, according to the observations recorded in § 401, and the laws there deduced from them. These same laws we shall find to prevail through the whole series of experiments contained in this section, when we compare them with the former.

418. We now come to the application of these laws, to the case of the *conversion of a horse-shoe bar into a magnet*, by stroking with a magnet.

I used as the magnet in these experiments, a powerful five-bar horse-shoe magnet, along which I drew a simple horse-shoe, which had lost almost entirely the magnetism it at one time possessed. I laid the five-bar magnet on the table, so that its poles projected over the edge, the curve being towards the middle of the table. It was heavy enough to remain firmly in this position while the stroking was performed, and so powerful, that it held fast the single bar in every position in which the latter was placed in contact with its poles. I held the single bar vertically, the poles upwards.

Originally, the single bar had, in the dark, and of course this applies only to the then actual perceptive power of Mlle. ZINKEL, on its northward pole, a small blue flame of about 0.4 of an inch, and on its southward pole only a reddish smoke. The five-bar magnet had a blue northward flame as long as a hand, a red southward flame of a finger-length. Both horse-shoes had on one side a bluish, on the other a reddish odylic glow, duller on the single bar, very much brighter in the five-bar magnet. This glow was hardly visible at the curve, and increased gradually towards the poles.

I began by placing the curve of the single-bar on the poles of the five-bar magnet, the southward limb of the former on the northward pole of the latter. The single bar adhered firmly, in a vertical position, with its limbs and poles projecting above the poles of the five-bar magnet. The result was, that,—

419. A. *The odylic glow of the single bar* became brighter at the curve, dividing into two halves, red and blue, the former on the part attached to the northward, the latter on that attached to the southward pole of the five-bar magnet. At the same time, both limbs of the single-bar acquired a brighter glow, the colours being on the same sides as before, but more intense. When I now slowly drew the single bar downwards, till the poles of the two magnets came in contact, the glow of the curve in the single bar increased, the two colours meeting and mingling in the middle; while in both limbs the glow diminished, becoming duller and more grey; but the colours still retained their places. The northward limb of the single bar had always a blue, the opposite limb a red glow, and the *quality* of the glow remained unchanged; that is, blue on the northward, red on the southward side, whether the single-bar were free, or in contact at its curve, with the poles of the five-bar magnet; or finally, when the poles of both were in contact. But the distribution of magnetic and odylic force, in regard to amount and intensity, was not, in these three positions, the same.

420. B. *The odylic flames of the single bar*, when its curve closed the five-bar magnet like an armature, rose from its original size of 0.4 inch on the northward pole rapidly to 4.8 inches; while the reddish smoke of the southward pole changed into a red flame of 1.2 to 1.6 inches long, with a thick yellowish grey smoke above it. In later experiments, when the single bar had become pretty strongly magnetised by the earlier ones, and had, in the free state, a blue flame 2.4 inches long, it rose, in this position, to 4.8 inches on the blue side, and to 2.4 inches on the red side.

On drawing the single bar slowly downwards, its polar flames gradually diminished; the southward flame soon vanished, the northward flame became shorter and duller, while the glow at the curve became brighter. When the poles of both magnets were brought in contact, all flames disappeared.

In this experiment, and in every phase of it, the colours of each limb remained constantly the same in every position. The two lateral portions of the curve, the limbs, above or below the points of contact, the flames over the poles, the sparks, the fiery threads, and the luminous downy flames round the limbs, continued throughout blue on the northward, red on the southward half of the single bar horse-shoe. At no point did the quality of the odylic light vary, but only its intensity, corresponding to the amount of odylic charge.

421. C. *Odylic glow of the five-bar horse-shoe.* As soon as the poles were joined by the curve of the single bar, the glow at its own curve became brighter, while that of the limbs, and still more of the poles, became duller, the northward limb acquiring a greyish blue, the southward a yellowish red colour. Here also, both colours met and passed into each other on the curve.—On drawing the single bar slowly downwards, the poles of the other became gradually brighter in their glow. They were brightest when the poles came into contact, but still inferior in brightness to the curve.—The glow, therefore, in the five-bar magnet, pursued a course nearly the reverse of that observed in the single bar; which is in accordance with the general laws already ascertained.

When I removed the single bar, the curve and limbs of the other horse-shoe became less bright, but its poles again became brighter. We cannot here consider the odylic flames of the five-bar magnet, because, as the poles were never free, no such flames could appear.

422. An experiment with horse-shoes, in which like poles were used, as was done with straight bars (see § 403), gave similar results, confirming those there mentioned. I laid the northward *pole* of the single bar on the northward pole of the five-bar magnet, and the southward pole also together, and drew the single bar *upwards* till the curve reached the poles

of the other. In this way, the poles were reversed, but the direction of the stroke was also reversed, and the result must correspond to the former. It was as follows. As soon as the corners of the poles touched each other, there appeared at the blue pole of the single bar a short red portion, 0.4 inch long, and at the red pole a similar blue portion, also 0.4 inch long. But when the single bar was moved on about 0.4 inch, both these portions disappeared, the northward pole of the single bar became entirely blue, the southward pole entirely red, exactly according to the usual law, which they then followed as before. These spots of reversed glow are the same as we have seen above in bar magnets on the contact of the two negative poles at their corners; a reversal of the polarity by the influence of a like pole. The single bar horse-shoe, therefore, was here, in the first moment of contact, multipolar, 0.4 of an inch red (at the northward pole), then the limb blue, the other limb red, and at the southward pole, again a streak of blue 0.4 inch long.

423. *The force*, existing in these bars, *is*, *therefore*, *never uniformly distributed*, *not even when the magnetic circuit appears to be closed.* This closure or completion of the circuit, as we have seen in §§ 404, *b.* to 407, is probably never perfect, and hence, even without reference to the disturbance caused by the earth's magnetism, these constant inequalities in the distribution of the forces. We find new examples of this in some investigations into the

424. *Effects of the armature on the odylic flames.* We have seen that the armature, when attached to a horse-shoe, diminishes the glow of its poles, and increases that of the curve. But all the sensitives saw the odylic flames instantly disappear, when the armature was attached to a horse-shoe. I find the following persons named in my journals as witnesses of the fact; Mme. KIENESBERGER, Mlles. WINTER and DORFER, FRIEDRICH WEIDLICH, Mlle. ZINKEL, Mme. FENZL, Baron VON OBERLAENDER, Prof. ENDLICHER, Baroness VON AUGUSTIN, and many others.

425. When I placed before Mme. KIENESBERGER, the armature on the curve of a horse-shoe, this had, as might be

expected, no sensible effect on the polar flames. But when I moved the armature slowly towards the poles, transversely over the limbs, the polar flames steadily diminished, and disappeared when the armature reached the poles. On moving it in the opposite direction, a similar gradual increase of the flames was observed. The details were these. When the armature was moved from the curve towards the poles, the flames became duller, the blue greyish, the red muddy yellowish, while their length gradually diminished. At the middle, the red flame disappeared, and only reddish smoke remained over the southward pole; the diminished bluish flame became quite grey. When the armature approached the poles, it also was extinguished, leaving a grey smoke, which disappeared when the armature had reached the poles. All these phenomena, in the reverse order, appeared when the armature was drawn from the poles to the curve; first, over the northward poles, grey smoke, then grey flame; then, at the southward pole, reddish smoke; at the other pole, the blue flame; then, at its own pole, the red flame, both increasing till they attained their original size. The glow was at first very weak at the curve, strong at the poles; on moving the armature towards the poles, the curve became more visible; and as the poles gradually lost in glow, the glow of the curve gradually increased; till at last, when the armature had reached the poles, it nearly equalled the limbs and the poles in brightness.

426. But the armature itself varied in its glow. When at the curve, it was grey and hardly visible, at the poles it was brighter, and red on the side next the northward pole of the horse-shoe, blue on the other. In all the intermediate positions, it exhibited corresponding degrees of light and tints or colour. It was changed into an induction magnet, and exhibited the regular colours of one, although somewhat duller than those of the inducing magnet. The above phenomena are explained by considering the part of the horse-shoe at any time lying between the curve and the armature to be cut off from the whole. The remaining part is so much shorter, and hence its polar flames are of less size, and *vice versa.*

427. I endeavoured to ascertain *the state and the action of*

the armature and the single horse-shoe during the stroking. This was done on the occasion of the last-mentioned experiment of drawing a single-bar horse-shoe along a five-bar magnet, in the presence of Mlle. ZINKEL. When, under the circumstances mentioned in the latter part of § 418, and when the five-bar magnet was closed by the curve of the single bar, as by an armature, and the flames of the latter were at their highest, I attached to the poles of the single bar its own armature, it adhered firmly, and was converted into an induction magnet, with strong glow, the colours opposite to those of the poles of the single bar. The limbs lost much of their glow, the blue became bluish grey, the red a muddy yellowish red.

428. When I drew the single-bar downwards, till the poles of both horse-shoes touched, as in Fig. 9. and again attached the armature to the single bar, it again adhered firmly, and was converted into an induction magnet, with strong glow, *but now the colours of the glow were reversed*, being no longer opposite to those of the poles of the single bar, *but to those of the five-bar magnet.* From this it follows, that the armature was no longer under the influence of the single bar, but had come under that of the five-bar magnet; and this, not directly, but through the medium of the thickness of the steel forming the limbs of the single bar. These must first have been converted into short induction magnets; and then from them, the armature was converted into an induction magnet, formed by the five-bar magnet. All these things are clearly visible to the sensitive eye, while science, which has hitherto refused to derive any benefit from the perceptive powers of the sensitive, could only hitherto attain to them with difficulty and uncertainty.

Fig. 9.

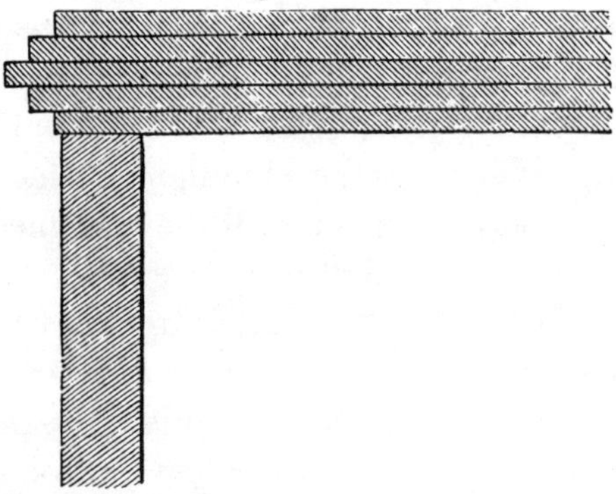

429. The next question was, *when and where does this*

striking reversal in the poles of the armature of the single bar take place? To ascertain this, I closed the five-bar magnet with the curve of the single bar, and placed the armature on the poles of the latter projecting upwards. I now drew the single bar slowly downwards, and begged the observer to describe the results from time to time. When the single bar had been drawn down to two-thirds of its length, the armature which had at first adhered firmly, fell off. I again applied it, but found that it was no longer attracted in the least. This continued, till I had moved down the single horse-shoe to three fourths of its length, when the armature began to be again attracted. The attraction increased as I proceeded, and was strongest when the poles of the single bar reached those of the other. The course of the colours in the glow of the armature was as follows. The armature had at first colours opposed to those of the single bar, from which it received magnetism by induction. As I drew the horse-shoe downwards, the brightness of glow and of colour in the armature diminished, the light became duller, the colours less distinct, till, at two-thirds, all colour in the armature was gone, and it retained only a dull uniform grey, like that of any other piece of soft iron. This lasted till I drew the bar down to three-fourths, when the armature again began to shine more brightly, and to exhibit colours in its two halves, but now reversed. Its blue and red were no longer opposed to those of the single bar, to which it adhered, but to those of the five-bar magnet, to whose poles it was now brought near. The seat of indifference, in relation to colour, coincided therefore with that of indifference in relation to attractive force; and these observations furnish a beautiful proof of the parallelism between the two phenomena, under ordinary circumstances.

430. It was also observed, that when the armature was in the position in which it had no attraction, and exhibited no colour, but only a grey glow, the curve of the single bar also exhibited only a grey glow, although tolerably bright. It would appear, then, that in this position, *both the armature above, and the curve below, were in a state of neutralisation or indifference,* when the single bar was drawn down to from two-

thirds to three-fourths of its length, and in that position adhered to the larger magnet.

431. Once, when, from the presence of the catamenia, the perceptive powers of Mlle. ZINKEL were three or four times greater than usual, she saw the single bar horse-shoe, already magnetised by frequent stroking with the five-bar magnet, with a blue flame on its northward pole six inches long, and a red flame on its southward pole of 2.4 inches. When I now attached its curve, as armature, to the five-bar magnet, unlike limbs and poles together, both flames increased in size about one-half. But on drawing the single bar downwards, the flames diminished, and when they had fallen to their original size of six inches and 2.4 inches, I found that I had drawn down the bar to one-third. In this position *the effect of the five-bar magnet on the flame of the single bar was = O;* and the forces had come into a certain equilibrium, so that the influence of the larger magnet on the poles of the other, was, at least *as far as concerns odyle*, suspended. When I moved the bar further down, to two-thirds of its length, I came to the *point of magnetical indifference*, where the *magnetic* effect of the larger magnet on the poles of the smaller was also suspended.

432. In this observation, *the divergence between odylic and magnetic influences* was strikingly seen. In the experiment just described, which was often repeated, and was also tried in the ordinary state of the observer, when the magnetism of the single bar was = O, and the armature fell off, this was far from being the case with the odylic flame. It was, indeed, smaller, having fallen, on the blue side to 1.2 inches, and on the red side having changed into reddish smoke, thus being reduced to one-fourth or one-fifth of its proper size; but it was not destroyed, as the magnetic attraction was. Nay more, when I drew the single bar down to three-fourths of its length, and its armature had already reversed the colours of its glow, so that the half resting on the negative end of the single bar became blue (§ 427), yet the remains of the flame continued always blue. It diminished, as I drew the poles of the single bar towards the poles of the five-bar magnet, from 1.2 inches to 0.8, — 0.4, — 0.2, and 0.04 of an inch, then to the thick-

ness of paper, and, at last, to a blue film playing over the negative end, but still blue and bluish grey, till, when the poles of the bar came into contact with those of the five-bar magnet, it entirely disappeared.

433. Here *the difference in the courses followed by odyle and by magnetism* becomes very striking. While the — M of the northward limb of the single bar is converted above into + M, the flame over that pole continues blue, and persists in — O; from a steel bar, which gives out + M, nay brings the part of the armature on it to the state of — M by induction, the blue flame, indicating — O, flows uninterruptedly, as long as the armature of the single bar and its poles remain at any perceptible distance from the poles of the five-bar magnet. Therefore, *positive magnetism and negative odyle can exist at the same time in one and the same limb of a horse-shoe.*

434. As *electro-magnetism* is a means of exciting the odylic glow in iron (§ 374), so also does it produce odylic flames, and that in a very high degree (§ 12). In the experiment mentioned in § 374, to which, for the sake of brevity, I refer, Baroness PAULINE VON NATORP, a healthy subject, and possessing little power of seeing the odylic light, saw, in the large horse-shoe of soft iron, when converted into a magnet by a current of electricity, not indeed a flame, but a nebulous light rising to the height of a hand.—Baroness ISABELLE VON TESSEDIK saw it in about the same degree. M. PAUER saw a flame, from 12 to 16 inches long, resting like a mass of luminous vapour on both poles. Mme. FENZL saw lights over the poles, the longer one ten inches high and greyish, the other somewhat shorter and yellowish. Baroness VON AUGUSTIN saw, on both poles, a flaming light, about twenty inches broad, and as high as a man, rising, united, to the ceiling. The upper part was blue, the middle brighter and whitish, the lower part again darker. They were unsteady, and ended, above, in thick clouds of smoke.—STEPHAN KOLLAR saw, when one pair of SMEE'S plates was used, only on the negative pole a flame; with two pairs, both poles had flames nearly twenty inches high, with smoke rising from them. The much more sensitive Mme. KIENESBERGER saw a blue flame from

the negative pole, eighteen to twenty inches long, and from the positive pole a red flame about 6.5 inches long, both ending in smoke. She described both as more unsteady than in the case of a permanent steel magnet, and as leaping up, like the flame of a tallow candle. This leaping, and continual change of size, and variation from short to long, is no doubt caused by the varying amount of electricity at each moment generated in the battery, and passing through the coil, by a flow and ebb in the inducing agent, arising from inequalities in the chemical action between the elements of the battery. I must here also mention a circumstance, to which I shall return at a future period; namely, that both the polar conducting wires and the coil round the horse-shoe were surrounded by flaming lights. These flames were brighter near the poles, duller at the middle of the wires, and about the curve of the horse-shoe. Mme. Kienesberger first perceived the odylic glow; then, after a minute or two, the flames appearing over the poles at first greyish and feebly luminous, then increasing in intensity and developing colours. And when the current was interrupted, the colours of the flames first disappeared, the grey flames vanished, and the glow continued longest. Prof. Dr. Endlicher saw, on the poles of an electro-magnet, flames forty inches high, unsteady, exhibiting a rich play of colours, and ending above in a luminous smoke, which rose to the ceiling and illuminated it. M. Delhez saw the flames of the same size, but did not distinguish the colours. The flames appeared to him darker below (red), brightest in the middle (yellow), and darker again above (blue). Mlle. Glaser saw, over the poles of the same electro-magnet, flames five feet high, and smoke rising from them to the ceiling. The flames exhibited the most beautiful and varied play of colours, blue predominating over the northward, reddish yellow over the southward pole. Mlle. Zinkel studied very often and with great care the flames of the electro-magnet. Under the circumstances above stated, she saw the flame of the northward pole forty inches high, that of the southward pole upwards of one foot in height. Both were coloured, blue predominating in the former, red in the latter. The appearance and order of the phenomena, including the flames surrounding the coil, the conducting wires, and the battery, were always described by her exactly as by Mme. KIENESBERGER. But she observed that, on one occasion, five or six minutes, on another ten minutes, were required to bring the flames to their full stationary size. In this last case, I had used a SMEE'S battery with five square feet surface of zinc and silver. (No doubt the battery did not, in these cases, and probably on account of the surface of the zinc not being quite clean, reach at once its full power.—W. G.) The columns of flame from each pole astonished her by their size and beauty. The negative flame was more than as high as a man, the positive as long as an arm. Adorned with the most beautiful rainbow colours, they constantly flickered and leaped up and down like the flame of a gigantic candle. Streams of sparks flowed into the smoke, and were scattered on all sides. The smoke reached the ceiling, and was broken or flattened by it.

I showed to Mlle. REICHEL in the dark a SCHWEIGGER'S multiplicator, and caused a weak voltaic current to pass through the coil of wire. She saw the coil and the needle immediately assume the glow, and the latter then gave out at its points fine streams of flame, flowing out in the line of its length. But as the containing case was too narrow to allow them to extend in a straight line to their full length, they struck on its sides, were then deflected upwards, struck again on the glass cover, were again deflected along its surface, and proceeded along it till they were lost to the eye of the observer.

435. We have already seen the relation of the electric atmosphere to the odylic phenomena, in its effects on the odylic glow (§ 370). But this influence is especially powerful in regard to the flames. In the beginning of my researches, Mlle. REICHEL had already observed, that the flames of a magnet grew larger as often as it was brought into the sphere of any kind of electrical action. Mlle. ATZMANNSDORFER also often made the same observation; both have been often mentioned. M. HOCHSTETTER saw the flame of the nine-bar horse-shoe, to him usually twelve inches high, increase to twenty-four inches. Mlle. PAUER observed, that the flame of

the same magnet increased to three times its former height, and illuminated the roof. M. Fernolendt saw the flame of this magnet, to him sixteen inches long, rise to thirty-six inches, and illuminate the ceiling. Mlle. Glaser saw the flame of a twenty-five inch bar magnet, when I brought the negative pole near to the positively electrified conductor of the machine, rise to four times its former height, namely, from about 3.2 inches to thirteen inches. She also saw the flame of the nine-bar horse-shoe, to her sixteen inches long, rise, when brought near the conductor, to sixty-four inches, or about the height of a man, casting light on the ceiling. Mme. Bauer saw nearly the same phenomena. Baroness Von Augustin saw the blue flame of the nine-bar magnet, when its negative pole was brought to within twenty inches of the conductor, increase to from twice to thrice its former size, and illuminate the ceiling. I placed before Mlle. Zinkel, the nine-bar magnet at forty inches from the conductor of a powerful electric machine, which I set in motion. The magnet stood with its poles upwards, the limbs equidistant from the conductor. The flames were at first four inches high on the negative, 1.6 inches on the positive pole. The former rose to twenty inches, the latter to six inches, both thus increasing to four or five times their former height. The colours became brighter on the blue side, more turbid on the red side. As soon as I ceased turning the machine, the flames sank, and soon recovered their original size. This was often repeated at different times.

436. With Baroness Von Augustin as observer, after she had seen the great increase of the blue flame on the negative pole when near the conductor, I reversed the position of the magnet, so that the poles being still upwards, the positive pole was now nearest the conductor. Very soon the reddish light over it disappeared, and was replaced by blue. In this position, therefore, the odylic polarities were reversed. When, in other experiments with Mlle. Zinkel, I turned the blue negative pole of the horse-shoe to the conductor, while the opposite pole was turned away from it, the flame of the latter became greyish red, that of the former pure and bright blue, with a play of rainbow colours, above which, smoke and sparks rose

nearly to the ceiling. The flame was so brilliant, that she could not conceive how I did not see it, at least in part. But when I reversed the position of the poles of the magnet, the red flame being now next the conductor, it rapidly became discoloured and turbid, then grey, and in half a minute passed into blue; while the blue flame became violet grey. The blue flame on the positive pole now rose high, the violet grey flame on the negative pole shrank in dimensions. Here then, as in the case of the glow (§ 371), *a reversal of the odylic polarity had taken place,* without any such reversal of the proper magnetic polarity. The electrical polarity in this case, dictated the law, and reversed the odylic polarity, but was not able to reverse the magnetic polarity. When I now reversed the electric polarity by charging the conductor negatively, I obtained the same odylo-luminous phenomena, but of course reversed. Now the red flame of the positive pole, when brought near the conductor, increased in size, brightness, and purity of colour; but the blue negative flame, when brought near, was soon changed into a red one. The odyle passed into + O, and that at the negative magnetic pole.

437. To bring out these facts still more distinctly, I repeated them, using bar magnets instead of horse-shoes. Two bars, one twenty-six inches long, the other five feet long, were placed on a table, unisolated, in the magnetic parallels, and at forty inches from the conductor. When their blue negative flames were turned towards it, they lengthened out towards the conductor to three and four times their former length. But when their red positive flames were placed nearest the conductor, the flames were at first rendered turbid and less luminous, and then, after a struggle between the forces here in action, manifested by the playing of small flames backwards and forwards on the bars, they were changed into blue, and increased within one or two minutes to considerable length and breadth. As often as I stopped the rotation of the machine, or discharged the conductor, the flames were again reversed, and returned, in all respects, to their former state. The small struggling flames played over the surface as before, till all was again brought under the dominion of the magnetism of the bars.

438. The appearances were still more beautiful, when I brought the bars nearer to the positively electrified conductor. I tried this with bars both of square and circular section, grasped by a GUIDO's support of wood. At a certain distance from the conductor, towards which the northward or negative poles pointed, the colours of the flames were simply blue and red at the opposite poles. But on bringing them gradually nearer to the conductor, not only did the vividness, intensity, and brilliancy of the flames increase, but other colours were by degrees developed from these two, till at last, at about twelve inches from the conductor, all the rainbow colours were presented to the eyes of the sensitive observers in full splendour. I shall return to this further on, when treating in detail of the colours of the odylic light.

438. *b.* We see here the odylic flames undergoing the same changes in the electric atmosphere as we formerly saw in the odylic glow; and if we have hitherto found, in most of the phenomena, *that magnetism proper to a certain extent predominated over odyle, we now find electricity, with superior power, assuming the entire control over it.* It appears that odyle is even more intimately connected with electricity than with magnetism; that its dependence on magnetic poles is not a necessary one, but variable and conditional only; that odyle partakes more of the mobility of electricity than of the rigidity of magnetism, which latter quality it only possesses at the poles in so far as the rigidity belongs to magnetism, to which odyle is, up to a certain point, attached. Odyle therefore appears here, even more distinctly than elsewhere, *to hold a kind of middle place between electricity and magnetism.*

439. *The influence of the earth's magnetism* on the size and brightness of the odylic flames of the magnet is pretty strongly marked, and keeps pace with its effect on the odylic glow. When a bar-magnet lies conformably in the magnetic meridian, its flames are largest and brightest. But when reversed, the positive or southward pole pointing to the north, the blue flame of the negative pole became dull and compressed, tending to grey, turbid, vaporous, less distinctly visible, and smaller; the positive red flame of the southward pole became

also feebler, more yellowish grey than red. Mlle. ZINKEL saw a difference in the flames in every direction in which the bars were placed. When I placed a horse-shoe magnet so, that, with its poles upwards, the northward pole was next the north, both flames were vivid and bright; but if the southward pole were next the north, and the northward pole of course next the south, the observer only saw dull, turbid, and feebly coloured flames, both being sometimes only grey, or passing into grey vapour.

440. The same thing happened, when I brought the magnets, on the vertical plane of the meridian, into different angles with the magnetic dip or inclination. In every position, the size and colour of the flames were different. This also happened (with the glow ?—ED.) when horse-shoes were closed with their armature.

441. Even *masses of unmagnetic soft iron*, when made to revolve on their centres, in the vertical plane of the meridian, exhibited different aspects of their odylic flames in different positions. I shall enter into this more minutely in its proper place. On one occasion Mme. KIENESBERGER, awaking in a very dark night, saw at the window a bright flame. She sprang out of bed, terrified, in order to extinguish the fire which she supposed had broken out. But when close to the window, she saw nothing more; the flame was no longer visible to her eyes. On returning to bed, she again saw it, and again got up and approached it; but it again vanished. (We shall hereafter see that the flame is only distinctly seen at a certain distance.—See §§ 504-506.) It was a window-bolt of iron which stood vertically, and gave out flame from its upper end.

442. Not only do magnetic and electric influences affect the odylic flame of the magnet, but also purely odylic influences do so. This is the case with *crystals* of all kinds, including earths and saline salts, in which we are not acquainted with any magnetic properties, in the strict sense of the term. We have seen, that when a bar magnet was placed in contact with another of the same size and strength, in a straight line, with unlike poles touching, the flames in the middle disap-

peared, while those at the ends became nearly twice as large as before. A crystal produces the same effect. When I placed the negative pole of a crystal, of rock-crystal, gypsum, heavy spar, &c. in contact with the positive pole of a needle of equal length, both being in one straight line, Mdle. ZINKEL, Mme. KIENESBERGER, Mlle. ATZMANNSDORFER, Mlle. GLASER, and FRIEDRICH WEIDLICH, saw the blue negative flame of the magnet increase to about twice its former size. When I applied the other end of the crystal to the negative pole of the magnet, they saw the red positive flame of the latter become longer. *The crystal therefore acted on the odylic flame of the magnet as powerfully as another magnet, although it did not add in the least to the supporting power of the magnet.* The peculiarity and independence of the odylic influence is here obvious.

443. If this view were correct, then we should obtain, from an experiment of control, a confirmation of it. The effect of the flames of magnets and crystals, when opposed to each other, ought to concide with what we have seen of the flames of two magnets, when made to approach each other, in § 401. To test this, I placed before Mlle. ZINKEL a bar-magnet of six inches conformably in the meridian, and approached to its southward pole the negative pole of a crystal of gypsum, which had a blue flame about two inches long. As soon as the poles were about one foot asunder, the magnet and the crystal being in a straight line, both flames became narrower and longer, and seemed to struggle to meet each other. But when near enough for this meeting to have occurred, it did not take place; they became broader and shorter, and contracted round their own poles, and finally disappeared on contact. But when hostile, that is, like poles were approached, no lengthening of the flames towards each other was observed. They became compressed, and on contact were turned back over their own poles; all in perfect agreement with the phenomena exhibited when two magnets were made to approach each other.

444. A still more powerful means of strengthening the odylic flames is found *in animal organs, such as the human hand.* I placed a bar magnet, eight inches long, in a wooden GUIDO'S

holder, conformably in the meridian, and brought Mlle. Atzmannsdorfer into the dark chamber where it stood. When she took hold of its southward pole with the fingers of her right hand, she saw the blue flame at its northward pole increase both in size and brightness. When she took the negative pole in her left hand, the red positive flame on the positive pole increased. When she reversed the hands, applying, for example, the right hand to the negative pole, the flame at the opposite pole became feeble and turbid. I repeated the same experiments a year later with Mlle. Zinkel, and with the same results. *Their hands therefore acted on the flame exactly like magnets.* The addition of their right or negative hand to the positive pole, increased the odylic current at the opposite pole considerably, but without in the least increasing its power of attracting iron. There was an addition of vital odyle made to the magnetic odyle, and the same produced a flame of double size. *And what crystals did in the last experiments, was done by the human hand on this.*

445. I made afterwards a similar experiment with Mlle. Glaser. I took hold of the southward pole of an eight-inch bar magnet with my right hand; the blue flame at the northward pole became twice as large as before. With my left hand on the northward pole, the red flame of the southward pole was doubled in size. But when I held both poles alternately with reversed hands, the flames at the projecting poles became turbid. In these experiments, only weak magnets must be used, otherwise the results are not sufficiently obvious; since the power of the hand is equal only to that of a small magnet. These results confirm in every respect those of the preceding paragraph.

446. A variation of the experiment, but essentially of the same nature, was often tried with Mlle. Zinkel. I made her bring her fingers near to bar-magnets. When she directed, at a certain distance, the fingers of her left hand, joined together, towards the negative pole of the magnet, the flames on both sides lengthened out towards one another; but on a nearer approach, they became broader and shorter, retreated and disappeared as soon as her finger touched the bar. This

did not always occur equally. When the magnet was very small or weak, its flame was extinguished, but not that of the fingers; when it was strong and large, the flames of the fingers were extinguished, while a residue of magnetic flame was left. Bars of four or six inches were found generally to yield the most distinct results, coming more easily into equilibrium with the force of the hand. But this varied in different persons, and no doubt, in the same person in different states of odylic power. My own fingers often had their flames sooner extinguished than those of Mlle. ZINKEL. This always happened when the flames were brighter on her fingers than on mine, which was frequently the case. She is tall, and has small hands, in which the odylic force of her person is concentrated, and thus rendered more intense. When she directed the fingers of her right hand to the negative or northward pole of the magnet, there was no lengthening of the flames, nor tendency to meet each other. But on a near approach, the flames contracted, each round its own pole, became flattened and at last inverted backwards, when the fingers touched the pole; but both flames were then smaller and duller than before. The same results, *mutatis mutandis*, were obtained at the other pole. These experiments all testify to the same truth, *that the points of the fingers act on flames of magnetic poles just as crystals and as magnets do, and may be substituted for them.*

446. *b.* The variations just mentioned, produced by unequal power in the odylic poles, which act on each other, went so far, that in one case, when powerful hands acted on a feeble magnet, its polar flames were actually reversed. This deserves to be noticed, because it is *another proof of the independent nature of odyle as compared with magnetism*, similar to that which we have noticed in § 436. I caused Mlle. ZINKEL to hold between two fingers, and conformably in the meridian, a four-inch needle, not strongly magnetic. When I held the southward pole in my right finger-points, the blue northward flame became three times as long as before. This showed the feebleness of the needle in comparison with my hand. But when I held the same pole with the fingers of

my left hand, the blue flame disappeared, and a red flame took its place. When I made the experiment at the other end of the needle, with my left fingers on the negative poles, the red flame of the opposite pole became brighter, and three times longer than before. But when I applied the fingers of my right hand to the same negative pole, the red flame disappeared, and was replaced by a blue one. It clearly appears from this, that the odylic force of my hands so far outweighed that of this weak magnet, in reference to the quality or kind of odyle, as indicated by the flame, that it changed the blue flame of the magnet into red, and the red into blue, when its odyle was opposed to that of the flame at the further end of the magnet; that is, when it was the same as that of the end grasped in the fingers; *but without disturbing, in the slightest degree, the proper magnetic polarities of the ends of the needle.* My hand, which has of course no magnetic power, had so strong an odylic effect on the magnet, that the magnetic odyle of the latter was overpowered by the vital odyle of the former, and the colours of the flames were reversed (just as we have seen them to be by electricity).

447. In contrast to this means of increasing the odylic flame, I have to mention *those which diminish it;* of which, up to this time, I have detected two. The first is *heat.*—Mlle. PAUER saw the flame of a horse-shoe, when cold, two inches and more in length; but when I had given it, by placing it on a stove, the heat of the hand, the flame had sunk it to 0.8 inch. On cooling, the flame recovered its former size. I repeated the experiment next day with a larger magnet, and heated it more strongly. The flame diminished as the heat increased, and returned as cooling took place. Mlle. GLASER saw on a single horse-shoe ten inches long, while it was cold, a flame 4.8 inches high. When heated to about 100° F., the flames appeared only 3.2 inches high. At 145°, when I could hardly hold it in my naked hand, the flame had sunk to 1.6 inches. But when, after allowing it to cool with the armature attached, I removed the armature, the flame had recovered its former size. Mlle. ZINKEL observed the same in the same magnet. She found the effect of heat still more striking in bar-magnets, in which

the flame almost disappeared. According to these experiments, just as the odylic glow was diminished by heating (§ 369), so also is the odylic flame, and even in a greater degree. It is probable that at a temperature little exceeding that of boiling water, all flames would vanish from steel magnets. The flame consequently decreases with the rise of temperature at a much more rapid rate than their magnetism does; which indeed also decreases when they are heated, but very much more slowly.

448. A second cause of diminution of the odylic flames is very important to the investigator. This is *the proximity of surrounding objects.* We have seen the effects produced by crystals, hands, &c., and I shall treat of the action of metals, metalloids, alkalies, acids, &c. when brought, in considerable quantity, near the magnetic flames, in the section on the light of these bodies. I shall here only allude to *the striking effect produced by the approach of persons.* Mlle. ATZMANNSDORFER often told me that she only saw the flames of magnets well, when I was at a certain distance from them, and when they were not too near to herself. Mme. KIENESBERGER observed that when, in preparing or arranging a strong electro-magnet, I came very near it, its polar flames became smaller and more turbid. M. HOCHSTETTER saw the flames of a bar, six inches high, entirely disappear when I accidentally came near it, and reappear as I removed further off. I placed, as a control for this observation, the nine-bar horse-shoe conformably with its poles upwards on a table. He saw the flames rising a foot high. As I approached it the flame diminished, and when I was so close as to touch it in one part, the flame disappeared, and only the glow remained visible. But when I retired the flame returned, and I could repeat this as often as I chose. Dr. NIED, before whom the magnet was placed on a chair, saw the same thing, as did also M. DELHEZ and Baroness VON AUGUSTIN. M. PAUER saw the magnet flaming in the dark; when I had approached to within about twelve or thirteen inches, the flame became dull, and when I stood close to the magnet, it had disappeared. Mlle. PAUER also often saw this. The flames of all magnets first became duller,

and then were extinguished when I came near them, and again appeared when I went further off. Mlle. GLASER saw the same thing often in the flame of the nine-bar magnet, and this when it had recently been stroked with another magnet, and its magnetic intensity thus brought to the highest point. When I removed to a very short distance, the flames immediately began to appear as before, and were perfectly restored when I was a pace distant from it. Mme. BAUER often accidentally made the same observation. When I happened to ask about the flames of magnets, she complained that I came too near them, that I thus rendered the flames turbid or smaller, or even extinguished them, which prevented her from observing the points on which I questioned her. She desired me, on these occasions, to go further off. Mlle. ZINKEL explained that, when I retreated, after having by my approach extinguished the flame of the nine-bar magnet, it was not instantly, but only gradually restored; and that I must be about 40 inches from the magnet before the influence of my vicinity became imperceptible.

449. In order to become master of these deceptive appearances, I made the following experiments. I placed the nine-bar horse-shoe on a table, with its poles directed upwards. At one time the northward limb was turned to the east, the southward to the west; at another, this was reversed. In the presence of Mlle. ZINKEL, and in the dark, I now approached and retreated alternately to the poles at the same time, from the north, with my front towards them. Both flames disappeared when I came close, and re-appeared when I retreated. But when I approached both poles with my right side only, only one flame was extinguished, namely, the blue northward flame; while the other became more luminous and more intensely red. With my left side, this was reversed. The southward flame vanished, the northward flame increased remarkably in luminousness and intensity of blue colour. This cleared up the matter, and brought it under the regular law.

450. It is therefore the odylic state of the whole person which acts on a magnet, even a very powerful one, and as strongly

as magnets act on magnets, when like poles are brought together. We have seen how the flames were mutually forced back and extinguished; and just so do the odylic poles of the body act on magnetic poles, even when we cannot always, or, at least, in every case, clearly point out where the animal pole lies, and which of the many interwoven odylic axes in the body are those which have caused the effects we perceive. But a cautious investigation will see in this an inexhaustible source of numberless blunders, misapprehensions, incomprehensibilities, and enigmatic appearances, which have up to this time pressed, like a night-mare, on what is called animal and mineral magnetism; and, without clearing these away, it was absolutely impossible to obtain harmonious or trustworthy results. Every observer, in animal magnetism, or in the effects of magnets on the system, saw a different phenomenon, and every day a new one, differing from that of the day before. Nay, during one experiment, the results changed under the hands of the experimenter, according as he changed his position, or as the number of his assistants and companions changed. The causes of this were unknown to these observers, who could nowhere find a firm footing in their experience, the scientific ground gave way under their feet, and the confusion became at last endless and hopeless.*

* My worthy friends, the Doctors and Professors of the self-styled Committee, should take this doctrine to heart, and learn not only to recognise the weak points of their experiments, but to feel the disgrace they have brought on themselves by their groundless accusations of the poor girl, Mlle. Reichel. They prepared a dark chamber, of which they themselves tell us, that the chinks of the door were hurriedly stopped with handkerchiefs to keep out the light. (See Journal of the Society of Physicians in Vienna. Year III., p. 138.) He who has worked long in the dark can easily imagine what sort of darkness that was, which was thus extemporised! In the confined space of a small room, the sensitive patient was always crowded along with ten to fifteen young men; and sometimes one, sometimes another, *went in and out.* (Journal, pp. 119 and 139.) But we know, that by the momentary admission of light, even through the smallest chink, the eye becomes, *as far as concerns odylic light, dazzled and almost insensible for more than half an hour.* What then must happen when such prodigious blunders are committed as opening a door to let people out and in? This alone is more than enough to make the results of such gropings in the dark, under the name of experiment, utterly useless and worthless. If Mlle. Reichel, under such circumstances, saw nothing, as, according to the pro-

The odylic flame, as we see, gives us no information as to the direction of the magnetic force; no lines of magnetic force,

tocol, she often declared, she spoke the truth, for she *could* see nothing, under arrangements so perverted, not because she had not the power to do so, but because experiments made blunderingly, and without knowledge of the subject, made it impossible for her to exercise that power. But, at last, she now and then saw something, gave confused statements, and was therefore called a liar and impostor. Let us examine the statements of these gentlemen, and see where lies and imposture can be found. Mlle. Reichel was from three to six paces from the magnet, and had generally on each side a doctor to watch her; and these guards often had hold of both her hands. Opposite to them sat another doctor, holding the heavy horse-shoe magnet in his lap, and moving it backwards and forwards. Close to him, on each side, stood a fourth and fifth doctor, who were to control the motions of the magnet. The sensitive girl was now required to tell how and where the magnet emitted light, when it was shifted, now here, now there. She was urged to show her art; she was laughed at and treated with scorn when she failed. She was in this way irritated and exasperated, insomuch that she struck out at the doctors, and was on the point of administering to one of them a box on the ear, &c. Now, in the first place, it is quite impossible that a sensitive, of moderate power, as Mlle. Reichel at that time was, could see any magnetic light in a darkness often interrupted by the going out and coming in of spectators. In the second place, three to six paces is far too great a distance from the magnet. Such a sensitive cannot with certainty perceive magnetic light at a greater distance than forty inches. In the third place, the magnet, a heavy nine-bar horse-shoe, was between the hands of the doctor, and held close to his person. The light was consequently utterly extinguished to the eye of the observer. *All this made it a threefold impossibility for her to see any luminous emanations from the magnet.* Three enormous blunders were therefore made, each of which was alone sufficient to render impossible the sensitive perception of light. In fact, the girl saw nothing but scattered and uncertain gleams of light, here and there, no doubt from a hand, a head, or two heads laid together, or hands and heads in contact, or hands touching the magnet and strengthened by it, or eyes, or breath, or the pits of stomachs, or the knees of the doctors sitting close together, &c. &c., where these things accidentally met and mutually strengthened each other's odylic power. Although, in my work, I had fully and minutely detailed how much light was emitted by all these living organs, yet, for the most part, no attention whatever was paid to this by the experimenters. They shifted and hid the magnet in all directions, and when she saw a light, it occurred to none of them that they had faces, eyes, hands, hypochondria, &c. &c., all of which emit light, and, in general, more strongly than the magnet. In this utter confusion of all the conditions of scientific research, the poor tormented girl, as may be easily understood, knew not what to do, or how to satisfy the demands with which she was overwhelmed; and when the end of such ill-made experiments was incongruous answers, the gentlemen, instead of confessing their own ignorance, had no hesitation in protecting themselves by the unconscientious and calumnious assertion, that the sensitive patient was an impostor. They were not ashamed to adorn this barefaced assertion by to use the words of Dr. FARADAY, appear in any odylic phenomena. In order to learn, if possible, something on this point, I made an experiment with the magnetic curves, obtained by means of iron filings. I laid on the floor of a darkened room a twenty-five and half-inch bar magnet, and covered it with a plate of glass of fifty-two inches, or more than twice as long. On this I strewed iron filings, and tapped it gently with my finger on all sides, so as to produce the well-

stories, which bear on their face the obvious marks of improbability. Mlle. REICHEL, for example, is said to have secretly crept towards the magnet held by one of the doctors, and to have felt for it with her hands, in order to discover its position. This story carries its own contradiction with it. For, *if she saw no light*, she could not find the magnet, because it might be held above the head, or under the feet, or to the side of, or behind, the person who had it. She could not therefore know even where to feel for it, but was sure to lay hold of the experimenter, on his body, head, or feet, or to thrust her hand into his eyes. This accusation is therefore absurd in itself, and it is in vain to try, by means of such absurdities, to throw dust in the eyes of a reflecting reader. The poor girl herself was moved to tears, when I read to her this passage from the journal, and could find no words to express the pain it gave her, or her indignation at such a calumny. The inaccuracy, obvious in these experiments, cannot be imputed to the calumniated girl, but only to her calumniators, and demands a serious rebuke. No one has a right to set himself up as a judge in a matter in which, at the same time, he so thoroughly exposes his own ignorance. But to cloak this ignorance at the expense of a defenceless girl, and by false accusations, is a line of conduct which can excite only abhorrence and indignation.—R.

It is painful to think that parallel cases have not been wanting in England. The spontaneous somnambulism, and apparent transference of the senses, in Miss M'AVOY, met with precisely similar treatment; as did the very interesting facts which occurred in the case of Dr. ELLIOTSON's patients, the OKEYS. There was the same predetermination to find the patient an impostor, the same utter absence of all cogency in the evidence adduced, and the same rash and unjustifiable, as well as unmanly, accusation of imposture, brought against persons of whom no evil was known, apparently because the authorities chose to assume the facts to be impossible. The still more recent case of Miss MARTINEAU's servant girl is another instance in point. Having seen that girl, and made observations on her, I can speak with confidence of her honesty and truthfulness. It is the duty of every lover of truth and of science, to protest energetically against the system of reckless accusations of imposture preferred against persons of blameless character, because their statements appear to us incredible, or, as has often happened, because we are unable, from want of knowledge of the rules of scientific research, to form a clear distinction in our minds between what is real, and what may be imaginary or delusive, in the results obtained.—W. G.

known figures. I now introduced Mlle. REICHEL, who had seen, and knew, nothing of the arrangements. She saw no longer any flame from the bars, but the magnetic curves appeared brilliantly illuminated. Every particle was luminous, and, connecting itself with the others, formed luminous lines, precisely in the direction of the magnetic curves. The whole image sparkled with all the hues of the rainbow at every point, and drew from her the exclamation, that she had never in her life seen any thing so charming.

452. Two years later, I repeated the experiment with the healthy sensitive Mlle. ZINKEL in the properly prepared dark chamber. The result was, the same description, and the same delighted surprise at the beauty of the sight. She saw on the glass plate millions of little brilliant stars, all arranged in curve lines. She testified the greatest pleasure when, by gently tapping the plate, I caused the stars to move and leap about. The whole of the northward half had a predominating blue light, beautifully variegated with all other colours; on the southward half an equally variegated and beautiful red light prevailed. She added, that the whole circumference of the plate *was set in a luminous border*, also exhibiting all the rainbow colours. It was narrow, only from one-eighth to one-twelfth of an inch broad, and in it the colours formed parallel threads. Neither the discased nor the healthy sensitive saw any other curves whatever, but the known magnetic curves. Nothing, therefore, had happened beyond this, that each particle of the filings had been, by induction, converted into a magnet which gave out odylic flame.

453. In concluding these details concerning the odylic flame, I shall make one more practical application. It is a fable, widely spread in Germany, and which has been often made, by our dramatic poets, the ground-work of the most striking scenes, that ghosts, witches, and devils, assemble for their hellish dance by night on the Blocksberg. Everything in the world, even such a fable as this, has a cause or origin in nature; and we can now see that this myth is not destitute of a natural foundation. It has long been known, that high on the Brocken, there are rocky summits which are strongly mag-

netic, and cause the needle to deviate. More minute investigation has proved, that these rocks contain disseminated magnetic iron ore or lead-stone; as on the Ilsenstein, the Schnarcher (Snorer), &c. The necessary consequence is, that they send out odylic flames. Now when persons of high perceptive powers for odylic light happened to come on such places in a dark night, as must often have been the case with hunters, charcoal-burners, poachers, wood-cutters, &c. they necessarily saw, on all sides, delicate flames of different sizes and colours, flaming up from the rocks, and in the currents of air flickering hither and thither. Who could blame these persons, imbued no doubt with the superstitious feelings of their age, if they saw, under these circumstances, the devil dancing with his whole train of ghosts, demons, and witches? The revels of the Walpurgisnacht (the night which ushers in May-day), must now, alas! vanish, and give place to the sobrieties of science;—science, which, with her torch, dissipates one by one all the beautiful but dim forms evoked by phantasy.

453. *b.* A condensed retrospect of the observations in the flame teaches us:

a. Feebly magnetic steel exhibits glow without flame. When the intensity of the magnetism rises beyond a certain point, luminous emanations are added, which appear first nebulous, then vaporous, then as flames, more especially at the poles of magnets, and frequently appear, to the eyes even of healthy persons, five or six feet high.

b. The magnetism of the earth has great influence on the size and colour of the odylic flames. According as magnets are placed with one or the other pole towards north, west, south, or east, upwards or downwards, in the magnetic inclination, or in any other position, the flames in each case have a more or less different aspect.

c. Bars of soft iron, under the influence of the earth's magnetism, behave like weak magnets.

d. In compound magnets, the odylic flames are found in colour, alternately stratified, just as the glow is.

e. The odylic flame has a tendency to rise in the air.

f. Magnetic flames, from unlike poles, when placed opposite each other, and gradually brought nearer, exhibit little or no mutual attraction, but mutually force each other back, collecting each round his own pole, and become inverted. On contact of the poles the inverted flames disappear, and a more delicate or less luminous flame appears, that from each pole enveloping the opposite pole.

g. Flames which cross each other act so, that one is carried along by the superior force of the other; the stronger being that which, at the crossing, is nearest its source.

h. The flame follows the movements of the air.

i. All the manifold effects produced on one magnet by another, are either directly mirrored in the associated phenomena of odylic light, or else, peculiarities in these luminous phenomena are observed. The stroking of magnets fnrnishes numerous examples of this.

k. The same occurs, when an armature is regarded as an induction magnet, and is moved in different directions along its surface.

l. In such experiments, the divergence between odyle and magnetism not unfrequently reaches a point of such contrast, that + O and — M are present at the same time in one and the same pole of a magnet.

m. Electro-magnets exhibit the odylic flames in the same way as ordinary steel magnets.

n. The electric atmosphere strengthens the flame, and, in some circumstances, reverses the odylic, but not the magnetic poles.

o. Heat enfeebles the odylic flame.

p. Crystals and animals (the human hand) act on the odylic flame as magnets do; they strengthen or weaken it, reverse its colours, or extinguish it, both on contact, and merely by a near approach.

q. The magnetic curves, produced in iron filings when over a bar magnet, exhibit to the sensitive eye a starry host of miniature magnets, giving out odylic flames, and glittering in rainbow hues.

r. The odylic flame is a material something; most probably a body rendered luminous; but it is not magnetism.

III. ODYLIC THREADS, ODYLIC FIBRES, ODYLIC DOWN.

454. Even from the earlier observations of Mlles. REICHEL, NOWOTNY, and MAIX, related in Part I., we know that magnets, besides the polar flames at their edges and solid angles, emit lights *in the form of a fibrous down.* These observations have since been confirmed by those often repeated of Baron VON OBERLAENDER, who described the lateral flames as fibres and bundles;—of Mlle. ATZMANNSDORFER, who saw, both in a nine-bar and in a seven-bar horse-shoe, the space between the limbs full of fibrous flames, and the whole surface of the magnets clothed in a fine fiery down; both of which she often described, without my having in any way drawn her attention to them;—of Mlle. ZINKEL, who saw, not only nine and seven-bar horse-shoes, but also a single horse-shoe, while drawn along a five-bar magnet of the same form, covered, within the limbs, with a luminous down, 2.4 inches broad, and externally with one 0.4 inch broad. Besides this, she saw the light on wires conducting odyle, the flames at the ends of wires, or those of plates, discs and balls of iron, when under the induction of magnets, in the first at its edges, in the two last on their whole surface, and many other luminous emanations, assume a fibrous and downy character; to which I shall return with more detail in the proper places. These downy lights were strongest in open horse-shoes; when closed, they sometimes disappeared, but most commonly became only smaller and less luminous:—of Mlle. WINTER, who described a three-bar horse-shoe as on all sides clothed in a delicate luminous down:—of Dr. NIED, who saw a simple horse-shoe, when closed, enveloped in a luminous down, and so also with one of seven bars:—of Mlle. PAUER, who saw iron discs and balls, under induced magnetism, as Mlle. ZINKEL had done, clothed in a luminous downy nebula:—of Mlle. WEIGAND, who observed a small but powerful horse-shoe, when lying on her hand, at night, to be enveloped in fiery down:—of Mme. BAUER, who saw, surrounded with luminous down, all magnets, from the smallest pocket horse-shoe to that of nine bars:

—of the young STEPHAN KOLLAR, of the aged SEBASTIAN ZINKEL, and of Mlle. DORFAR, who observed, in all horse-shoes, the downy flame, as it were, licking the surface, both within and without, like tongues, or playing backwards and forwards over it:—lastly, of Mme. KIENESBERGER, who, especially during the catamenia, saw all horse-shoes surrounded with a luminous fibrous down, feeblest at the curve, and sometimes imperceptible there; but increasing in size and brightness towards the poles, just as Mlle. REICHEL had formerly described it.

455. These delicate emanations also exhibited colours. Mlle. ZINKEL saw the down between the limbs of simple horse-shoes, red on the one side and blue on the other, playing into one another, so that the interior space had a variegated appearance. With compound horse-shoes, this was still more strongly marked, because each limb, from the stratification of colours in it, had both red and blue down, extending, within and without, from the poles to the curve.

456. This luminous down was very vivid along bar magnets, *when placed in the electric atmosphere.* Single bars, straight or horse-shoe form, were clothed throughout in down one and a half to two inches long. The space between the limbs of horse-shoes was quite filled with it, and had the appearance of masses of luminous threads, variegated in red and blue, sometimes even in rainbow colours.

457. These emanations are always stronger from edges and solid angles than from flat surfaces. In bar magnets, the down is arranged symmetrically on both ends; it is almost null at the axis or middle point of the bar, and increases towards the poles, where it is strongest at their extremity, not at the magnetic foci which lie about a seventh of the half length of the bar within the poles. In horse-shoes it is not symmetrically arranged, and was seen by Mme. BAUER, Mme. KIENESBERGER, and Mlles. ATZMANNSDORFER and ZINKEL, much stronger between the limbs externally, so that the fibrous light often filled nearly or entirely the whole middle space with a fiery down. M. PAUER only saw it distinctly between the limbs.

458. When I *attached the armature* to horse-shoes, which, when open, were clothed in the downy light, it instantly diminished, but did not disappear. According to Mlles. ATZMANNSDORFER and ZINKEL, it became less than half of its former length and size, and especially duller and more turbid. Where it had been yellow or red, it was now greyish yellow or quite colourless and grey. This also happened when I closed a horse-shoe, not with its armature, but with another horse-shoe. Dr. NIED, who saw on a single bar horse-shoe a strong downy light, lost this, and saw only the dull glow of the steel when the armature was attached. His sensitiveness is not great. In all cases, therefore, it would appear that the conduction through the armature is imperfect, and consequently a part of the current of imponderables residing in the magnets continually passed off into the air.—From all these observations we may be more than sufficiently satisfied of the accuracy of the original observations of Mlle. REICHEL (the impostor of the Vienna Doctors, W. G.) on this point, as described in Part I.

459. These downy envelopes are very probably constant but weak odylic emanations, extending over the whole magnet, which, from the low intensity of their light, are generally visible only to the most highly sensitive, hardly or not at all visible to those of lower sensitiveness.

460. Almost all the most sensitive observers described also *scattered coloured threads of fire, which they saw in the odylic flame.* I have spoken of them before, as seen by Mlle. ATZMANNSDORFER and others. Mme. BAUER also described them to me. Baroness VON AUGUSTIN often saw them rising unto the flame of the nine-bar and of a five-bar horse-shoe, as well as in that of the electro-magnet. I studied this point most minutely with a healthy subject, Mlle. ZINKEL. In the nine-bar horse-shoe, she saw many single threads like strips of fire, of greater brightness, rising from the poles into the flame. They had the thickness of a knitting needle, and did not actually originate in the steel of the poles, but in the flame above it, where they became distinct. They were always blue or red. They flowed on steadily through the flame, especially

the upper part of it. She saw them most distinctly when observing the stratified alternation in the colours of the odylic flames over the plates of the nine-bar horse-shoe, as mentioned in § 396. Here she clearly distinguished that from the plates, in red odylic glow, red, and only red threads arose, and from the blue glowing plates, blue, and only blue threads, ascended into the flame, which had always the same colour.

460. I do not regard these appearances as of a peculiar kind, but rather consider them as of the same nature with odylic flame generally. The cause of them probably lies in inequalities in the efflux from the minutest points of the surface of the magnets. Since we know that edges and solid angles give rise to stronger emanations, I think we are justified in concluding, that finer inequalities may serve as points of efflux, and thus give rise to the formation of threads, fibres, and a downy form of light. They are locally concentrated currents of odyle in the general odylic current, blue from the negative, and red from the positive pole.

IV. ODYLIC SMOKE.

462. The next phenomenon, after the odylic flame, is a peculiar luminous vapour or smoke which rises from the magnet, and is seen in the dark by sensitives. I have said little of it in the earlier part of this work, but as, in pursuing my researches, it was unavoidably forced on my attention, as appearing always in the same way and under similar circumstances, I was compelled to regard it as an essential part of the phenomena, and to direct my attention to it.

We shall first hear the testimony of the numerous witnesses whom I have examined on this point, and afterwards make a comparative examination of their statements.

463. And first, those of healthy sensitives.

M. EDUARD HUETTER saw, over a pocket horse-shoe, a feeble nebulous gleam, which, when he moved the magnet about in the dark, moved with it, and therefore belonged certainly to the magnet. Baroness PAULINE VON NATORP saw luminous nebulæ only on the northward poles of five-bar and seven-bar horse-shoes. On the nine-bar magnet she saw traces of a luminous vapour, and from a very strong electro-magnet she perceived grey clouds rising to the height of a hand. Prof. RAGSKY saw the northward pole of a simple horse-shoe giving out, with intermissions, a bluish, feeble, vaporous light. Dr. HUSS of Stockholm, body physician to the King of Sweden, saw over a strong electro-magnet abundant smoke rising in the form of clouds. M. TIRKA observed that the nine-bar magnet was clothed in a nebulous light, casting a feeble illumination around. M. PAUER saw all large horse-shoes exhaling luminous vapour. In a strong electro-magnet he saw a luminous nebula over each pole. M. HOCHSTETTER saw, over the nine-bar magnet, smoke upwards of forty inches high; and when the magnet was placed in the electrical atmosphere, he saw above the flame, smoke rising five or six feet, and casting light on the ceiling. M. SEBASTIAN ZINKEL saw, over the flame of a simple horse-shoe, smoke three times as high as the flame. M. FERNOLENDT saw, above several horse-shoes when in the electrical atmosphere of the positively electrified conductor, a luminous turbid smoke rising to the roof. Mme. FENZL saw, over the flame of the nine-bar magnet, at different times, nebulous light five or six feet high, resembling thin smoke. Baroness VON TESSEDIK and the young STEPHAN KOLLAR saw bars, horse-shoes, and the electro-magnet, covered with vaporous lights, sometimes only on one pole, sometimes on both. In the electro-magnet it rose whirling in the form of a cloud. M. KOTSCHY saw, above the nine-bar magnet, a wide-spreading light like a vapour floating, about forty inches high, and feebly luminous. Baroness VON VARADY and Mme. VON PEICHICH saw the same. JOHANN KLAIBER saw, above the flame of a three-bar horse-shoe, a luminous smoke gradually losing itself above in the air. M. ANSCHUETZ saw vaporous lights playing like tongues over a three-bar horse-shoe, and alternately appearing and disappearing; they were only on the northward pole, sometimes paler, sometimes brighter. At another time he saw unipolar luminous vapours over several horse-shoes, which did not intermit. On the nine-bar magnet he saw, in imperfect darkness in his own house, a ball of luminous vapour undulating over one pole. M. DELHEZ perceived, over the flames of a large electro-magnet, a column of grey smoke rising to the roof, and there causing an illuminated space larger than was formed by the nine-bar magnet. The smoke was checked by the ceiling, and visibly flowed along its surface. Mlle. ERNESTINE ANSCHUETZ saw the vapour over the nine-bar and a five-bar horse-shoe most distinctly, when I brought near to them the unlike poles of two small bar-magnets. M. NIKOLAUS RABE saw every flame from a negative pole end in a luminous vapour; in the nine-bar magnet this vapour was five feet high. On a large bar-magnet, four feet long, he saw flames, that on the northward pole twenty inches, that on the southward pole twelve inches long; both flames passed into smoke, light and thin on the northward pole, denser and heavier on the southward pole. Mlle. PAUER saw, on the nine-bar magnet, a flame from twelve to sixteen inches high, surrounded above with a smoky light which extended nearly to the ceiling, which it reached entirely, spreading out on it, when the magnet was in the electric atmosphere. Baron VON OBERLAENDER saw many of the magnetic emanations like fine vapour, and the flame of a large nine-bar horse-shoe seemed to him above, where it nearly touched the roof, to pass into a kind of fine smoke. Prof. ENDLICHER observed, over the flames of a strong electro-magnet, which were forty inches high, a smoky, dull column of vapour, which rose to the ceiling, flowed horizontally along it, and illuminated it. Mlle. GLASER saw the same bar-magnet alone, but still more when in the electric atmosphere, as well as large bars, sending out over the flame, smoke up to the ceiling. So also with the electro-magnet.—Mme. BAUER saw this in a still higher degree. On all magnets she saw smoke over the flame, in strong magnets sometimes variegated in colour, iridescent, and exhibiting sparks. It was always stronger, thicker, and duller over red flames, fine, thinner, and lighter over blue flames. Dr. NIED observed vaporous exhalations over every flame from any form of magnet. They were always stronger on the southward than on the northward pole, especially in horse-shoes of one and of seven bars, in the latter case as long as an arm. From the nine-bar magnet he saw a column of vapour rise up to the roof, on which it cast a light.—Baroness VON AUGUSTIN saw over the nine-bar horse-shoe, and still better over the electro-magnet, smoke ascending in heavy clouds to the ceiling.—Mlle. ZINKEL saw, in a series of experiments, to enumerate which would be tedious, on every magnet of any power, the red flame passing into a thick, heavy, feebly luminous, reddish yellow smoke, and the blue flame ending in a fine, etherial, greyish blue vapour. On smaller and simple horse-shoes, these cloudy emanations were from four inches to a foot or more in length; on the nine-bar magnet often as long as an arm, and when it had been strengthened, six and a half feet long. Magnets of great intensity appeared to her, especially during the catamenia, covered near the poles with a thin vapour. Above the stratified flames of compound horse-shoes, she saw the smoke rise to the length of an arm. During the stroking of one horse-shoe with another, she saw a single bar horse-shoe pouring out thick clouds of smoke, especially at its southward pole. In most cases she saw only a blue flame on the northward pole, and no flame on the southward pole, but only heavy smoke. When the curve of the single bar lay on the poles of the five-bar horse-shoe, whereby the flames of the former became twice as large as before, the smoke at the positive and the vapour at the negative pole always increased greatly. But when the single bar adhered to the five-bar magnet, at two-thirds of its own length, so as to attract no longer its own armature, and its poles therefore were magnetically indifferent, there appeared a small blue flame at the negative pole, but at the positive pole only reddish smoke. A large electro-magnet, of horse-shoe form, had on both poles masses of smoke, rolling rapidly upwards into the air, in which the observer could always distinguish one mass of cloud or smoke from another, as KLAIBER and others often did likewise. In the electrical atmosphere she saw smoke pouring in masses from the nine-bar magnet (§ 436); and this occurred even when a near approach to the conductor had reversed the odylic poles. These masses of smoke rose to the ceiling, and sometimes enabled her to distinguish the lines of the design painted there, by the light they shed. They were always more abundant, especially in the electro-magnet, over the positive than over the negative pole. She sometimes saw no flames on small bars or four-inch needles, but a blue vapour over the negative pole. In bars of twenty-four inches and of forty-eight inches, the flames passed into smoke, thinner and bluish on the negative, thicker and yellowish red on the positive pole. If I placed on the bars caps of different forms, so that the poles ended in two, three, or four points, each of these, over its flame, exhibited a corresponding column of smoke. When I brought near to each other the unlike poles of two eight-inch bars, she saw, at a certain distance, the same apparent tendency to unite in the smokes as in the flames; but on a nearer approach, the smokes retreated with their flames, collected with them round the poles, and became inverted, as I have explained of the flames. The same thing happened with two horse-shoes. If they lay on the table (§ 394), the smoke appeared between them at the ends of the flames. When the poles were so close that the flames were inverted, the smoke no longer appeared before the poles, but flowed to and beyond the curve, as already explained in that paragraph. The course of the smoke, in size and brightness, was always parallel to that of the flames.

464. Now let us turn to the delicate and diseased sensitives.

Mlle. Dorfer saw, in various magnets, the flame passing into smoke. Mlle. Winter saw, in a three-bar and in a five-bar horse-shoe, emanations of vapour, not only on the poles, but over all the surface. Mlle. Weigand saw a small pocket horse-shoe, beyond the luminous down, enveloped in luminous vapour. Mlle. von Weigelsberg saw the same magnet giving out at its poles unsteady nebulous lights, from 1.2 to 1.6 inches long, longer on one pole than on the other. This light seemed to grow and to fade. At another time, she saw a nebulous light from all magnets, especially at the poles. Mme. Johanna Anschuetz saw a luminous vapour rising from the poles of a five-bar horse-shoe, the height of a hand; from the nine-bar horse-shoe, twice as high. She also saw nebulous light moving between the limbs and on the poles of a large simple horse-shoe. Mme. KIENESBERGER saw, over all compound horse-shoes, the flame passing into vapour, which rose, and gradually lost itself. This was most distinct over the nine-bar magnet, where the vapour was five or six feet high. She even distinguished between the lower and the upper parts of these emanations, the lower part being more fine and ethereal, the upper more smoky, but gradually growing thinner as it rose, and disappearing at last. No doubt she saw better, in the lower part, the cloud from the negative pole, which all describe as shorter, and over it that of the positive pole, which rose over the other, and reached even the ceiling. Her account of the vapour and smoke over a large electro-magnet agreed with that of Mlle. ZINKEL so exactly that I may omit its further description; but she only saw the vapours of such size as to reach the ceiling, when a strong SMEE'S battery was employed. With a less powerful current, both the flame and smoke were shorter, the former only a foot, or upwards, in length. Even the earth's magnetism sufficed to produce similar effects. A bar of soft iron, held in the middle by a wooden support, and brought into the meridian, showed at both poles small flames and short smoke. When now placed in the line of dip or inclination, the remarkable fact was again observed, that the magnetic and odylic polarities became, at each pole, opposed, the latter being reversed, the former unaffected. A reddish grey flame appeared at the negative pole, which pointed, of course, downwards, and a bluish flame at the positive pole. The vapour on both had the same colours; the reddish grey smoke from the lower pole reached the floor, two feet distant, and there spread in all directions. FRIEDRICH WEIDLICH at first did not see the numerous magnets I placed before him; but after having been more than an hour in the dark, he clearly saw both the flames, and the smoke in which they ended. The flame passed so gradually into the smoke, that it was difficult, at any spot, to distinguish between them. He saw a peculiarly strong smoke from a freshly magnetised and powerful five-bar horse-shoe. From the flame of the nine-bar magnet, which was five or six feet high, he saw thick reddish clouds rolling up in separate masses or cumuli, with a continual revolving motion, and reaching the ceiling. When I blew into this, it was displaced, but soon recovered its position. In another series of experiments with the nine-bar magnet, he again saw the smoke rolling up to the ceiling in heavy masses which followed one another. JOHANNA KYNAST saw luminous smoke five or six feet high over the flame of the nine-bar magnet. Mlle. ATZMANNSDORFER saw this smoke on so many occasions, that I ceased to make notes of it at last. I shall only state, that when her vision was in any way rendered less acute, or when she had not been long enough in the dark, she always perceived the smoke before she saw the flame. As her vision improved, the smoke appeared to give place to the flame, and to assume its usual position over it, and as the flame became brighter, the smoke became apparently paler. She often saw the smoke from the nine-bar magnet reach and spread over the ceiling, which it illuminated for some minutes.

465. Odylic smoke is given out, not only by permanent magnets and induced magnets of all kinds, but also by unmagnetic bodies, such as copper or silver, when they become conductors of odyle. When I held a hollow roll of copper wire over the pole of a magnet, Mlles. ZINKEL and GLASER and Mme. KIENESBERGER saw the flame which appeared at the end of the wire ending in thick smoke, which rose from it.

466. The general tendency of the odylic flame to rise has been already mentioned. The same is true of the odylic smoke. When I laid a magnet so that one of its poles, or both, in the case of a horse-shoe, projected over the edge of the table, Mlles. ATZMANNSDORFER, ZINKEL, and REICHEL, saw the flame proceeding horizontally outwards, then curving upwards so that the smoke at last rose vertically.

467. In the presence of Mlle. ATZMANNSDORFER, and in the dark, I drew lines with phosphorus on paper, and showed her the luminous vapour arising from it, and also the effect of blowing on it. She assured me, that it had the greatest resemblance to the odylic smoke, except in the intensity of its light, that of the odylic smoke being beyond comparison paler and feebler; it was also not so green, but more blue and reddish.

468. *How feeble the light* of the odylic smoke must be, is shewn by some experiments I made on the reflection of odylic light by mirrors. Neither Mlle. ATZMANNSDORFER nor Mlle. ZINKEL could see the image of the smoke. Both were in their ordinary state. Only Mme. BAUER, during pregnancy, was able to perceive it. In the case of the two girls, the loss of light by absorption had sufficed to render the image invisible to them.

469. The odylic smoke appears stronger in an exhausted receiver. This was seen by many sensitives; among others, by Mlle. PAUER, Mlle. ZINKEL, M. HOCHSTETTER, Mlle. GLASER, &c. We shall return to this in § 480.

470. In the early part of my researches, both Mlle. STURMANN and Mlle. REICHIEL told me much about the smoke which rose from magnets and other bodies giving out odylic light. But I was absorbed at that time by so great a number of other surprising phenomena, which met me at every step, that I had no ear for these statements, which I regarded as variable, and thus I neglected them. It was only afterwards, when I found all sensitives, including the healthy, repeating these statements so decidedly and unanimously,* that

* Here, then, in addition to all the proofs already given, we have the testimony of a dozen of new witnesses to the exactness of all the statements of Mlle. REICHEL. It would be utterly inconceivable how it was possible for the doctors to come to the monstrous assertion, that Mlle. REICHEL had never seen any magnetic light whatever, and was therefore an impostor, if a glance at their protocol did not show the reader how their experiments absolutely swarm with mistakes and blunders. One of these, and not the least important, was that, among other things, there were always ten or fifteen young men crowded round the sensitive in a small room. He who has the slightest knowledge of the subject is well aware that human beings have a very strong mutual odylic influence on each other; and in my work it had been very minutely explained, that a human body is a constant source of magnetic, or, more correctly, odylic force, radiating on all sides from it. Under such circumstances, how could a dozen of doctors and professors expect or insist that an ignorant girl, driven in the midst of them, should find her way among the confused influences acting on her from all sides, and give clear and scientifically available answers to questions, which they themselves did not understand how to put rightly? Every single person is a far stronger source of odyle than a magnet is. The close proximity of one man is, in many cases, sufficient, on the one hand, to annihilate the perceptive power of the observer; and, on the other, to extinguish the magnetic light. When I make experiments on this light in the dark, the first thing I

I felt the necessity of testing them more minutely; and I then recognized their weighty significance in the series of phenomena connected with odylic light.

471. That which is here testified by more than thirty witnesses, not only agrees essentially with itself in all details, but also harmonises with what we already know of the odylic flame, with which the odylic smoke is in the most intimate connection. It therefore bears the obvious impress of truth so certainly, that it can only be denied by those who talk recklessly and illogically, as unfortunately many do, who would fain be

do, before asking a single question, is to remove to a certain distance both from the person and from the object, in order to avoid disturbing the result by the odylic influence of my person, which would confuse them and render them utterly worthless. Instead of observing similar precautions, these gentlemen actually placed a doctor on each side of Mlle. Reichel, and they generally had hold of her hands—an arrangement, besides, which no sensitive can endure—and then placed the magnet in the lap of a doctor opposite, &c. &c. It is impossible to resist smiling when we read of researches made in such a style. The girl, thus tortured, was expected to justify the reckless statements of the ignorant person who had placed her in so false a position, and whom every negative answer from her exposed to shame. She was also expected to satisfy the excited expectations of those assembled, who incessantly irritated her by expressions of abuse and scorn, and by their contemptuous treatment drove her into ebullitions of anger. In this general confusion and restlessness, she was expected to solve with precision the most delicate problems which can be proposed to the senses of touch and sight; and that, under physica.a nd moral conditions, which rendered their solution absolutely impossible !......What could possibly be the result of such a mode of investigation, but the pitiful hotch-potch presented by the answers of the persecuted girl in the protocol! In such circumstances, it is not worth while to enter on the individual statements, whether they be truly or falsely recorded, understood or not understood, by the investigators. The whole resolves itself into a hopeless mass of confusion and misapprehension.

Mlle. Reichel was, in her time, (that is, in her then state of health,) an admirable sensitive; the best that could be desired for scientific purposes. The senses of touch and sight were wonderfully acute; she was very obliging, persevering, exact, and truthful in her statements, very modest and humble in her behaviour, and intelligent in apprehending questions rightly framed. But we must not trample with jack-boots on so delicate an instrument for delicate researches. Neither did the doctors understand their own object, nor the girl and her helpless conductor understand what they had to do. Science cannot be thus promoted: but folly may be shown in making such experiments, and then it is cloaked by the contemptuous calumniation of a defenceless woman.

called natural philosophers, but who often have no tinge either of logic or of natural science. If we now compare all the above observations, scattered over a period of more than three years, and made by so many different persons, we arrive at the following propositions :

a. All magnets, whether permanent magnets of hardened steel, or induction magnets of soft steel or iron, magnets produced by the induction of the Earth's magnetism, or electro-magnets, give out, in profound darkness, along with the odylic flames, at the poles, a luminous, nebulous, or *smoky thin vapour*, which is also emitted by their surface generally, although much weaker and often imperceptible. The strength of this emanation diminishes from the poles to the axis, where it is relatively small, but still discoverable.

b. The size of this smoky light is directly proportional to that of the associated flame. When the latter is only an inch or two in height, the length of the smoke does not much exceed this. When the flame rises to three or four feet, the smoke extends to five or six feet or more. Its size is also directly proportional to the size, as well as to the intensity, of the magnet from which it flows. Large magnets of small intensity, (such my nine-bar horse-shoe was, when weak,) gave relatively long flames with abundant vapour. It will not be expected that I should produce exact measurements at this early stage of such an investigation.

c. The odylic smoke is *propelled from the magnet with a certain force* which gives to it its first direction, but it exhibits, after that, *a constant tendency* to flow up. When it reaches the ceiling, it spreads out, flows over it, illuminates the painting on it, and lasts for a short time, not exceeding a few minutes. Whatever material substratum may lie at the foundation of this phenomenon, it is either lighter than air, or is in some way repelled from the earth, so as to be compelled to rise.

d. The smoke differs according to the pole which yields it. At the positive southward pole it is reddish grey or yellowish reddish grey, and thicker, inclined to form heavy masses of clouds ; the negative northward pole yields it blue grey and bluish grey, finer, lighter, more etherial. When its intensity falls, more of grey appears in both varieties, and finally it becomes colourless or pure grey. There are exceptional cases, in which the colour at the poles are reversed; as when bars are placed in the line of magnetic inclination, and in some other peculiar cases. In these, the odylic poles are reversed, while the magnetic poles remain as before.

e. Although it is always present over the odylic flames, yet it is sometimes visible without them. This happens in magnets of feeble intensity. In such cases, blue flame is often seen at the northward pole, but no flame on the southward pole, and instead of it, reddish thick smoke. In magnets of still lower intensity, no flame at all is seen, while smoke is observed on one or on both of the poles.

f. The odylic smoke is so far material that it is disturbed by blowing on it, when, after a short pause, new portions rise and arrange themselves as before. It somewhat resembles the luminous vapour arising from the slow combustion of phosphorus at ordinary temperatures, but has a much feebler light.

472. The question, in what way odylic flame and odylic smoke are related together, is an obvious one, but for the present difficult to answer. Whether they are two specifically different things, or the same thing under two varieties of form, is a point which I must here leave undecided. All the researches I have made with persons who saw both at once, lead to the conclusion, that they differ in aspect as an ordinary flame does from the luminous smoke which rises from it, and loses itself in the air. But when I consider, that sensitives of low power only see smoke, where those of higher sensitiveness see flames with smoke above them; that where the former see small flames and little smoke, the latter perceive much larger flames with much more smoke; that the same individuals, in a less sensitive state, see those things which formerly appeared large, now much smaller; that further, when blue flame and grey smoke are seen on the negative pole, only reddish grey smoke appears on the positive pole; that there are magnets which exhibit only smoke at both

poles; and lastly, that sensitives who have remained with me in total darkness for several hours, at first saw only smoke, after an hour flame and smoke, and after several hours large flames and great rolling clouds of smoke on all magnets;—I feel compelled to suspect, that the flame and smoke are perhaps only the same phenomenon, varying on the one hand in degrees of intensity, and on the other, differently perceived according to the sensitiveness of the observers, or in the same observer, according to the more or less perfect developement of his perceptive power, either from variations in his natural state, or from the effect of longer or shorter exposure of the eye to perfect darkness. The first degree of the phenomenon would then be a feebly luminous nebula; the next a more distinct vapour, duller or brighter, and first grey, then on the positive side yellowish reddish grey, on the negative bluish grey, then reddish on the positive, bluish on the negative side; a third or fourth degree would be the appearance of flame covered with vapour, first of the blue, then of the red flame; and lastly, over these, especially over the latter, the abundant rolling upwards of thick clouds of smoke to the ceiling of the room. I say I feel compelled to entertain this suspicion; because I wish to keep all my theoretical views every where quite distinct from the observed facts, which I have recorded from the concordant statements of so many sensitives, and which, at all events, are more certain and safe than any speculation of mine, however simple. So long as we do not know what is the nature of these odylic phenomena in general, (and, as far as I can judge, there is not much prospect of our soon penetrating very deeply into the interior essence of that nature, considering the prejudiced aversion to such enquiries which prevails in the minds of so many natural philosophers,) so long as the whole nature of these beautiful phenomena shall remain mysterious to us, so long shall we be unable to come to a settled decision as to the identity or diversity of their multifarious forms; and we must therefore, for the present, hold fast, in our description or in our apprehension of them, as well as in our nomenclature, to the shapes in which they present themselves to the senses.

V. ODYLIC SCINTILLATIONS.

473. Besides the four forms of odylic light just treated of, the glow, the flames, the fibrous or downy flames, and the smoke, there is another, of less extent indeed, but of lively intensity. This consists in *scattered sparks or scintillations*, which appear in the smoke, and move about separately in it. Mlle. REICHEL first noticed them, and saw them often, not only in the smoke from magnets, with which alone we are at present concerned, but in many other odylic lights from other sources, in treating of which I shall return to the subject. The first signs of them are very slightly indicated in Figs. 1. and 2. Plate I.; in the flame, there very inadequately figured, they were most distinct in the side view.—In this form they were very plainly seen by Baron VON OBERLAENDER, who compared them to scintillations of burning pine charcoal, (chiefly produced by the charcoal of the bark, as is well seen when charcoal burns in oxygen gas.—W. G.)—Several other persons, and particularly Mlle. ATZMANNSDORFER and JOHANN KLAIBER, often compared them to flying fire-flies.—Mlle. GIRTLER called them little scattered stars.—Mlle. WINTER saw them flying about abundantly, and near the wall, rising up with an angular or zig-zag motion. She has often seen them formerly, when attacked by severe nervous fits.—M. DELHEZ saw them in the smoke of the electro-magnet, jumping about separately, scattered and without arrangement.—Prof. Dr. HUSS saw a great number of them above a spherical electro-magnet (see § 587).—Baroness VON AUGUSTIN saw them in the smoke of the nine-bar magnet, and still more in that of the electro-magnet.—Mlle. NOWOTNY saw the greatest number of them proceeding from the magnet, and Mme. KIENESBERGER not only saw them rising nearly to the roof in the smoke of the nine-bar magnet, but also proceeding from the body of the electro-magnet.—FRIEDRICH WEIDLICH and Mlle. STURMANN saw them in the smoke of small as well as large magnets.—Dr. NIED and M. RABE saw them streaming upwards in the vaporous emanations of the nine-bar magnet,

many disappearing by the way, but some reaching the roof. —Mlle. VON WEIGELSBERG, M. ANSCHUETZ and his sister, Mlle. ERNESTINE ANSCHUETZ, compared them to fire-flies rising with the vapour, and then wandering about.—Mme. BAUER saw them rise to the ceiling in the smoke of the nine-bar magnet.—Professor ENDLICHER saw them rise singly with the smoke of a strong electro-magnet, and reach the roof; they flew about irregularly in the smoke, and even out of it, and resembled points of light, larger and smaller, more or less brilliant. Mlle. GLASER saw so many of them in the smoke of the nine-bar magnet, when in the electrical atmosphere, that they seemed to rush upwards in a stream. She also saw many of them in the smoke of a large electro-magnet. Mlle. ZINKEL described them as very luminous points, which moved irregularly upwards with the smoke, sometimes singly, sometimes more or fewer, but never in great numbers; and which sometimes fell down in the smoke, and immediately rose again. Sometimes none were visible for a time, then three, four, or perhaps, at another time, eight to twelve, would appear in different parts, sometimes several being grouped together. It even happened that single sparks fell on the table, or on the arm, or on the bed (when she observed a magnet in bed), and continued visible for some moments before they were extinguished. In a bar magnet, the negative pole of which was placed in the electrical atmosphere of the positively charged conductor, she saw their number much increased, not only on that, but on the opposite distant positive pole. All these persons expressed a lively pleasure at the sight, such as is heard from a company returning home at night through a wood, when they meet with fire-flies, the sight of which suddenly takes the attention of all from other things, to fix it on the beautiful appearance of these insects.

474. In an experiment with the healthy sensitive Mlle. ZINKEL on this subject, I used the nine-bar horse-shoe. Its poles were pointed upwards, and I brought them into different positions, sometimes in the meridian, sometimes at right angles to it in the parallels. In all of these she saw sparks rising,

singly, or by twos or threes, very brilliant and very small. She observed two colours in them, red and blue. The red chiefly flowed from the positive, the blue from the negative pole. But occasionally blue sparks were seen over the former, red over the latter. The cause of this apparent anomaly, which at first was enigmatical to me, is well explained by the observations recorded in § 396. The blue sparks were from the negative plates, the red from the alternating positive plates at the negative pole, formed by reversal of the polarities in the alternate plates. The same thing, reversed of course, occurred at the positive pole. They became abundant in the electric atmosphere, in which all odylo-luminous emanations are increased and intensified. They were more numerous and brighter, the nearer the magnet was to the conductor. See § 436.

475. The finest developement of this phenomenon was seen by Mlle. ZINKEL in a large electro-magnet, excited by the current from a powerful SMEE'S battery. Not only did the sparks fly out on all sides from the large variegated flame, but they formed a shower, or rather a stream, which rose constantly to the ceiling. They were so bright that she could not conceal her astonishment at my not being able also to see them.

476. Finally, something similar was observed when I suddenly detached the armature of a horse-shoe, and this occurred the more strongly, the greater the intensity of its magnetism. At the moment of separation, Mme. BAUER, Mlle. ZINKEL, Mlle. REICHEL, Mlle. DORFER, and others, saw numerous sparks flash forth and instantly disappear; after which the flame began to appear and to develope itself.

477. The occurrence of this phenomenon, established as it is by the unanimous and uniform testimony of so many sensitives, both diseased and healthy, and confirmed by countless repetitions, admits of no doubt. I do not allow myself for the present to form any conjectures as to its nature, or even the relation which it bears to the other phenomena of magnetism or of odyle with which it is associated. I can here only esta-

blish the physical fact, as it actually presents itself to the eyes of the sensitive.

* * * * * *

478. Having now become in some measure acquainted with *the various kinds* of luminous emanation from the magnet, let us turn to *differences in the circumstances* in which the magnet, while emitting them, may be placed, and strive to investigate the influence which such differences may exert.

LIGHT FROM THE MAGNET IN DIFFERENT EXTERNAL CIRCUMSTANCES.

ODYLIC LIGHT IN DIFFERENT MEDIA.

479. We know how different are the luminous phenomena of electricity in air, in vacuo, or even under diminished atmospheric pressure. In order to test the effect of the air on odylo-luminous emanations, I frequently placed magnets *under the exhausted receiver of the air-pump* in the presence of sensitive persons. I used small and large horse-shoes, the poles upwards, standing in large tumblers under the bell-jar, so that every change was easily visible, as well as small bar-magnets, such as could lie horizontally under the bell-jar.

480. Even the blind man BOLLMANN solved the question. When all was ready, but before commencing the exhaustion, I led him to the air-pump. He saw nothing. To be certain that he looked in the right quarter, I placed his hand on the bell-jar, but he still saw nothing. The magnet sent out too little light to affect his vision through the glass. I now began the exhaustion. Very soon, when it was about half completed, he saw light. And as the rarefaction increased, this light also increased, and reached its maximum of size and brightness for his mutilated organs of vision, when the mercury in the gauge stood at 0.12 to 0.16 of an inch, the utmost degree of exhaustion, unfortunately, which could be obtained with my air-pump. When the air was rapidly re-admitted, un-

known to him, he was disagreeably surprised by the sudden extinction of the light and return of darkness. Mlle. KRUEGER observed, after some exhaustion, a small flame only on the negative pole becoming brighter as the exhaustion advanced, but she only saw it distinctly during the action of the piston; when it moved in the opposite direction the flame became pale, and nearly disappeared to her eyes. M. TIRKA, JOHANN KLAIBER, and Mme. KIENESBERGER, also saw nothing at first; but when the air was half removed, they saw the contents of the bell-jar become luminous, the magnet in the odylic glow. On further exhaustion, KLAIBER saw the flame appear on both poles, first dull, then brighter as the air was removed more completely, increasing in vividness at every stroke of the piston, so that at last very bright flames flowed about under the bell-jar. When the air was admitted, all light suddenly disappeared to the three observers, and it returned as soon as the pump had again been worked for a time. M. HOCHSTETTER also could not at first see the magnet under the bell-jar; but when the air was partially extracted, the magnet, the jar, and its whole interior, became luminous, ceasing to be so when the air was admitted. Baroness VON AUGUSTIN did not see the magnet under the full jar; but when I had rarefied the air sensibly, she perceived it, and as I went on, the light increased till the whole jar was filled with light, of which the magnet formed the central point. Mlle. DORFER, after a few strokes of the piston, saw little flames on the poles varying with each motion of the piston-rod. When the rarefaction was great, she saw the flame bent back by the roof of the jar, and flowing down its sides. She compared its form to that of water flowing in a curve out of the mouth of an inclined pitcher. The admission of the air extinguished all light. Mme. JOHANNA ANSCHUETZ and her husband M. ANSCHUETZ, his sister Mlle. ERNESTINE ANSCHUETZ, and Mlle. VON WEIGELSBERG, all of nearly equal sensitiveness, saw, with slight variations of size and brightness, the magnets at every stroke acquire a brighter glow, then flames appearing on the poles, larger on the negative, playing on the roof of the jar and flowing down from it, and all disappearing at the moment when I opened the air cock, and again step by step re-appearing as the exhaustion was renewed. Baron VON OBERLAENDER saw precisely the same kind and order of appearances. The statements of FRIEDRICH WEIDLICH, in various experiments made at different and distant times, also exactly coincide with the preceding. Mlle. PAUER, Mme. BAUER, and Mlle. GLASER, were all, at first, unable to see a bar magnet under the jar; but it was seen in odylic glow, when half the air had been extracted. On further exhaustion, first flame, then smoke appeared, filling the whole jar; the jar itself and its knob acquired at last odylic glow. The two latter of these observers saw the flame blue on the northward, yellowish red on the southward pole, as in the open air, and it rose obliquely upwards along the sides of the jar. On admitting the air, all light disappeared. These experiments were often repeated at different times. Mlle. ZINKEL added, that when the experiment lasted long, the whole vacuum was filled with odylic vapour, and the jar itself became luminous. Its knob acquired a white glow, which lasted for a time after the air had been admitted. I showed the experiment for the first time to Mlle. ATZMANNSDORFER, when she was in the state of somnambulism (spontaneous), in which I do not generally or willingly try experiments on odylic light. She described the succession of the phenomena in complete accordance with other sensitives;* after some strokes, increasing glow, flame rising to the roof of the jar, illumination of the whole jar, and sudden extinction when the air was re-admitted. She added, that the glowing steel was transparent, almost like glass; a statement which we have formerly met with. Two months later, when in her ordinary state, and entirely ignorant of the former experiments, I repeated the same trial several times. But she gave always the same account; and she, like Mlle. ZINKEL, always saw the jar and its knob glowing at last. She saw fine flames between the limbs of the magnet, filling

* This result proves, as some former ones also do, that there is no essential objection to observations made in the sleep-waking state in themselves. Perhaps the Author has found it difficult, in that state, to obtain answers to his questions.—W. G.

the whole space. The luminous down from its outside filled the jar to its sides. The flames were bluish and reddish as usual, but mixed with rainbow colours. In regard to the smoke, she added the peculiar remark, that it increased in size and brightness during the first part of the exhaustion, but only up to a certain degree of rarefaction; that it then became duller, diminished, and when the exhaustion was complete, nearly disappeared, while the flame shone beautifully, and flowed down the sides of the jar in variegated colours. The smoke rolled round within the jar as long as it was visible. These last statements, for which I have as yet no confirmation from other observers, and the conclusions from which I therefore reserve, are obviously of great interest, in reference to the distinction between odylic flame and smoke.

481. These observations, collected and compared, yield the following results. *The odylo-luminous phenomena are affected by changes in the pressure of the atmosphere. Under diminished pressure they increase considerably in brightness.* Magnets, invisible in the open air or under the full jar, acquired a bright odylic glow, and very distinctly visible flames on their poles and between their limbs, when the air was half exhausted. These emanations, bluish and reddish on the opposite poles, and mixed with rainbow colours, increased as the exhaustion became more complete. Then the jar became so charged with odyle, that it, as well as the external knob, acquired the glow. The flame did not pass through the glass, but was deflected by it as an ordinary flame would have been, and as Mlle. Reichel often described the odylic flame to have been, as in the glass case of Schweigger's multiplicator, § 434, and by the approach to a large lens, § 20, like fire on the bottom of a pan held over it. The smoke also appears and increases, but only up to a certain point of rarefaction, beyond which it diminishes, and probably disappears in a perfect vacuum. The glowing jar produces no smoke externally.

482. It appears from this, that the pressure of the air opposes the developement of odylic glow and flame, and that these increase as the pressure is diminished or removed. In this respect, both shew analogy with electric light, but it is

analogy only, not identity; odylic smoke appears only to derive advantage from rarefaction of the air to a certain point, but not to exist at all when air is absent, and consequently its appearance depends on the presence of air. The sensitives did not speak of the occurrence of sparks in the exhausted receiver. The glass of the jar seems to be a means of confining odylic lights, especially flame and smoke, as it turns them back; while the odyle itself penetrates into it, and excites in it the odylic glow.

483. As a medium presenting *a greater density* than air, I selected *water*. I showed to the following four sensitives a small and freshly-magnetised horse-shoe, whose flames flowed brightly in the dark, alternately in air and in water at different times. Mme. KIENESBERGER saw the flame in air 0.8 inch long. When I immersed it in water, the flame and smoke quickly vanished, but the glow remained, and the steel lay in the water a luminous mass. She also spoke of a bright point of light remaining at one pole. FREIDRICH WEIDLICH saw the flame in air two inches long. I then sank the magnet, lying in a glass basin, into water. The flame instantly disappeared, but he saw the magnet glowing and translucent almost like the glass itself. But he declared that the flame had not entirely vanished, but that one small point, as if on a corner, was left. As often as the magnet was taken out of the water, it exhibited, although dripping wet, the flame two inches long. The same appearances were observed at every repetition. Mlle. ATZMANNSDORFER, in whose presence I made the same experiment, saw the magnet instantly lose its flames, but retain its glow brightest towards the poles. Mlle. ZINKEL laid in water a horse-shoe with both poles pointing northward. Flame and smoke disappeared, and they returned as soon as it was taken out of the water, and while it was still wet. The glow continued in the water without diminution. A small point at one pole remained brighter, like a residue of flame. This point was on the side next the opposite pole, and from it there proceeded an exceedingly slender thread of light, which was found to be the cross edge of the pole, next to the other pole. I marked, in the dark, the pole on which this spot appeared.

When brought to the light, it appeared, contrary to my expectation, to be the *positive pole.* It was no doubt the same on which Mme. KIENESBERGER and WEIDLICH had seen a similar bright spot. In many repetitions of this experiment, Mlle. ZINKEL saw, especially during the catamania, traces of similar light, but less bright and visible, at the negative pole. They were less striking, because they were blue and greyish blue, with low intensity, while those on the positive pole were reddish yellow, and sometimes red like glowing charcoal.

484. A denser medium, therefore, such as water, destroys the flame and smoke, whether it absorb them and become thereby magnetised, or rather odylised water, or whether it prevents their developement farther than as a bright point and a bright thread on the inner corner and edge of a horse-shoe pole. Mlle. NOWOTNY had already pointed out such a thread as the lowest degree of odylic flame. § 3.

485. I have not tried solid bodies, as media of another kind. It would be well, in this trial, to use an induction magnet hermetically inclosed in glass of considerable thickness, or rather placed in close contact by melting the glass round it. It would be necessary, if a permanent magnet were used, to restore its magnetism from without, as the heat of melting glass would demagnetise it. But there would be no great difficulty in doing this. I have made some experiments with copper wire approaching to this, but essentially different. I wished to try whether the odylic charge which, as we know from § 45, can be given to other bodies by magnets, and which is recognised by its action on the nerves of touch, might not also be rendered sensible by luminous emanations, and thus lead to further insight into this difficult matter. I made with this view an irregular coil of copper-wire one-twelfth of an inch thick, of ten or fifteen turns, so as to form a kind of net or basket, which I pressed flat, and laid on the northward pole of the nine-bar magnet standing vertically. It fitted pretty closely on the pole, and the end of the wire projected about eight inches laterally towards the east. When this was done in the dark chamber before Mlle. ZINKEL, whose eyes were well prepared from her having been

long in the dark, the flame, formerly eight inches high, shrank together, and did not extend beyond the net of wire, which visibly absorbed it. The wire rapidly increased in intensity of glow, becoming brighter and as if transparent, and after a few seconds, this increased so much that it cast a bright light round it, from a luminous atmosphere, or nebulous veil, which covered it to the depth of nearly an inch. The wire was seen glowing within this, and immediately afterwards a bright flame flowed from the end of the wire, much more intensely luminous than the blue flame of the magnet had been, so that it illuminated the floor to more than forty inches. It was eight inches long, and showed a fine fibrous structure, such as was often seen in the flame of the magnet, and which had therefore been communicated to it by the magnet. The magnet lost none of its glow, but only its blue flame. I made a similar experiment, with the same results, with Mlle. Atzmannsdorfer, but as I do not find it written down in the journal of experiments made with her, I cannot give further details of it. The experiments with Mme. Kienesberger, however, are minutely recorded. I brought one end of a long wire, from a room in which an assistant was, under double doors and carpets, into the dark room, and placed the sensitive before it. I now directed the assistant to lay the negative pole of a five-bar horse-shoe on the other end of the wire. After about a minute, the glow of the wire began to increase, and it became gradually brighter till, after four or five minutes, it reached its maximum, and appeared transparent. There also appeared here and there along the wire, single bright nebulous points, almost like sparks, but less vivid, larger, and permanent. Blowing on them extinguished them for a moment, but they immediately returned. On the point of the wire appeared a similar blue spot of light, larger than the others, and not so easily extinguished by blowing on it. When the opposite pole of the magnet was laid on the end of the wire, the same appearances were seen, but smaller and duller. They correspond entirely to those observed by Mlle. Zinkel, only they were feebler, because the magnet was smaller, and a much smaller surface of wire was in contact with it. The nebulous spots were obviously the commencement of the luminous veil, appearing here and there. We shall find that Mme. KIENESBERGER also saw that appearance fully developed in another experiment. With FRIEDRICH WEIDLICH I had the end of the wire, in the other room, coiled up so as to bring a larger surface in contact with the magnet, the force of which was also increased by induction from the coil. He saw the glow of the wire increase, and a flame eight inches long rise from its end. With the positive pole the flame was shorter, broader, and more turbid; all in correspondence with the other observations above mentioned. With Mme. KIENESBERGER I laid the end of a copper wire, not coiled, on the nine-bar magnet in the dark chamber. At the negative pole the wire, forty inches long, became brighter in glow, shining all round, and a flame, in size, form, and aspect, like that of a wax candle, small, slender, yellow below, blue above, conical, and with a smoke four inches high, appeared at the point. With the positive pole the appearances were similar, only feebler, smaller, and less luminous, the flame red and smoky. The succession of the phenomena in arising was slow, as before, and they also disappeared slowly when the magnet was removed. In a later experiment I used the coiled up wire, as with Mlle. ZINKEL, only making the projecting portion forty inches long. When the coil was put on the northward pole of the magnet, the blue flame instantly sank, and only a small residue remained playing on the wire, which to all appearance had absorbed the rest, that is, had absorbed the force that caused it. The glow of the wire became more intense; it soon after acquired a veil of luminous vapour along its whole length, which was tranquil, bluish, and nearly as thick as a finger. At last a flame, four inches long, pale yellow below, blue above, rose from the end, and over it was a stream of fine luminous vapour. On removing the wire from the negative pole, the wire soon lost its increased light and its flame, while the flame of the pole blazed up, every thing returning to the former state. With the coil on the southward pole the results were similar. The red flame was instantly absorbed by the wire, the natural pale glow of the wire became dark red; a bright red luminous veil appeared over its surface, 0.6 of an inch thick, and at last a flame of two inches broke from the point, red below, yellow above, pointed, and ending in a thick dull smoke which rose from it.

486. These experiments, to which, if it were necessary, I could add many similar ones, which entirely confirm them, teach us with consistency and gradually increasing clearness, *that the conduction of odyle through other bodies*, which we have learned, in the seventh treatise, to recognise by its effects on the sense of touch in sensitive persons, *is also accompanied by corresponding luminous phenomena;* that as the force, which the magnet emits, and with which it saturates all other bodies, so also the flames proceeding from it, may be absorbed by other substances, and again poured out by them, exactly as from the magnet itself. The odyle, or its essence, or its substratum, therefore, when transferred, carries with it not only its power of acting on the animal nerve, *but also its property of emitting light into other solid media; its glow, its flames, its smoke, with their fluidity and mobility, their light and their colours;* and, as we already know, their power of exciting the peculiar sensations of coolness, warmth, oppression, and refreshing lightness.

487. A retrospect of the contents of the present section tells us,

a. The luminous emanations from the magnet, the glow and the flame, are strongest, largest, and brightest in highly rarefied air, and would probably, in vacuo, become even a little more intense.

b. That the odylic smoke, in its greatest intensity, seems to be connected with a certain definite density of the air, on either side of which it diminishes.

c. That the ordinary atmospheric pressure considerably lowers the intensity of the odylic luminous phenomena.

d. That the density of water so much limits the developement of the odylic flame as almost to destroy it entirely, without, however, affecting the odylic glow.

e. That the density of solid bodies, such as glass, absorbs and to a certain extent retains and gives out odylic light;

while others, such as metals, especially in the form of wire, along their surface, but especially at their points, easily pour forth the absorbed odyle in the form of light and flames.

f. In a word, *that odylic light is differently affected by different media.*

488. As far as we have hitherto penetrated into the subject, it appears in some degree probable, that odylic smoke is *odylised air*, exactly as what is called magnetised water is *odylised water*, that is, air and water charged with odyle. For water also, when odylised, whether by the magnet, by crystals, by the hand, or by chemical action, acquires the odylic glow, and becomes visible in the dark, where it was before invisible, exactly as odylised metals acquire the glow, or exhibit an increase of their natural feebler glow, when odylised in the same way.

THE COLOURS OF ODYLIC LIGHT.

489. The character and succession of the colours exhibited by the different forms of odylic light, when closely studied, becomes a highly important branch of the subject. From my further researches it appears to be, not, as it seemed at the beginning, accidental and irregular, but subject to very regular physical laws; and thus its intensity, its tints and its shades, serve as a kind of measure, or means of estimating, on the one hand the polar quality and the strength of the odylic emanations, on the other for the degree of sensitiveness in the observer. This investigation also leads, or promises to lead, to most interesting disclosures in regard to magnetism proper, and its internal distribution.—Odylic light, in the lowest degree, appears on the magnet in the form of a feeble grey, that is, colourless nebula, only visible after the observer has been for hours in absolute darkness, and of the reality of which he can only convince himself when the magnet emitting it is slowly moved to and fro. Thus M. EDUARD HUETTER, in a pocket horse-shoe magnet of great intensity, saw the gleam of light from the northward pole so feebly grey in the blackness of the deepest night, that for a moment he doubted whether

what he saw were a reality, or the effect of self-deception. But when the magnet was moved about, he saw the grey nebula moving also, and then satisfied himself of the accuracy of his observation. He is perfectly healthy in all respects. The same thing often occurred with many of the less acutely sensitive, when commencing experiments with them before they had been long enough in the dark; as with the healthy sensitives, Baroness von Natorp, Mme. Fenzl, M. Tirka, M. Kotschy, M. Schuh, M. Delhez, and many others. I omit adducing more examples, as we have already, on previous occasions, become familiar with this phenomenon.

490. This feeble dawn of grey light, at first only visible over the northward pole, becomes stronger as the intensity of the odyle rises in the scale. It first becomes more distinctly visible, and then by degrees denser, more like vapour, and more concrete. It next appears, in the nebulous form, on the southward pole; and now on both poles it increases in luminousness and in density, till it resembles luminous smoke.

491. Then comes a point, when colour begins to appear, at first dull, and only slightly tinging the general grey. Mme. Fenzl saw over the electro-magnet only nebulous masses of light; but that over the negative pole, compared with that over the positive pole, was more of a bluish grey, while the other had a yellowish grey colour. Prof. Endlicher saw the same in several horse-shoes. This is the very beginning of the perception of colour. Proceeding higher in the scale of objective intensity (or of subjective sensitiveness), the part of the smoke nearest the steel acquires the aspect of flame, the upper part remaining smoky. M. Sebastian Zinkel (aged 77), saw on the northward pole of a single bar horseshoe a bluish light rising, to which he was doubtful whether to apply the term flame or smoke; on the opposite pole he saw a similar, but smaller and smoky emanation of light. The smoke is always thickest at the point of the flame, and decreases further off, passing gradually into vapour and nebula, thus becoming constantly feebler and more delicate till it is lost at its upper extremity. The flame below now acquires more colour. It is first noticed, when the poles are pointed

upwards, in the northward flame, that its grey becomes yellowish or bluish, and passes gradually through greyish blue into yellow or blue. The southward pole often exhibits only smoke, long after the northward pole has a blue flame. At last the southward smoke passes also into flame below; the grey tends to whitish grey, then to yellowish grey, and rises through yellow and orange to red. The smoke above it is now very dense, rising in heavy masses, in which at last scattered scintillations, like fire-flies, are seen. (If the poles point downwards, the appearances are different; I shall very soon speak of these changes.) The red flame of the positive pole, although later in appearing, has yet the greatest intensity of light. The blue negative flame shines more feebly, and where both appear in close proximity, and of nearly the same size, the blue flame is always the less luminous, the yellowish red or red flame the brightest. When this appears reversed, as in *horse-shoe* magnets it generally does, it is simply because, in our latitude of 48° N. (and still more in that of London or Edinburgh, W. G.) the northward flame is larger than the southward, and therefore appears relatively more luminous.—The succession of appearances just described was seen, in all its stages, by the delicate sensitives, such as Mme. KIENESBERGER, Mlles. WINTER, DORFER, KYNAST, WEIGAND, KRUEGER, Mlle. VON WEIGELBERG, Mme. JOHANNA ANSCHUETZ, FRIEDRICH WEIDLICH, and others; and by the healthy sensitives, M. PAUER, M. ANSCHUETZ, M. TIRKA, M. SCHUH, M. KOTSCHY, M. RABE, DrNIED, STEPHAN KOLLAR, Baron VON OBERLAENDER, JOHANN KLAIBER, Mlle. PAUER, Prof. ENDLICHER, Mlle. ERNESTINE ANSCHUETZ, Baroness VON NATORP, Mme. FENZL, Baroness VON TESSEDIK, Mme. VON VARADY, Mme. VON PEICHICH-ZIMANYI, Mme. BAUER, Baroness VON AUGUSTIN, and many others. It would be superfluous to enter on the individual details of these very obvious phenomena, which may easily be observed and confirmed by means of any sensitive person, and which I have everywhere mentioned as opportunity occurred.

491. *b.* That in proceeding further, these two primary odylic colours became associated with others, namely, with

green, orange, and violet, producing a variegated, and apparently confused play of colour on the odylic flame, I have already mentioned, on the authority of Mlles. Nowotny, Reichel, Sturmann, Atzmannsdorfer, and Maix.—Prof. Endlicher saw in the flames over a strong electro-magnet, an irregular mixture of different colours, playing about. The same appearance was described by Baroness von Augustin, by Mme. Kienesberger, Stephan Kollar, Mme. von Varady, Friedrich Weidlich, Dr. Nied, Mlles. Winter, Girtler, Zinkel, and others. The last named saw it very often both in horse-shoe and bar magnets. It attracts the attention of all highly sensitive observers in the very first trials with odylic light, and they always express lively astonishment, satisfaction, and delight at the spectacle. *Ceteris paribus*, it was best seen in highly rarefied air. Mlle. Atzmannsdorfer, for example, saw the variegated play of colours over a horseshoe, which was hardly seen by her when it was in the open air, become brighter and more intense in its rainbow hues with every stroke of the piston.

492. But there is a still higher degree of these luminous phenomena, which deserves the most exact detail of the experiments, and of the various peculiarities connected with it. *This is a perfectly regular Iris or rainbow*, the occurrence of which astonished me, and will astonish every one, who may take the trouble to investigate these very remarkable phenomena. The variegated play of moving colours, when all things combine to permit its tranquil developement, arranges itself in determinate forms, and follows fixed laws. Even in 1844, Mlle. Reichel had told me, that the magnetic flame often looked like a rainbow. I paid no attention to this, supposing that she understood by rainbow merely the play of various colours in the motion of the flame, as we are accustomed to see it in electrical light, in the sparks and luminous bundles of rays. But the statements of Mlle. Reichel have always proved accurate in the end.* Besides her, Friedrich Weid-

* *Avis au lecteur* for the gentlemen of the self-styled committee, so often alluded to. It occurs to me, that I may take this opportunity of giving, and that it may not be quite superfluous to give, to such readers as are not likely to see

LICH distinctly stated, that when the air was tranquil, and the flame not agitated by the breath of those who were near it, so as to mix up its different parts, its colours arranged themselves in a regular Iris. He gave me a general account

the journal in which their protocol (under an assumed title) appears, a small specimen of that production, directed as it is against my researches. At page 50, for example, occurs the following passage. "[Dr. VON EISENSTEIN led her, (Mlle. REICHEL) in this state (the supposed magnetic sleep) into a large room, where he made her sit down on a sofa, and tried, by passes with his hands, and with four bar-magnets, to raise her state to that of clairvoyance, and, at the same time, to destroy the influence of the sun upon her, and give the preponderance to the magnets. When he brought the magnets into the region of the heart, and Mlle. REICHEL, as if involuntarily, shuddered, (or was affected with slight spasm,) he exclaimed—" Aha! here, then, resides this filthy sun?! thou hast him in thy heart? Wait a moment; I shall soon expel him"—and now he made spiral *tours* near the heart with considerable energy.—The same scene followed when he magnetised her over the back, and on the pit of the stomach. The sun was remorselessly pursued, and driven out of every lurking-place. At one of these operations, Mlle. REICHEL sprang up, and struck at her magnetiser, who forced her down on her seat, and magnetised her lips with circular tours. When she offered to resist this, and put her hands before her face, he removed them, and reproached her " because she would not kiss the magnet, her benefactor which cured her. The abominable sun must be driven away from her lips, and its place taken by the magnet, &c."] On turning the leaf, we find the acconnt of an experiment, in which, in a room *by day-light*, Mlle. REICHEL was expected to see magnetic flames on the magnets presented to her, and, in addition to this, her eyes were bandaged with handkerchiefs! This ends with the following words :—[" Dr. VON EISENSTEIN, who conducted the experiment, gave us no explanation of its tendency. Baron VON REICHENBACH always made his experiments on the luminous emanations from magnets in darkened rooms, and found that they were seen the more distinctly, the more perfect was the darkness. Why Dr. VON EISENSTEIN tried this experiment in a room brightly illuminated by reflected daylight, why he chose the time when her eyes were blindfolded; whether he wished to test her power of divination, or whether he wished to prove something else, we know not. He gave us no explanation of the experiment just described."] Similar drivelling is not unfrequently met with in the course of the report. Who would have the patience to wade through 200 pages of it?—R.

I have waded through it, and experienced nothing short of disgust in doing so. Dr. VON EISENSTEIN would indeed find it difficult to explain his most absurd experiment, which could not, by any possibility, have any bearing on the researches of the author, inasmuch as he had expressly declared that he avoided making experiments on persons in the magnetic sleep. But even if he had made experiments on persons in that state, it is inconceivable how a man of ordinary education could imagine that such experiments as those of Dr. E. could prove any thing, or how he could make them in the manner described in the report.—W. G.

of the order of the colours, and of their relative extent. He saw them best in a three-bar horse-shoe of high magnetic intensity.—Then came Baron VON OBERLAENDER, who saw a regular Iris over the same magnet.—Dr. NIED also saw it, but with intermissions.—Mlle. ATZMANNSDORFER often described the beauty of the rainbow which she saw over magnets of every form.—Mme. BAUER gave me the most lively descriptions of the rainbow-like stratification of the colours, of which red was always lowest, then yellow, green, &c. followed in ascending order.

493. One experiment made with Mlle. ZINKEL may serve as a specimen of the whole. When in her ordinary state of good health, she usually saw the polar flames of the nine-bar horse-shoe simply blue and red. But during the catamenia, she saw them not only larger, but both assuming the form of an Iris, blue *predominating* at the negative, red at the positive pole. This was when the poles were upwards, and conformably placed in the meridian. But when I turned *both* poles towards the north, laying the magnet in the meridian, the Iris disappeared from the southward or positive pole, leaving only a bluish greyish red flame, while the rainbow on the negative pole increased to twice its former size, and reached 20 inches in length. With both poles towards the south, the reverse took place, the northward pole lost its Iris, retaining only a dull reddish greyish blue flame, while the southward pole exhibited a beautiful Iris, nearly 20 inches long.

494. Even the magnetism of the earth sufficed to produce the Iris, visible to highly sensitive persons. Mme. KIENESBERGER saw, during the catamenia, an unmagnetic soft iron bar twenty-four inches long, lying in the meridian, pour out a blue flame on the north end, a red on the south. But the former was only blue in general; for besides that colour she saw all the other rainbow colours, into which the flame divided itself, although they had less intensity of light. They were stratified horizontally, reddish below, then yellow, then green, and above, predominating blue with violet.

495. But this phenomenon was exhibited in greater beauty and purity on electro-magnets. Here I was not only able to

intensify the light and render it more distinctly visible, but I had also the advantage of using only a single horse-shoe bar, and thus avoiding the various disturbances which arise from the mutual action of the individual bars of a strong compound horse-shoe. When I had produced on such a bar, by means of a SMEE'S arrangement of one-seventh of a square foot, a flame of a hand in height, it was Mme. KIENESBERGER who first told me, that the negative flame was not blue only, but *yellow* and *blue*, the former nearest the magnetic pole, the latter *horizontally stratified over it*, passing above into grey vapour. At the positive pole she saw only red flame, with copious smoke. I now added a second SMEE'S pair, of one square foot. The flame became more than thrice its former size, rising on the negative pole to nineteen or twenty inches, on the positive pole to eight inches. The former now exhibited the interesting appearance of a perfect Iris. Close to the pole, which stood vertically, appeared a red stratum, next to that a stratum of orange, then one of yellow, then one of green, one of light blue, one of dark blue, and lastly, one of violet blue, above which rose a grey vapour. At the same time, the positive pole exhibited close to the iron a blood-red stratum, then light red, and above this, orange, from which a thick heavy turbid smoke rose to the ceiling. She described the appearance as one of extraordinary delicacy and splendour, and was filled with the same delight and astonishment as all the preceding observers of similar appearances on permanent magnets. The intensity of the colours was greater than in any permanent magnets. Some weeks later, I made the same experiments with Mlle. ZINKEL. She described the appearances in the same way as Mme. KIENESBERGER, being about equally sensitive, and added, that each coloured stratum was not uniform, but subdivided into smaller strata of different shades of colour, so that the whole Iris had the appearance of a great number of coloured bands overlying each other. Beyond the violet she observed *a narrow streak of pure red*, in which the violet ended, after becoming gradually redder, and which passed above into the smoke. She even saw the Iris with the small pair of one-seventh of a square foot; but the colours were dull and indistinct, so that she could not with certainty name them. Thus she thought she saw pale blue between yellow and green; but when I added a battery of six pairs, all the colours became incomparably brighter and purer, and the supposed blue was found to be a transition colour from pale yellow to pale green. In order, therefore, that this experiment should succeed, it must be made with large electro-magnets. Mine has limbs a foot in length, and is about one and a quarter inch thick. These experiments were repeated, three months later, with the same results. Some months later still, I used a battery of two and a half square feet. The Iris was splendidly developed, with smoke which rose to and illuminated the ceiling. The appearances on the negative pole were again the same, only larger and brighter; but the Iris was also more brightly developed at the positive pole; to its former red and yellow colours was now added blue. By a further increase of power in the electro-magnet, the green and violet would no doubt also be brought out. This experiment also was repeated after some months.

496. The young KOLLAR also saw the rainbow colours of the flame on the electro-magnet. As he had not the least idea of what was to happen when I made strong batteries act on the horse-shoe, he was much astounded by the succession of the phenomena, from the glow to the Iris, and their growth to a flame of twenty inches in height, variegated in colour, and rolling clouds of smoke up to the ceiling.

497. Professor ENDLICHER saw no regular Iris arrange itself over the electro-magnet; for which, in this particular experiment, there was perhaps neither time nor leisure (that is, to seeure perfect stillness of the air.—W. G.); but he recognised different colours in the flame. Below, where red lies, he only saw a dark and doubtful colour; above this, yellow; then green, and lastly, above, blue mixed with violet. They were unsteady; but yet, like the others, he saw the rainbow colours in their order, although less perfectly developed.

498. Mlle. GLASER saw, on the same electro-magnet, excited by two SMEE's batteries, a flame five feet high over the negative pole, and one of half that height on the positive

pole. The smoke rose to the ceiling; but both flames again showed the beautiful Iris, the negative having all the colours, blue predominating, *with a narrow streak of red above the blue* at the top, ending in smoke, the positive exhibiting only red and yellow, the latter passing into thick smoke.

The chlorotic girl, ANKA HETMANEK, saw on the electro-magnet a flame larger than that of the nine-bar horse-shoe, and next to the negative pole red, then the series of colours up to blue and bluish red, passing into smoke up to the ceiling.

499. Mme. BAUER, in the pregnant state, described still more vividly than the others, the appearance on the large electro-magnet. Before the current passed through it, she saw only the whitish glow of all iron. But as soon as I connected the wires with the thick coil, the flames appeared at first small, growing gradually to half the height of the room, and ending in smoke, which rolled along the ceiling. (The two batteries were not even in good order, and acted but feebly.) Blue predominated at the northward, red at the southward pole. But still the flames arranged themselves into a most beautiful Iris on each pole. Rainbow colours also appeared on other parts of the apparatus, on the limbs of the horse-shoe, the voltaic plates, &c. I shall return to this in its proper place.

499. *b.* I placed at different times a seven-bar horse-shoe, the poles upwards, in the vicinity of the conductor of the machine, and showed it to Mlle. ZINKEL. She saw the usual blue and red flames. But as soon as the conductor was electrified, the flames not only increased *in the electrical atmosphere*, both in size and brightness, but formed an Iris, blue only predominating at the negative, red at the positive pole, but mixed with all the other colours. The upper part of the northward flame was the brightest of all; the dullest was the lower part of the red.

500. In all these cases *the Iris was horizontally stratified; the red was always lowest, the violet blue at the top.* This relation to terrestrial magnetism must be borne in mind. In the first few minutes of the developement of the colours, they move unsteadily through each other, but gradually arrange

themselves to a floating Iris. This takes place with singular slowness, generally requiring, for the complete developement of the Iris, four, five, or six minutes.

501. From the above observations it appears, *that the odylic flames on both poles do not consist of blue and red alone, but of an Iris on each side, in which blue predominates on the negative, red on the positive side.* The Iris becomes visible where magnetism and odyle possess a certain intensity, and is invisible, that is, it is confined to the prevalent colour, when that intensity is low, or the sensitiveness of the observer is less acute.

502. But besides this simple Iris, we have to consider a complex kind of compound Iris.

There had occurred, as has been formerly alluded to, a certain *variableness of the colours of odylic light*, which often threatened to introduce confusion into the results. This induced me to undertake a long special investigation of the point, consisting of numberless experiments made with the utmost attention, and from which I endeavoured to deduce the laws regulating the phenomenon. I shall here detail the most essential parts of this investigation.

503. A magnetic bar, with the poles in the line of dip or inclination, always emitted colours different from those seen when it lay horizontally in that of the meridian; and when a northward pole was turned towards north or towards south, its flame was at one time more blue, at another more red and grey. Another uncertainty consisted in this, that when I placed a northward pole vertically upwards, the observers generally described the flame as bright blue, but sometimes as grey, occasionally even yellow, with other similar incongruities.

504. To reach the cause of all this, I began by supposing the possibility of *subjective variations in perceptive power.* I placed a two feet bar-magnet vertically, the northward pole upwards. If feebly magnetic, Mlle. ZINKEL, when looking at it from a distance of eight inches, saw the flame of a turbid and doubtful greyish yellow. I made her try the distance at which its colour was most distinct, and its form best defined, which, *in every experiment with her*, was from seventeen to

twenty inches. It then appeared pure yellow. At a still greater distance it again became indistinct, tinged with light grey, not unlike light blue; still further off it became unequivocally grey, which became at greater distances duller, difficult to see, and at forty inches or more it nearly disappeared.

505. Mlle. PAUER, in repeating this experiment, saw the same appearances, except that in her case the distance of clearest vision for the yellow flame was only about eight inches. As soon as she removed further, it became bluish. Now Mlle. PAUER is short-sighted. This peculiarity had therefore an influence on her perception of odylic light, and it was thus proved how much the apparent variation of colour at different distances depends on subjective causes.

506. Similar observations made with Mlle. ZINKEL at different times, and in all different directions, in the same way, yielded proof that there is only one fixed distance, differing for each individual, at which the colour of the odylic flame is pure and distinctly visible; *but that at this distance it remains constant for the same individual, and does not vary at all; whereas, at other distances, less or greater, it exhibits different shades*, being dull yellow at small, bluish-grey and grey at greater distances. In order, therefore, to obtain unmixed results, we must always attend to the proper distance for the eye of each observer.

507. In fact I had sometimes, before I had been led to make the last-mentioned investigation, found myself in no small perplexity, since one observation had yielded an azure blue colour, while another, apparently made under precisely similar circumstances, had given a grey or a yellow colour of the odylic flame. As long as we do not see our way out of such a confusion, the researches, in which we cannot perceive the phenomena with our own senses, but must obtain them by asking questions of another person unacquainted with the subject, and this in the dark, become not unfrequently wearisome and exhausting to a degree that is indescribable, and occasionally it required all the charm of so interesting a subject to support my wavering patience.

508. Having once obtained this apparently trifling explanation, (the absence of which, however, would have opposed an invincible obstacle to further progress,) a part of the way was smoothed, and I began an extended study of the variations of the colours of the odylic flames from the *objective* point of view. I fixed the same two-feet bar-magnet by its middle in a GUIDO'S support of wood furnished with an universal joint, so that it could be placed in any direction. I placed it longitudinally in the magnetic meridian, and at first conformably pointed northwards, then in the line of the magnetic inclination of Vienna, at an angle of 65°. From this position I caused the northward pole to pass through the whole vertical circle described by the magnet round its short axis in the plane of the meridian.

I did this in the dark chamber first before Mlle. ZINKEL, who looked at the phenomena from the west. Beginning from the position in which the northward pole was pointed vertically downwards at 0°, she saw the flame pass through the following series of colours :—

At 25° (in the inclination)	pure grey.
„ 45° (rising towards north)	a narrow red streak.
„ 67°	violet blue.
„ 90° (horizontal)	dark blue.
„ 110°	light blue.
„ 127°	dark green.
„ 145°	light green.
„ 163°	canary yellow.
„ 180° (vertical,—M pole upwards)	bright yellow.
„ 200°	golden yellow.
„ 225°	orange.
„ 247°	flame red.
„ 270° (horizontal)	red.
„ 290°	deep vivid red.
„ 325°	greyish red.
„ 360° (vertical,—M. pole downwards)	reddish, whitish grey.

Fig. 10.

Above.

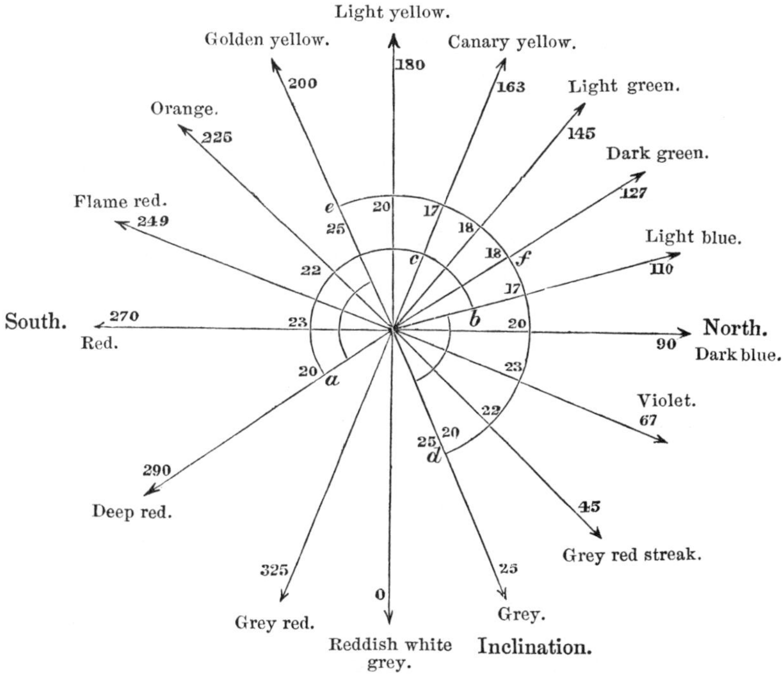

Below.

For the sake of greater distinctness, I represent it here graphically in Fig. 10, in which the *radii* represent the northward half of the bar in sixteen positions. The southward half is not here shown.

509. If we examine the colours of this circle, we there find a new Iris, the colours of which are arranged in a circular form, or rather follow each other round the circumference.

510. It is very remarkable that the position of greatest magnetic intensity, namely, that of the *inclination* at an angle of 65° with the horizon (in Vienna), is precisely that in which all the colours disappear, and nothing but *dark grey* is left. Does this grey represent white, the summation of all colours, or black, the absence of all colour? I have not yet been able with certainty to decide this, further researches will soon settle the point, but I am inclined to the latter opinion.

511. It is further remarkable, that the colours which stand diametrically opposite to one another *are not complementary colours.* For here we find opposite blue, not yellow, but red, opposite violet blue, not green, but flame red, opposite yellow, not blue, but grey, &c. The circle appears divided into sectors of 90°, the zero of which must be placed in the line of magnetic inclination. Opposite to this zero, at the distance of 180°, stands golden yellow; and on each side at 90° from these two points, we find red and blue opposite each other, that is, again, 180° apart. Under the magnetic equator, where the polar line or meridian and the inclination coincide, the direction and order of the colours would be different. I hope to see the time when some one may try these experiments there, which cannot be difficult, for certainly no ship ever crosses the line, in which one or more sensitive persons may not be found.

512. Another peculiarity appears in the circumstance, that the observer persisted in describing *a narrow streak of red between the grey* of the northward pole, pointing downwards in the line of inclination and the *violet blue.* This remarkable fact had occurred in former experiments; for we have seen that Mme. KIENESBERGER often spoke of a narrow red streak above the blue in the beautiful Iris of the electro-magnet, passing into the grey of the smoke. Mlle. GLASER, STEPHAN KOLLAR, Mme. BAUER, and Mlle. PAUER, all observed the same thing. It would appear, therefore, that from the blue or the violet, the red is again developed so strongly as to appear as an independent red at the other (the violet) end of the Iris; and consequently *red closes the spectrum at both ends,* at least certainly the odylic spectrum or Iris.

513. *Those colours which lie in the lower semicircle,* that is, for 90° on each side of the inclination, *differ strikingly in intensity of light from those of the upper half, or 90° on each side of the golden yellow.* The upper half, occupying the space parallel to the arc a c b, is bright, strongly luminous, lively, and brilliant; the lower half dull, turbid, and feebly luminous. *The greatest intensity of light lies in the golden yellow, the greatest darkness in the grey of the inclination.* We might call

the upper half *the side of day*, the lower in which all colours are reited and loaded with grey, *the side of night. The yellow*, which we already know to be the most luminous part of the spectrum, *represents, therefore, noon, blue and red the morning and evening twilights, grey the night. The earth is opposed to the heavens in its action on odylic flames.* §§ 536, 356.

514. When I used, instead of the northward pole, the southward pole of the same magnet, I prepared myself for a great difference in the results. But my expectation was not fulfilled. In the general result, it was nearly indifferent which pole was made to revolve in the vertical circle in the plane of the meridian. There were, however, some trifling modifications. The colours, which were almost precisely the same from both poles when in the same direction, always appeared somewhat later when the southward pole moved from the downward position in the inclination towards north, and so on round the circle. It was necessary always to advance the pole a few degrees further than in the case of the northward pole, in order to obtain the announcement of the same colour from the observer. This retardation of the colours on the southward pole was the same throughout the circle, so that the grey of the inclination was seen in a position of the bar forming a less angle with the horizon than that which, with the north pole, gave the same result, an angle, as above stated, of 65°, being that of the dip in Vienna. See § 534.

515. Moreover, the two poles each again divide the circle into two halves, those of greater and less purity of colour. *From the inclination by north to the gold yellow, on the arc parallel to e f d, the colours, in the case of the northward pole, were purer and more distinct, but from golden yellow over south back to the inclination, they were less pure, less distinct, and as it were veiled. Exactly the reverse occurred with the southward pole. From the inclination northward, over the same arc, parallel to e f d,* Mlle. ZINKEL *described the colours of the southward polar flame as turbid and veiled; on the opposite half they were pure and clear.* The two halves of this phenomenon passed into each other above, in the golden yellow, and were most strongly differentiated in the line between the

terrestrial poles. It is obvious, that conformable and unconformable positions of the poles of the magnet here made their influence felt as they had always done; and that the northward pole, exerting its full power when conformably placed, was, when unconformably situated, that is, when pointing to the south, enfeebled by the direct opposition of terrestrial magnetism. The red flame which it had in the latter position, was therefore turbid and veiled by an admixture of blue, belonging to its own magnetic polarity, which was opposite to that of the earth then acting on it. Hence a red, which was sometimes described as bluish red, sometimes as greyish red. The same is true of the southward pole, in the reversed position and with reversed terms.

516. *The circular Iris is therefore twice divided, by external influences, into two halves, in reference to intensity of light and purity of colour. In one case this division is that of a horizontal, in the other, that of a (nearly) vertical section. In the one case, the influence of terrestrial polarity, in the other that of the polarity of the bar itself, predominates.*

517. The odylic flame, in any one position, did not exhibit the Iris, but always appeared to Mlle. ZINKEL monochromatic. But the chief part of these experiments, when repeated with Mme. BAUER, yielded still more complete results. When I showed her the same magnet in the meridian, and held as before in the middle by the moveable arm of the support, she saw flames one-half longer than Mlle. ZINKEL had seen, and *not monochromatic, but always in the form of an Iris, except in the inclination*, where the northward pole gave out only a dark colourless grey. On moving this pole upwards towards north, she saw the flame form an Iris, first dull, then gradually brighter. But within this, in every position, one of the various colours always exhibited a size, a strength, and an intensity of light, such as to predominate over all the others, and this to such a degree, that, unless examined with attention, only the predominating colour was seen. Without reference to the constant presence of this Iris in the flame, she gave me the succession of the (predominating) colours as follows: (compare Fig. 10.) —First, on moving the pole from the line of the inclination a

little towards north, a short space of red, soon after which appeared violet blue; and then, as it approached nearer to north and reached it, dark blue, and light blue. From north upwards, various shades of bluish green, sap green, and canary yellow appeared. In the vertical position pure yellow appeared; golden yellow was diametrically opposite the grey of the inclination. On passing towards south, red appeared in the yellow, increasing through orange and flame red till, at south, red appeared nearly pure, with a trifling tinge of blue. Proceeding farther, grey became mixed with the red, the former increasing as the latter diminished, till at the inclination red, and all the subordinate tints of the Iris in the flame, totally disappeared, and gave place to the pure colourless grey from which the circle had been commenced. All this exactly agrees with the account given by Mlle. ZINKEL. The only difference lies in this, that Mlle. ZINKEL saw, in all positions, only one colour, Mme. BAUER a prevailing colour, accompanied by all the other colours in subordinate size and brightness. This difference certainly depends on a subjective difference of perceptive powers. Mlle. ZINKEL, less sensitive, saw only the predominating colour in each position; Mme. BAUER, more sensitive, saw the flame larger, and with rainbow colours. (We shall see hereafter, in treating of the odylic light of crystals, that Mlle. ZINKEL, in some instances, stood in the same relation to Mlle. GLASER as Mme. BAUER here did to Mlle. ZINKEL.) With this explanation, the observations of Mme. BAUER confirm entirely those of Mlle. ZINKEL.

518. I went through the same experiments with Mlle. GLASER, who is less sensitive than Mlle. ZINKEL. In her ordinary state the flames appeared to her only blue, yellowish, or red. But during the catamenia, she saw distinctly the colours of the circle in Fig. 10. She was sometimes doubtful as to the transition colours, and it was necessary strictly to observe the proper distance from her eye to the pole. For example, she sometimes called light blue, grey, pale yellow, pale reddish, and *vice versa*, but in the end, after careful repetitions, her final decision in every case agreed with the colours named by the two other sensitives. She had no hesitation at

all as to the grey of the inclination, the yellow opposite to this, the blue in the north, or the red in the south. She also perceived little difference between the two poles in the same positions; there was a general faint tinge of blue in the colours of the northward, of reddish yellow in those of the southward pole.

519. Mlle. Pauer saw, in the same experiment, the following appearances. The northward pole, pointing downwards in the line of the inclination, gave a pure dark grey vapour; rising towards north, it became at the top, first reddish, then reddish grey, violet blue, dark blue, dark green, bright green, at the top yellow, then turning towards south, orange, in the south red, below that reddish grey, to the pure grey in the inclination. This was early in the day, when she was fasting; after dinner, her powers of vision became less acute, so that she made mistakes in the grey tints.

520. I tried these experiments with persons still less sensitive. Mme. Fenzl had the kindness and patience to devote herself to this experiment, and she, with her husband Dr. Fenzl, were so obliging as to allow themselves to be shut up for half a day with me in the dark chamber for this purpose. The result was the circle in Fig. 11, representing the positions of the northward pole when the magnet was made to revolve vertically in the plane of the meridian.

It agrees perfectly with that in Fig. 10. What is here called the brightest part, is the bright yellow of the former circle; the brownish red here, is the dark greyish red of the former, &c. The correspondence could not be more complete.

521. I now tried M. Hochstetter, who was still less sensitive, under the same circumstances. He saw only dark smoke in the north position; brighter and stronger in the vertical position upwards; thicker and duller in the south; darkest of all in the inclination, the pole pointing downwards. Although he could distinguish no colours, yet the intensities of light exactly corresponded to those formerly described. Thus he, the weakest of the sensitives tried in this way, yielded proofs of the accuracy of the observations made with those who were most sensitive.

Fig. 11.

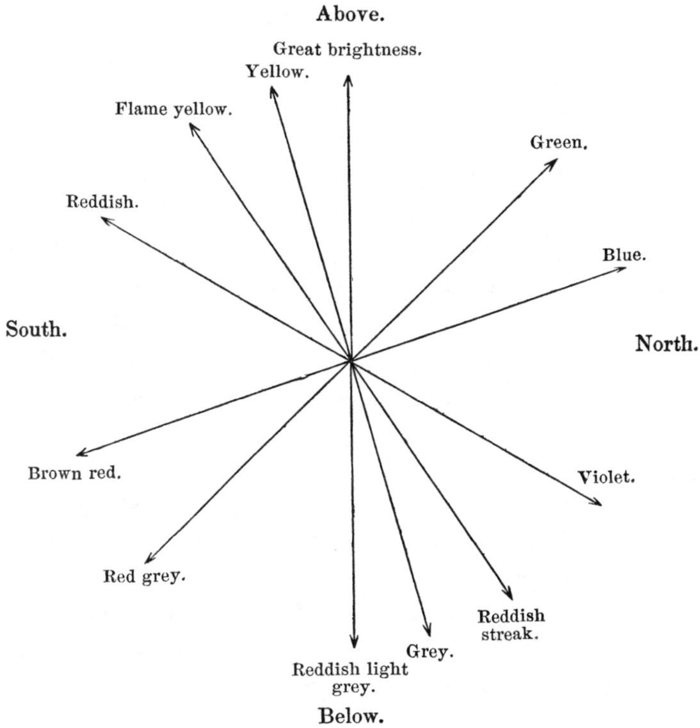

522. As a control, we may refer to the results obtained with a single bar horse-shoe (the form of which does not admit of all the same experiments being tried), when it lies in the meridian, with both poles turned either north or south. When to the north, Mlle. ZINKEL, at two spans distance (twelve or thirteen inches), saw the flame on the northward pole blue, on the southward greyish violet, or reddish blue, with a veil of grey. If to the south, the southward pole had a red flame; the northward a bluish, or greyish red, and darkened flame. All four cases agreed with the colours of the bar magnet on the same positions. Only single horse-shoes must be used here, because the results are confused, in the case of compound magnets, by the mixture of flames from alternately reversed odylic poles of the plates, as appears from § 396.

523. When I tried the experiment of directing the horse-

shoe poles vertically upwards or downwards, I obtained the results formerly stated (§§ 393, 439), which must, however, be here repeated, for the sake of comparison. *When both poles were downwards, and conformably placed*, the flames were grey on the northward, reddish whitish grey on the southward pole. In the inclination the grey of the former was darker, that of the latter reddish dark grey. Rising with the poles from the inclination northwards, the northward flame was grey with a tinge of red, the southward dark grey. Here, therefore, we again meet with the retardation in the southward pole. When the northward flame is dark grey, the other has not yet reached that point, and when it does reach it, the northward one is already reddish grey, and has arrived at the beginning of the short streak of red before the violet. When I now, in the same position, reversed the limbs, so that they lay unconformably, there were exactly such variations of colour as the previous experiments might lead us to expect, and which need not be detailed.

524. Lastly, *a horse-shoe, with its poles vertically upwards*, always gave at the — M (northward) pole, a flame which, according to the distance of the observer, was described as chiefly grey, bluish, or yellowish, while the flame of the + M pole was chiefly yellowish, or reddish grey. Mlle. ZINKEL, at her normal distance of two spans, and when all objects that could exert odylic influence were removed, saw, as did many others, the flame of the — M pole always chiefly light yellow, but blue and bluish grey further off. I often heard it described by Mme. KIENESBERGER, Mlles. REICHEL, ATZMANNSDORFER, STURMANN, by JOHANN KLAIBER, FRIEDRICH WEIDLICH, Baron VON OBERLAENDER, and others, as pale or light yellow below, and blue above, the reverse of a candle. This was probably the commencement of an Iris. The + M pole appeared reddish yellow below, greyish yellow above. All these things are merely confirmations and applications of the results obtained by causing a bar magnet to revolve vertically in the plane of the meridian.

525. In order to see how much of all this was due to the action of the terrestrial magnetic poles, and how much to that

of the poles of the magnet, I made the same experiment *with an unmagnetic soft iron bar*, and first with Mlle. ZINKEL. It had a small flame at each end, about a quarter or one-fifth of the size of those of the magnet, also duller, and the colours less easily observed. Nevertheless, the results were precisely the same, except that the angles varied from those observed in the magnet by a few degrees. This trifling difference was certainly due to the imperfections of the observation; for although I left it, as before, to the sensitive to state for each colour the point where it was most pure and intense, yet in so delicate a matter, and with the very feeble light, particularly of the unmagnetic bar, the angles cannot possibly be always correctly ascertained, so that they do not vary by a few degrees, until we have obtained means or instruments expressly adapted to the purpose. It is enough that the results in this case perfectly coincided in all essential points with those obtained from a magnetic steel bar.

On repeating the same experiment afterwards with Mme. BAUER, she saw the colours with perfect distinctness, but duller and smaller than in the magnetic bar. The order was the same; grey in the inclination below, yellow vertically above, blue to the north, red to the south.

Mlle. PAUER also went through this experiment. She saw the soft iron bar, in the inclination, give out below grey vapour, and the same when vertically downwards; to the south yellowish red, vertically upwards pale yellow (at a certain distance pale bluish), to the north blue. All the colours were dull, small, and feeble, so that she found some difficulty in recognising the tints where grey prevailed.

Mlle. GLASER also confirmed the results of this less distinct experiment. I held before her a soft iron bar in my left thumb and finger. When I turned it round, she saw the nebulous light appear grey, blue, yellow, and red, in the corresponding positions, and even pointed out green between blue and yellow.

526. It appears, then, that the circular iris, observed in a magnetic bar revolving vertically in the meridian, is also seen, although less bright, in an unmagnetic iron bar, and may

consequently be produced by the action of the terrestrial poles alone. This explains the effect on the light from magnets of the unconformable position.

527. I now proceeded to examine the appearance of the flames of a bar magnet revolving vertically in the plane of the magnetic parallel of my residence, a league north from Vienna. The experiment was performed like the former, and first with Mlle. ZINKEL. For the — M or northward pole, I obtained the following diagram, Fig. 12.

Fig. 12.

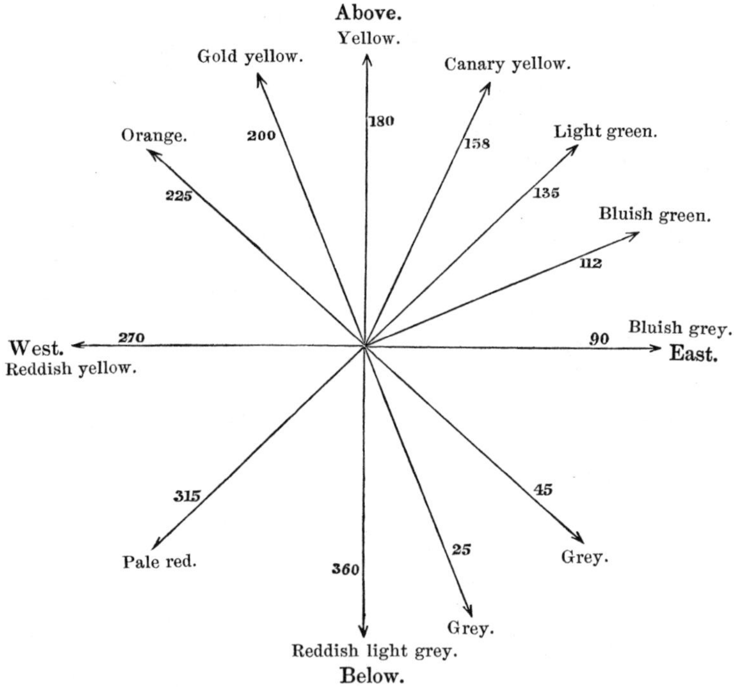

528. The + M or southward pole gave the diagram, Fig. 13.

529. The next sensitive tried was Mlle. GLASER. The results of her observations, often repeated, were as follows :— In the east grey, above yellow, in the west yellow, below again grey. Between east and above, traces of green, between west and below, traces of red.—Soon after this I tried Mme. BAUER. This highly sensitive lady gave the same statements

Fig. 13.

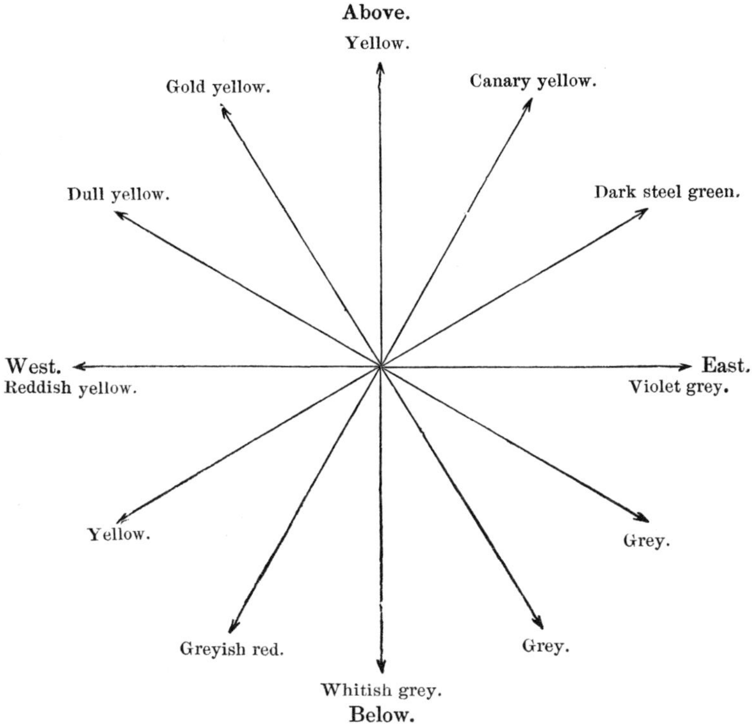

in all essential points. To the east, below and between them, grey; to the west, above and between them, yellow. The slight variations of shade in numerous repetitions of the experiment, when closely tested, were always found to arise from variations in the distance of the observer from the flame, or in the position of the bar in reference to the magnetic parallel. Mlle. PAUER saw the northward pole grey below, yellow towards west, yellow above, grey towards east. M. HOCHSTETTER saw only vapour, darker below and towards east, brighter above and towards west. This is the only perception of the grey and yellow in persons of low sensitiveness.

530. It is here distinctly seen that, in the parallels, the two poles differ very slightly in the colours of their flames. The general result is, that in both, *when pointed towards the east, grey is the prevailing colour, and towards the west yellow.* East, as essentially grey, corresponds, therefore, with the inclina-

tion; west, as essentially yellow, lies diametrically opposite to the inclination in the vertical circle in the plane of the parallel, as yellow was in that in the plane of the meridian. The northward pole of the bar appears, in the east more bluish grey, the southward more mixed with red, or violet grey. The former, towards west, has more reddish yellow, the latter more pure yellow. These shades are difficult to determine, and are therefore only approximative.

531. It was necessary, in order to complete this part of the investigation, to make the experiment in the parallel with a soft iron bar. This was done with Mlle. ZINKEL, and yielded the diagram Fig. 14.

Fig. 14.

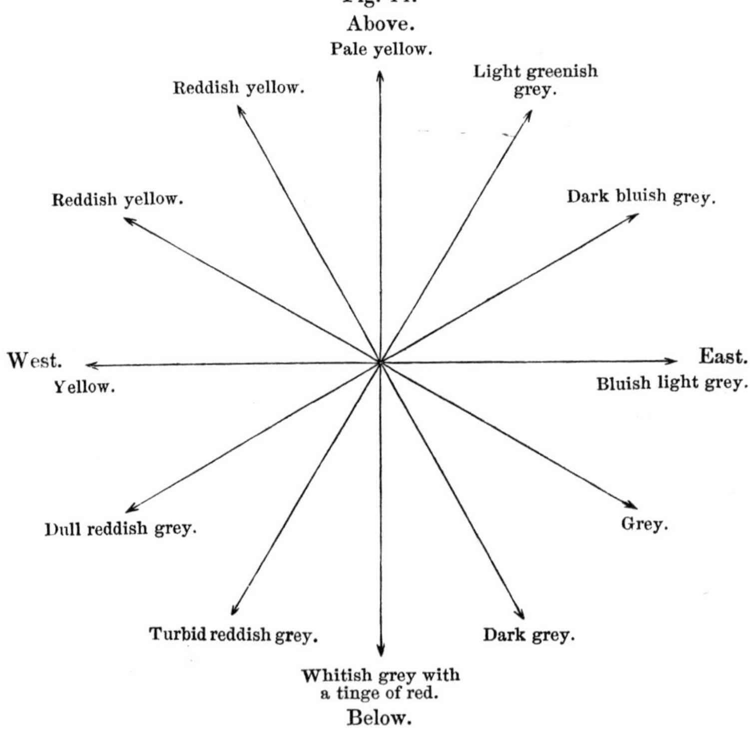

532. All the colours agree pretty closely with those of the magnetic bar; the flames, however, being so much smaller and duller, that it required great attention to distinguish the colours.

533. Yellow, with traces of red, always appeared on the west side ; grey, with traces of blue, on the east side. Here also grey and yellow are opposed, as are the beginnings of blue to those of red.

534. The third direction which I had to study, and which might in some degree serve as control to the results of the two former (vertically in the meridian, and vertically in the parallel), *was the horizontal.* A bar magnet, made to revolve horizontally, gave, with Mlle. ZINKEL, the following diagram, Fig. 15., for the northward pole.

Fig. 15.

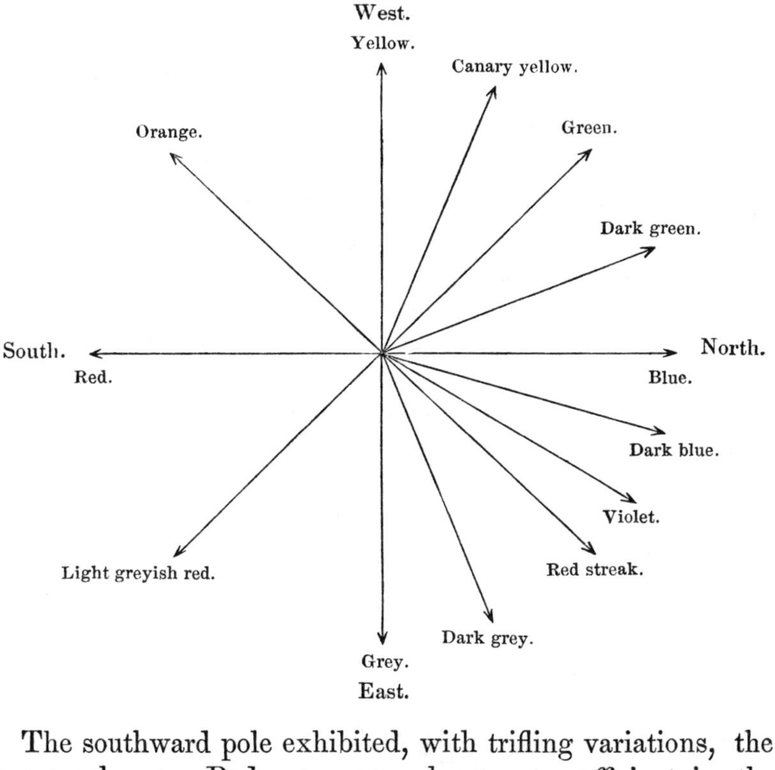

The southward pole exhibited, with trifling variations, the same colours. Red was everywhere more efficient in the latter, blue in the former, and both strengthened and vivified the colours on the conformable semicircle, enfeebled and troubled them on the unconformable half. Finally, here also was observed the retardation formerly mentioned (§ 514,) in the appearance of the colours of the southward flame, which was now noticed when the southward pole was moving from east to west.

535. The agreement which might be expected *in those points where the vertical circles cut the horizontal circle*, and thus form nodes, comes out in a most satisfactory manner. The coincidence in the line of north gave blue, in that of south red; of west yellow, and of east grey, exactly as they occurred in the same positions on the two former circles. The three observations, therefore, mutually control and confirm each other.

536. The observations made in the vertical circle of the parallel, and those made in the horizontal circle, are more difficult than in the meridional vertical circle. This is chiefly because, in the two former, north and south, which have the most powerful influence on the magnet, act during the whole revolution equally on its poles, while east and west, which represent much feebler points of the compass, have to maintain the conflict alone against above and below, which correspond with south and north. The size of the bar, its intensity, &c., all exert a greater influence on those colours which are produced by the feebler forces, and are more feebly supported. A slight difference, however, does occur between the colours in east and west; for in east, the grey of the northward pole had a tinge of blue, that of the southward pole, a tinge of violet; while in west the yellow of the northward pole was mixed with red, and that of the southward pole was there of the fullest and brightest yellow, with a shade of red. This indicates a similarity in the effect of west with that of south, of east with that of north. We thus arrive, by another path, at the same result which we met in § 391, namely, that generally, in reference to odyle, *east inclines to the side of north, west to that of south.* And as we already knew, (see § 356, and § 531,) that east and below, west and above, also harmonize in their effect, we arrive at the comprehensive result, *that north, east, and the earth* (*below*), *form a general contrast, in odylic polarity, to south, west, and the heavens* (*above*).

537. In all these experiments, in which the bars were

made to revolve in different planes, it appeared in several cases, as if the flames had not quite the same colours, at different corners of the same polar extremity. When I used an iron bar, twenty inches long, 1.4 inch broad (0.035 mètre), and 0.4 inch thick (0.01 mètre), Mlle. ZINKEL often observed, that the polar ends of the bar had flames of different colours on each end of their short sides. This was most observeable when the bar was so placed in the meridian, that its breadth lay in the parallels, and remained there during the revolution of the bar round its centre. The following representation of the end of the bar, or a cross section of it, will make this clear. It exhibited, when made to revolve in the meridian, always a darker greyish tint in the flame on its east edge, and a more yellowish, reddish, or greyish tinge in the flame of its west side. This phenomenon was not strikingly observed in a magnetic bar, especially if strong, when it was hardly perceptible; but it came out most distinctly with unmagnetic bars of soft iron.

Fig. 16.

West.

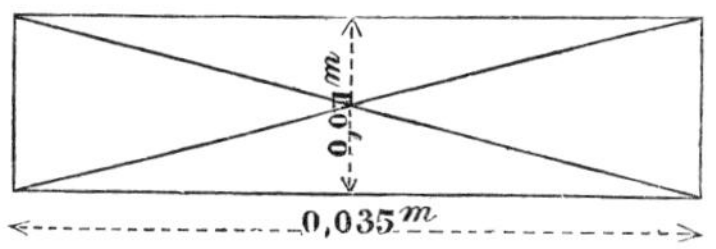

East.

538. I met with this also in experiments with Mlle. PAUER. In an iron bar standing vertically, she saw the flame only pine yellow on the west corner, but more of a bluish greyish yellow on the east corner. The southward end of a bar lying horizontally in the meridian, was nearly orange on the west corner, but greenish red on the east. And in all positions the corner towards east was duller and more veiled with grey; that towards the west brighter, clearer, and more luminous. When I placed it on the parallel, with the northward pole towards west, there always appeared on the north corner a greenish, on the south corner a flame-red tinge.

539. This led me to suspect *that the effect might be due to the transverse magnetism of the earth, or to an analogous odylic cause.* To investigate this, I took an iron plate one-fourth of an inch thick, forty inches long, and six inches broad, and placed it, supported freely in the middle horizontally, first in the meridian, and then in the parallels. In both of these

positions, the four corners exhibited dull odylic flames like those of a soft iron bar; but the colour was different at each corner.

With the length of the plate in the meridian, Mme. BAUER described the colours of the corners as they are represented in Fig. 17.

Fig. 17.

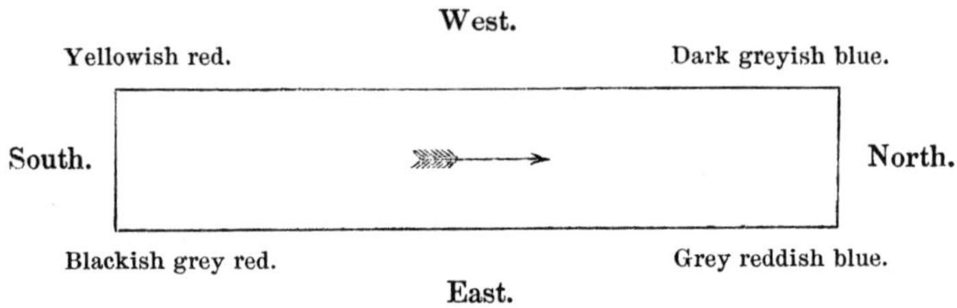

Here blue predominated at both north corners, red at both south corners, but mixed with yellow on the west, and with grey on the east side. *The two opposites, yellow and grey, appeared therefore again directly opposite one another in this experiment.* The dark green in the north-west was nearly black, and the blackish, greyish red in the south-east was like blackish, greenish red.

540. I made the same experiment with Mlle. GLASER. The result was that given in Fig. 18.

Fig. 18.

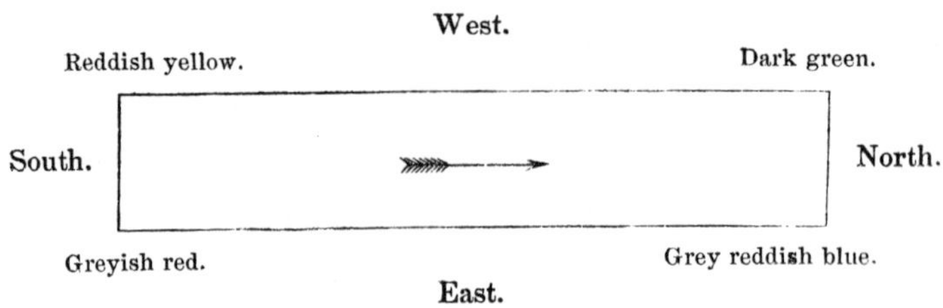

These statements agree in all essential points with those of Mme. BAUER, and harmonise, in the four leading colours, with the previous observations.

541. When the length of the plate lay in the parallel, Mlle. GLASER saw what is shown in Fig. 19.

Fig. 19.

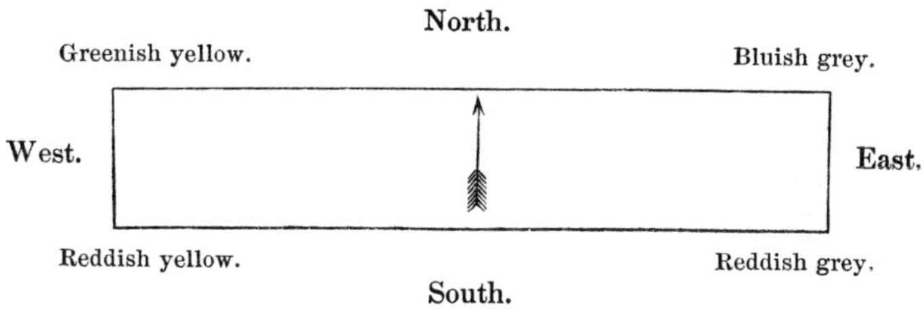

I made the same experiment many times, with all possible care and accuracy, with Mlle. ZINKEL. The result is shown in Fig. 20.

Fig. 20.

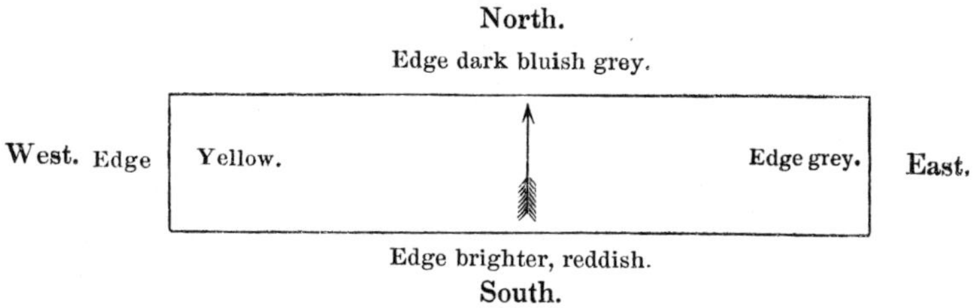

543. It thus appeared that the colours, although dull, followed decidedly the rule, that they appeared grey towards east, yellow towards west. But this was not all, for they again divided in each of these positions. In the east, the grey flame was bluish grey at the north corner, reddish grey at the south corner, and subtracting the grey common to both, blue is left on the former, red on the latter. In like manner, the yellow flame in the west appeared on the north corner canary yellow, that is, yellow with a tinge of blue, and on the south corner reddish yellow; and if we take away the common yellow, there is left blue on the north, red on the south corner. On this oblong parallelogram, then, we have at the same time all four primary colours, in east and west grey and yellow, and in north and south blue and red; or, accurately considered, all

four colours on one piece of iron, which is simply a broad bar influenced by terrestrial magnetism. In other words, *the polar pair of opposite colours, blue and red, reappears transversely in the other pair, grey and yellow.*

Fig. 20. shows that the edges agree with this. Of the two longer edges, one on the north, the other on the south side, the former is in general feebly greyish blue, the latter rather brighter and reddish grey corresponding exactly to their direction. Of the two shorter edges the eastern one is grey, the western yellow, in accordance with their direction.

544. To control these results, I placed, in an experiment with Mlle. ZINKEL, an iron-bar with a blunt point on the plate lengthways, and moved it into such a position that it projected about a handbreadth over a short edge. When it projected so as to form a continuation of either corner, it absorbed the odylic flame of that corner, and gave it out from its own point unchanged, but rendered stronger by concentration. But on moving it along a short edge from one corner to another, it acquired mixed colours on the way, composed of the two corner colours. Thus, between canary yellow and orange on the west side, it passed through all intermediate tints, while these two colours on the corners were reduced to a turbid residue. The flame at the point of the rod, therefore, was composed of the two colours lying at that side, into which the flame resolved itself transversely when space was afforded.

545. To strengthen these appearances, I tried the experiment of laying on the plate of iron, longitudinally, a bar magnet of half its length, and with its centre on the centre of the plates, so that its poles were equidistant from both sides. Intensity of light and distinctness of colour instantly increased in some parts, and diminished in others. See Fig. 21. The northward pole, directed towards the east side, there changed the bluish grey to blue grey, the reddish, whitish grey to dull grey; the southward pole, towards the west side, changed the canary yellow to gold yellow, the reddish yellow to orange. The northward pole therefore strengthened the blue tints and weakened the red; the southward pole strengthened the red

Fig. 21.

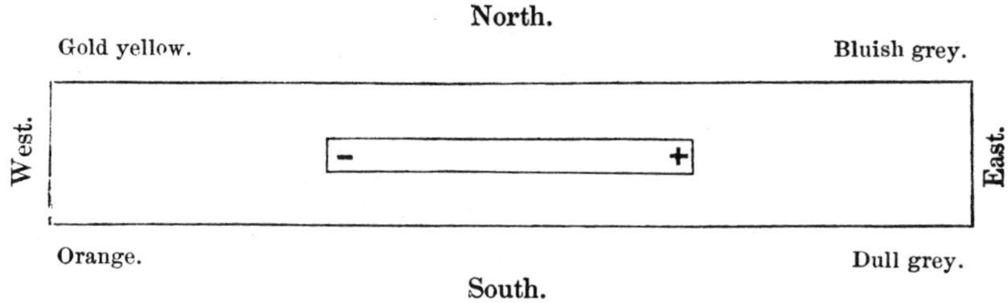

and weakened the blue. The results were analogous when I turned the northward pole to the west, the southward to the east. Now, the bluish-grey on the east side was changed to violet grey, and reddish-white grey to reddish-grey; and on the west side the canary yellow changed to dark greenish yellow, the reddish yellow into dull greyish yellow. All this corresponds to the action and mixture on one side of red, on the other of blue from the magnet, with the colours caused on the soft iron plate by the influence of the earth.

546. There remained the experiment, which I did not wish to omit, of placing the plate vertically in the meridian, with the long edges one over the other from north to south, and the short edges of its breadth vertical. Mlle. ZINKEL saw: the north edge darker, because blue, the south brighter, because reddish; the lower edge darker, because greyish; the upper edge brighter, because yellowish; the lower south corner darker, because tending to grey; the upper south corner brighter, because tending to yellow; the upper north corner darker, because tending to blue; the lower north corner brighter, because tending to whitish grey. All this distinctly and exactly corresponding in every respect to the principles so often explained.—To confirm this, I also tried the same experiment with Mme. BAUER. She, being more sensitive, saw the two corners towards the north with bluish flames, the two corners towards the south with yellowish red flames. Of the two former, the upper was mixed with blackish green, the lower with reddish grey, because, above, some yellow, below, some grey with the red streak formerly mentioned, had been added. Of the two latter, the upper was bright deep yellow,

the lower blackish grey red; again, above, yellow, and below, grey, had been added. To test both the observation and the observer, I several times turned the plates, so that the edge which was at the time below came to be above. But Mme. BAUER declared that nothing had changed but the position of the plate, that of the coloured flames remaining the same. This was inevitably the case.—Such a result, effected in the long horizontal by the short vertical, gave me true pleasure; it confirmed in the most satisfactory way the accuracy of the preceding observations.

547. When I turned the plate horizontally to the extent of 90°, so that it lay in the magnetic parallel, I obtained from Mlle. ZINKEL the following results:

Above and below, the colours as just mentioned.

The west edges brighter, because on the yellow side.

The east edges darker, because on the grey side—

All in agreement with the previously obtained results.

The whole of the plate, in such experiments, constantly exhibits a feeble odylic glow, increasing from the shorter axis towards the short sides. But here, as in bar magnets, the maximum of intensity is not at the edge, but a little within it. At the same spot, the odylic luminous down, which covers the whole surface as with fine fibres of light, is also strongest.

548. I united all these observations in one comprehensive experiment. I had a strong support made, with a universal joint attached to its arm. The arm grasped the middle of the heavy plate, which weighed near 15 lbs., and I could thus cause the plate to revolve on its short axis in any desired plane. I now placed it so that the long edges lay in the magnetic meridian, with the short edges running east and west, consequently so, that while the length of the plate revolved vertically in the plane of the meridian, the short edges always moved in the magnetic parallels. The results (obtained by Mlle. ZINKEL?) are represented in the diagram, Fig. 22. The roman letters mark the colours of the eastward corners, the italic letters those of the western corners of the short edges, in every position.

549. On examining the diagram, we find all round, the

Fig. 22.

whole of the colours, *in general*, exactly corresponding to the laws already ascertained. Towards north, blue prevails, towards south, red ; above, yellow, below, grey ; but *in the details* we find in each pair of corners, and in every position, the prevailing colour of the arc, transversely modified, and this uniformly in the sense of the side, east or west, towards which each corner lies. We thus see, all round the circle, the eastward corners invariably modified or discoloured by grey, the westward corners tinged with yellow, and these two influences affect all the colours.

550. I next made the same experiment, noting, however, fewer positions, in order to shorten it, with the length of the plate in the parallel. (It is not said whether the shorter edges *of the breadth* were vertical, or horizontally north and south, in which latter case the very short edges of the thickness would be vertical. Probably from the diagram, Fig. 23, the broad edges were horizontal, and ran north and south ;

but this, as we have seen in § 546, and as we shall see in § 568, *et seq.* could not materially alter the results, since a trifling thickness is sufficient decidedly to differentiate the colour.—W. G.) When the plate resolved vertically, the following results were obtained, (by Mlle. ZINKEL?) which I give in the diagram, Fig. 23. All the roman letters designate here the colours on the northern edges, the italics mark those of the southern edges, in each position.

Fig. 23.

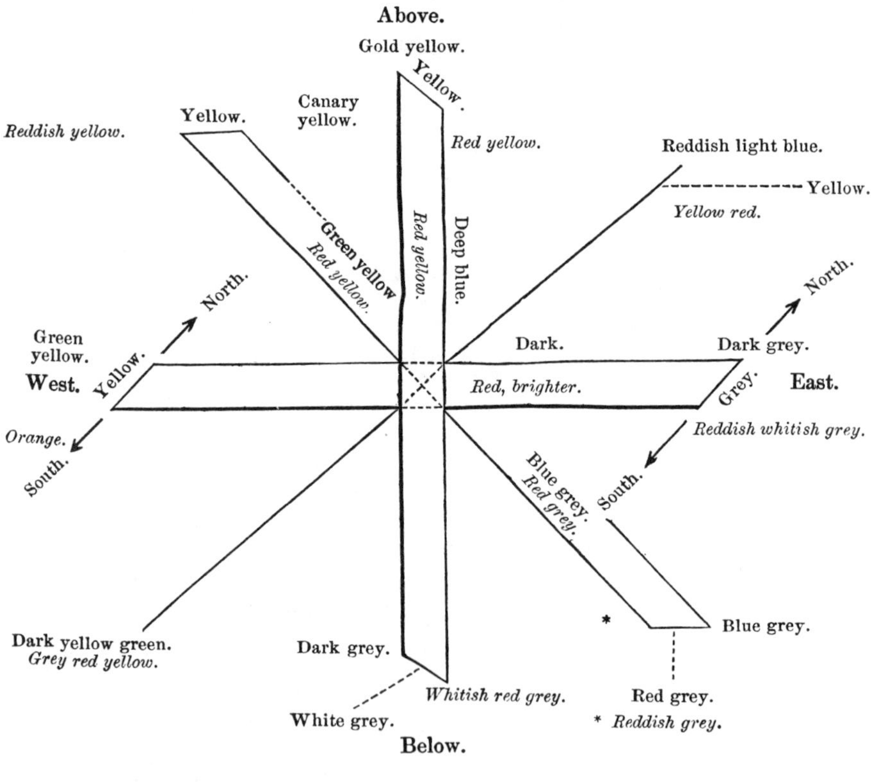

551. Here also, as in all the experiments detailed in § 513, and § 536, with soft iron bars, the *general* character of all colours above and to the west is yellow; that of all colours below and to the east is grey; but all round the circle, red on the south side, and blue on the north, make their influence felt transversely in the shade of the coloured flame, which

acquires from the mixture of blue sometimes a dark grey, sometimes a green tint.

552. The long edges correspond to this in their whole extent on both sides; nay, their two halves become polar and opposite to each other, at least in the line of the dip or inclination; thus :—

below,	Blue grey. Red grey.	above,	Green yellow. Red yellow.

All these contrasts could not possibly manifest, divide, and subdivide themselves more beautifully than they did in these multiplied experiments, with the most pleasing consistency. They exhibit the most immoveable constancy in the operations of the all-influencing natural forces which here operate, and which, in every variation of the experiments, always stand forth unchangeably the same.

553. And thus we have placed beyond a doubt, *the existence of transversality in the odylic phenomena*, as it has long been known, from the time of the first researches of M. PRECHTL, to exist in the magnetic phenomena. The difference, so far as my researches have yet extended, and as may be seen from the details I have given, consists chiefly in this, *that magnetic* transversality has greater independence, while odylic transversality exhibits more dependence on the threefold polarity of the earth.

554. After I had thus ascertained, under variously modified circumstances, that fixed points of the compass uniformly correspond to fixed odylic colours, the Iris itself, where all the colours appear at once towards any point of the compass, became only the more enigmatical. In this difficult and deep-lying subject, no doubt, we cannot for the present, with propriety, speak of solving enigmas; for here all is enigmatical; but we can collect facts, arrange them according to their analogies, and group those which are of one kind together in the meantime. With this view, I now wished to try whether the Iris, visible over the poles of strong magnets, has an independent existence of its own, neither bound down to nor

affected by the points of the compass, up to a certain point; or whether I might not perhaps succeed in isolating and exhibiting separately its component colours. We know, up to this time, that the Iris, on the magnetic poles, always appears horizontally stratified, so that the red lies lowest, then the yellow, and so on to the blue and violet, above which a streak of red occurs, followed by grey smoke. We know further, that when the flame, with its Iris colours, is directed towards any point of the compass, in each case a different colour, but always the same colour in the same position, becomes predominant, in size and brightness, over the others. I wished now to see, whether, if I gave to the flaming emanàtions at the poles, channels or rather points of egress of different forms, there would be developed, in all alike, the Iris corresponding to the direction of the pole?

555. For this purpose, I had a number of different caps of iron, made to fit on the end of the pole of a large bar magnet. They all consisted of a hollow four-cornered part below, which could be fitted on the end of the bar, but ending above in a variety of shapes, such as I thought it desirable to try. I could thus give to the pole, or end of the magnet, any shape or termination I pleased. I now fixed vertically my strongest bar magnet, and attached successively to the upward pole the different caps. This was done first with Mlle. ZINKEL, and some months later, with the less sensitive Baroness VON AUGUSTIN, and Dr. NIED.

556. The first cap tried was *hemispherical above*, and 1.2 inches in diameter. The magnet now showed to Mlle. ZINKEL a dull Iris, in which two colours, yellow and blue, predominated, the yellow below, the blue above; the other colours were turbid and indistinct. Here there was little change in the form of the pole, and the odylic flame was upwards of four inches long, as it had been before the cap was attached.

557. I now tried another, *hollowed out in the middle externally, and ending in a cutting edge all round.* It formed, in fact, a cup, in which the hemispherical end of the former cap would exactly fit, and was therefore the opposite to it. The raised edge now also gave out all the odylic light, while none

arose from the hollow part. The luminous wreath or circle again formed an Iris, but no longer, as in the last experiment, horizontally stratified, the colours being now arranged in vertical bands, the whole about two inches high. Both Mlle. ZINKEL and Dr. NIED saw this.

558. When I tried *a similar cup, with a free point projecting from its lowest point*, the coloured wreath remained unchanged, while the point gave out, not coloured flames, *but only grey smoke.* Dr. NIED only saw in the middle a bright luminous point.

559. With a cap, ending in a *four-sided pyramid*, and consequently in a single point, I obtained an elongated narrow flame, presenting *an Iris*, in which the colours were again horizontally stratified over one another. It differed from that of the hemisphere, in rising higher and exhibiting more colours. Baroness VON AUGUSTIN also saw it, and described it as narrowing above to a point.

560. A cap *with a simple straight edge*, running transversely across the pole, again gave an imperfect Iris, in which the colours formed vertical bands on the horizontal line (as in § 557). When the edge lay in the plane of the meridian, the colour on the north corner was violet and blue, that on the south red. Between these lay the other colours, but more dull. With the edge in the magnetic parallel, the east corner had a grey, the west corner a yellow flame, while the other colours appeared in irregular mixture along the edge between these.

561. When I took a cap with *two diagonally opposite points*, each of these showed a flame. When one was north, and the other south, the colours were blue and red. This was seen also by Dr. NIED and Baroness VON AUGUSTIN. When one lay east, and the other west, Mlle. ZINKEL saw the former grey, the latter yellow. In north-west and south-east, they were green and dark muddy red ; in north-east and south-west, reddish violet and orange.

562. I now tried a cap *with four points*, forming the corners of a square, and placed it so, that the points corresponded to north, south, east, and west. Flames, but of different colours, now appeared in all four points. Baroness VON AUGUSTIN saw that

of the east point grey ; of the south, red ; of the north, blue ; of the west, brightest, but doubtful in colour, sometimes pale blue, sometimes whitish.—Dr NIED gave the same account of them, with the addition, that the west point was decidedly yellow.—Mlle. ZINKEL saw all four colours with perfect distinctness.

563. On reversing the magnets and placing the caps on *the southward pole*, the results were the same in every case, but the colours were found by Mlle. ZINKEL to be duller and feebler than when the northward pole was upwards.

564. I had thus the satisfaction *of having decomposed or analyzed* the flame of both poles of a bar magnet (which originally formed each an Iris, with colours differing in intensity), *by putting on a cap with four points, into its four fundamental colours, each in a flame of equal size.* In this experiment with one magnetic pole, I reached the point which I had previously attained only by means of much more complex experiments, that, namely, of proving that every odylic flame of a magnet contains all the conditions necessary to produce all the colours of the spectrum ; which are always called into existence and made to appear separately, by the influence of the corresponding point of the compass, when the apparatus is so arranged that the colours can separate from one another, and each freely develope itself alone. We might say, in other words, that in every magnetic pole there is thus shown to exist a sort of double odylic transversality, depending on the polarities of the earth.

565. I now tried the effect of *a circular surface or disc.* A disc of iron plate, 13.2 inches in diameter, was well flattened, and an iron wire folded into its circumference, so that a smooth round clean border, one-twelfth of an inch thick, ran round it. It was suspended by a small hook in the middle, horizontally above the pole of the magnet, and could be fixed at any height. I could now let it down on the northward pole of the magnet, which stood vertically. In this way it formed a widely extended flat cap or termination, or in other words, an induction magnet of the small length of the thickness of the plate, but of the great thickness of the diameter of the disc.

566. I showed to Mlle. Glaser first the disc by itself. It had a pale whitish grey glow, but no colours could be observed. I now let it down on the pole, and it instantly acquired odylic light. In the middle point of the upper surface appeared a blue spot, the surface became bright, and the border coloured grey on the east, and blue, yellow, and red on the north, west, and south respectively. These colours passed into each other where they met.

567. Mme. Bauer saw the same; but the surface was not grey merely; the colours were seen not only in the border, but for a considerable distance inwards from it. When I used the southward pole of the magnet, the colours were the same, but feebler and less clear.

568. I showed the disc, resting on the northward pole, to Mlle. Pauer. The odylic glow instantly spread over it. The colours were developed, as might have been expected; on the upper centre a blue spot, on the lower, in contact with the magnet, a red spot, both upwards of two inches in diameter. They passed into a surrounding yellowish zone, faintly tinged with red on the under with green on the upper surface, and this again lost itself in a grey zone. This last continued to the border, where it was surrounded by a kind of downy fringe of light, 0.6 of an inch thick, and coloured, grey, blue, yellow, and red in east, north, west, and south respectively. In north-west, south-west, south-east, she saw respectively green, orange, and grey-red; in north-east violet, with a short patch of red. These colours formed a continuous wreath of tints passing into each other, and thus a kind of circular rainbow.

569. Mlle. Pauer saw a blue flame, two inches high, rising from the upper middle; it passed above into sulphur yellow, and that into grey; this of course could not be observed below, where the disc rested on the magnet. But the whole disc was surrounded by a downy nebulous light, only distinctly seen in profile, which was bluish above, and reddish grey below. It had a slow undulating movement, and flowed gently along both surfaces towards the centre. She could disturb it by blowing, when it became for a moment brighter, but soon recovered its former state.

570. I varied the experiment as follows with Mlle. ZINKEL; I connected with the poles of a SMEE'S voltaic battery of more than two and a half square feet of surface, the two surfaces of the disc; the wires being only separated by its thickness, about one twenty-fifth of an inch.

Immediately the observer saw round the upper centre of the disc, connected with the silver, a spot of blue glow, forming more than two inches in diameter. At the same time a similar red spot appeared on the under surface, connected with the zinc. No flame appeared. But the whole disc acquired a coloured glow, not merely on its border, but over its surface, blue, yellow, red, and grey, appearing respectively in the north, west, south, and east positions; green in north-west, &c. &c. as before. The blue and red central spots each formed a kind of star of innumerable points, or rather ray-like prolongations, stretching out towards the circumference, and uniformly exhibiting the colour corresponding to the point of the compass towards which they were directed. On the rest of the surface, the colours were arranged round the central spot in successive zones, so as to form a rainbow of parallel circles. A luminous web of fine downy fibres enveloped the border of the disc, like the down on closed horse-shoe magnets, and 0.6 of an inch thick. Besides this border, the whole surface was covered with a similar downy light or flame, rising as high as the thickness of a thin quill (about one-fifth of an inch), with a feeble undulating motion towards the middle, where it was rather stronger. It veiled the glow of the disc as with a luminous mist.

571. In order, if possible, to place the copestone on the results of the preceding experiments, I now had a hollow iron ball, without any projections, prepared. It was rather less than thirteen inches in diameter, and was turned in two halves, which could be fitted so closely (in the same way as the Magdeburg hemispheres), that the joining could hardly be seen. A small hole was bored through the junction, and a silk thread introduced, by which the hollow ball was freely suspended.

With this arrangement, and reflecting on all the facts above developed, I admitted the hope, that I might possibly produce something which might approach to an artificial polar light, a northern and southern light, or *aurora borealis* and *australis*. In order to supply my sphere, like that of the earth, with two magnetic poles of equal force, and to be able to regulate their intensity, an electro-magnet appeared to me the best means. I coiled a copper wire, one-twelfth of an inch thick and covered with silk, round a cylindrical iron bar, an inch in diameter, which I placed diametrically within the sphere. Its length was precisely that of the internal diameter of the hollow sphere, and it fitted into it so as not to fall out of its place, as long as the ball was not separated into its halves. I placed it vertically, touching the highest and lowest points of the hollow within the ball, and brought the two wires out through two small holes lined with quill, above and below. They were now connected with a SMEE'S battery, having an active silver surface of more than two and a half square feet. I showed the apparatus, thus arranged, in the dark chamber to Professor Dr. HUSS of Stockholm, M. HOCHSTETTER, Baroness VON AUGUSTIN, Mlle. PAUER, Mlle. GLASER, Mme. BAUER, and Mlle. ZINKEL, all perfectly healthy and vigorous persons.

572. Dr. HUSS, the least sensitive, saw the ball only for a moment or two, while the battery was acting most powerfully; but he soon lost sight of it, whether from weakness of the inducing current, or subjective intermission.

573. M. HOCHSTETTER saw a bright round spot appear round each pole, about two inches in diameter, without distinguishable colours, and in the central points of the outer surface exactly corresponding to the ends of the electro-magnet within. More he did not see.

574. Baroness VON AUGUSTIN saw the ball by itself feebly luminous. When the connection with the battery was established, the light increased, and the ball became enveloped in odylic smoke. At the poles she saw brighter portions, which cast a turbid but distinctly visible light on the ceiling and the floor. She observed differences of colour on its surface, feeble indeed, but on the north distinctly blue, on the east grey and dark, on the south reddish; the west side was brightest, but she could not with certainty define its colour.

Anka Hetmanek saw the ball in white odylic glow before it was connected with the battery. When connected, it became much brighter, with coloured streaks from one pole to another. The colours were on the north, blue; on the west, yellow; on the south, red; on the east, dark grey; on the north-west, green. The poles were brightest, the upper blue, the lower red.

Mme. Fenzl saw the ball become luminous as soon as it was connected with the battery. The south-west side was much the brightest, but could not distinguish colours. But from both poles she saw lights flowing, which, to use her expression, spread over the poles like expanded parasols; they were concentric with the polar surfaces, but at a short distance from them.

575. The phenomena appeared much better to the eyes of Mlle. Pauer. Before connection with the battery, she saw the ball like a pale ball of light in the air. When the current passed through the coil, the ball acquired a strong odylic glow, which soon became coloured. The north side, when she looked down on it, appeared blue; the west, yellow; the south, red; the east, grey; she saw green between north and west, orange between west and south, violet between east and north. The west side was brightest, the east darkest. The colours shaded insensibly into one another. But the most intense light appeared at the points corresponding to the poles of the inclosed electro-magnet, which were the poles of the ball. The upper had a blue, the lower a red spot, the former being connected with the silver, the latter with the zinc of the battery. Each spot was four inches in diameter, and passed at their outer border into yellow, which again was lost in grey. At both poles flames appeared, blue above, red below, and about four inches high. The whole surface was covered with a delicate grey nebulous veil, seen most distinctly in profile, and about 0.4 of an inch high. It appeared to her to flow gently from the equator to the poles, and was disturbed by blowing on it. Along the equator, all round, she saw a brighter ring, narrow and whitish yellow.

576. Mme. BAUER saw the ball by itself in a feeble yellow glow floating in the air. As soon as the current began to pass, she saw it assume a splendid variegation of rainbow colours. The light was most intense at the poles, blue above, red below. The polar spots were six inches in diameter, and divided externally into innumerable narrow meridional streaks of all the colours, running in straight lines towards, and at right angles to, the equator. These streaks were blue, yellow, red, and grey on the north, west, south, and east sides respectively, and between these appeared all the mixed colours and transition tints. The ball was darker on the east, the colours lighter and brighter on the west side. The intensity of light in the colours diminished towards the equator, where they disappeared. But here a new appearance presented itself, the equator itself formed a narrow luminous girdle round the ball. This last observation agrees with that made by Mlle. PAUER.

577. I was able to test these phenomena most accurately and minutely with Mlle. ZINKEL during the catamenia. Before the current passed, she saw the ball in white odylic glow. When the connection was made, it became very bright with the splendour of variegated colours, which always forced from the beholders expressions of admiration. She saw the appearances as the two preceding sensitives had done; the upper pole blue, the lower red; the light most intense at the poles, forming a large star of apparently countless points, or coloured radial prolongations, running over its surface towards the equator. They were blue, yellow, red, and grey. in north, west, south, and east, &c. &c. The whole east side was darker, the west brighter, and so forth.

578. She saw also the girdle round the ball at the equator. She described it as a luminous streak, dividing, at the greatest horizontal circle, the ball into two equal parts, upper and lower. It was one-fifth of an inch broad, and of a feeble whitish-yellow light, invisible to those of lower sensitiveness, and hence not seen by M. HOCHSTETTER, dimly seen by Mlle. PAUER, and only well observed by Mme. BAUER and Mlle. ZINKEL. Above and below it was not sharply defined, but rather

exhibited the appearance of a fine comb with numberless short teeth, pointing towards the poles, and at right angles to the equatorial girdle itself.

579. The light over the surface, its odylic glow, was described by her as by Mme. BAUER to be continuous, but within this continuity to be formed of an infinite number of distinguishable *fibres*, which appeared of the thickness of a small knitting needle, and ran perpendicularly from both poles to the girdle. She considered these not as independent threads of light, but rather as lines of greater intensity, alternating with lines of less intensity, and thus giving the appearance of threads running from the poles. They all had the colour of their position, blue on the north both above and below the equator, yellow on the west, and so forth.

The blue and red polar spots had the considerable diameter of six inches, or a radius of three inches from the poles. They then divided and formed the threads which passed over the remaining zones to the equator. This gave to the polar spots the aspect of stars, the more strongly marked from the different intensities of the different colours, which caused some points to project apparently further than others, and thus form a star.

580. The intensity of light in these phenomena diminished from the poles towards the equator, so that, in the tropical zones, they became gradually duller, and at last disappeared very near the luminous equator.

581. The colours round the sphere, which thus formed a spherical Iris, passed everywhere into each other, as blue through green into yellow, yellow through orange into red, &c. These transitions were formed in this way: each of the meridional threads had a colour very slightly different from that of the threads on each side of it; no two were perfectly alike, and the difference was marked enough to be distinctly seen by the observer. Thus the ball appeared as if composed of a thousand coloured threads, which gave to it so great a charm, that Mme. BAUER and Mlle. ZINKEL declared that they had never seen any thing so beautiful.

582. The chief colours also divided themselves each in its

own field into several chief shades, so that a certain number of threads, although no two were entirely of the same colour, yet formed bundles according to their resemblances. Thus there was a bundle of dark blue and a bundle of azure blue luminous threads. The green, which was developed from the azure blue, showed first a bundle of blackish green threads, then one of a grass green colour. The yellow, red, and grey, were also thus divided into principal shades. The grey had in the middle a dark grey bundle, 0.8 of an inch broad near the equator. On each side of this lay a bundle of whitish light grey threads, 1.2 to 1.6 inches broad. Before the violet was developed from the grey towards north, there appeared the often mentioned strange phenomenon of a narrow portion of red. It was pretty sharply separated from dark grey, and passed pretty rapidly into dark violet blue. Its breadth at the equator was not more than one-fifth of an inch. This remarkable red streak was very brightly luminous and strongly red, much brighter than any part of the red on the south side of the ball. *Red, therefore, occurred at both ends of the spectrum, beginning and closing the series of colours.* It was formed on one side from the yellow, on the other from the blue, and we thus perceive very distinctly *what is to be thought of the violet of the spectrum generally, which is not a peculiar colour, but in fact only a mixture of blue with the neighbouring pure and independent red of the streaks so often mentioned.* Why this red, which in the ordinary spectrum appears only as violet in a part of the blue, stands forth independently in the odylic spectrum—this is a fact, the causes of which can only be ascertained by further researches of another kind.

583. Besides the general difference in intensity of light, between the west and east sides of the ball, another was observed not less comprehensive, between the upper and lower halves. The upper half, the pole of which was blue, showed a very much stronger light on its red and yellow parts, but was duller on the blue and grey. The lower, with its red pole, was brighter in its blue and grey parts, and turbid in its yellow and red. This caused the appearance, that the red and yellow, when sent from the red pole, were dull, and be-

came brighter as they approached the other pole beyond the equator, while the blue and grey were sent off in dull threads from the star of the blue pole, which became more lively and bright beyond the equator, on the hemisphere of the red pole. Like colours, therefore, appeared in opposition, as to intensity of light, between the poles and the luminous threads, unlike colours harmonized in this respect.

584. Such were Mlle. ZINKEL'S descriptions of the *odylic glow* of the ball. Let us now attend to the *odylic flame*, as seen by her. At each pole a flame streamed out, perpendicular to the surface of the sphere, 2 to 2.5 inches high, 1.2 to 1.6 inches thick, *which then spread out and laid itself on all sides parallel to the spherical surface*, to the extent of a radius of 3.2 inches round the polar or central point. It divided into bundles of fibres or single fibres, and flowed, like the fibres of the odylic glow, and in the air, above and parallel to them, over the ball, to the extent just mentioned. Above each pole, therefore, there appeared a luminous mass of 6.4 inches in diameter. The observer compared the whole flame to a loosely bound sheaf of corn, standing on the ground, the ears and stems of which hang over on all sides, horizontally, or curved downwards. The bundles of flame were not tranquil, but flickered and scintillated continually backwards and forwards, shortening and lengthening, and shooting out rays like the bundle of electric light at the positively electrified conductor. *The resemblance of this phenomenon to the polar light of the earth or aurora borealis* (*and australis*) is so obvious, that it must occur to every one who may take the trouble to read these lines. The ball, in my experiment, is in fact a *terrelle* or miniature earth, analogous to the magnetic one of BARLOW.

585. Another luminous phenomenon, connected with the odylic flame over the ball, is *a luminous shell of vapour*, a kind of delicate *photosphere* surrounding the ball. It consists of a turbid veil of light, not lying on the ball, but at a small distance from it, and floating as a hollow sphere or shell in the air all round. Its distance from the surface is the thickness of a small finger, or about 0.4 inch, and its thickness about 0.08 inch. It resembles the luminous covering seen by

Mlle. ZINKEL and others over the surface of a ball connected with the electrified conductor, which I shall describe in one of the subsequent parts of this work; but it differs from this last in this, that while the ball on the conductor is entirely surrounded with the nebulous shell, our terrelle is only or chiefly covered with it on those parts where the blue and the red prevail. Over the yellow and the grey it is feebler, and in parts disappears. It therefore chiefly affects the line of the meridian in which the terrelle lies, keeping to north and south, and retreating from east and west. Its density was not inconsiderable, for it visibly veiled the threads of odylic glow beneath, and rendered their outlines less distinct.

586. *Odylic smoke* rose abundantly over both polar flames. From the blue, it rose vertically to the ceiling, inclosing the silk cord by which the ball was suspended. This cord acquired a golden yellow light, as did a small weight attached to it at a considerable height. The smoke formed on the ceiling a luminous spot, nearly thirty inches in diameter, so bright that the observer could, though with difficulty, see the pattern painted on the ceiling. The smoke was checked by the roof, and flowed along it for about forty inches; the cord also was luminous for a certain distance where it ran along the roof.

587. Six months later, I caused to be made a much larger ball of sheet iron. It had a diameter of 21.2 inches; the electro-magnet was one inch thick, and had a triple coil of copper wire one-eighth of an inch thick. The battery (one of SMEE's form) used was so powerful as to heat this wire. The ball was shown to Baroness VON AUGUSTIN and Mlle. GLASER. The former saw the coloured meridians running from pole to pole, as described in § 574. But she also saw short columns of luminous vapour rising from the poles, red and blue from the positive and negative poles. These stems of light, as she called them, spread out above, and turned over. She compared them to the aspect of a palm tree, the leaves of which are sent out from the summit horizontally in all directions. Mlle. GLASER saw the coloured streaks or meridional lines all over the ball, with the colours proper to every quarter of the

compass, as so often enumerated. The coloured bands were, at the equator, about a handbreadth across, and divided from each other by turbid, indistinct transition bands of the same width, in which the colours were fused together. Above was a mass of light, which she also compared to a tree, the stem of which rose from the pole, while its branches spread out from the summit on all sides. This is precisely the appearance which Mlle. ZINKEL described as an overhanging sheaf of corn, and which she also compared to a round painter's brush, held upwards, and spreading out at the top. A similar, but red mass, was seen at the lower pole, but less distinctly.—ANKA HETMANEK also saw columns of light from the poles, which spread out and bent over like a tree. Mlle. ZINKEL saw everything on the larger as on the smaller ball, except the equatorial girdle and the distinct threads in the meridional bundles of light. This probably depended on a somewhat less intensity of odylic force in the large ball.

588. The statements of Mlle. ZINKEL are controlled in these experiments by those of Mme. BAUER, of which they are only a more enlarged and detailed developement. Those of Mme. BAUER are again confirmed by those of Mlle. PAUER and of Mlle. GLASER; and these last have their foundation in those of the less sensitive Baroness VON AUGUSTIN, and of the still less sensitive M. HOCHSTETTER. All give us the same phenomena, only more detailed, in proportion to their higher sensitiveness. Mme. BAUER is indeed more sensitive than Mlle. ZINKEL, in her ordinary state, is; but during these investigations, the catamenia were present in the latter; and at such periods, she was not only as sensitive as Mme. BAUER, but even equalled in perceptive power persons subject to somnambulism, such as Mlle. ATZMANSDORFER and others. If Mme. BAUER did not describe some, and described less minutely others of the phenomena afterwards brought out by Mlle. ZINKEL, this was probably in consequence of the short time during which Mme. BAUER was able to try the experiments, and on the other hand, of the quiet and leisure with which I was able to experiment on Mlle. ZINKEL, to repeat the trials on different days, and thus bring out the details.

589. Such explanations, however, are not required for those who have read these researches with some attention. The whole course of them, continued as they have been for more than three years, shows, on every page, that, in about sixty persons included in the investigation, the same threads of physical law may be traced throughout; and that every new branching off or subdivision has only supplied new proofs of the accuracy of the observations, and new guarantees for the consistency which prevails in the connections between the facts observed. Thus we have seen, to take the example of the last-mentioned investigation (§ 525), that a yellow odylic flame flows from the end of a vertical bar of soft iron, but that from the pole of a magnet an Iris arises (§ 517), the colours of which lie in horizontal strata above each other, red appearing a second time above the violet. A horizontal line, forming the termination of a magnetic pole, vertically placed (§ 560), yielded an Iris arranged in vertical strata along the horizontal line. Bar magnets showed (§ 517,) coloured flames at both poles, with certain principal colours in uniform contrast, one such contrast in the meridian, another in the parallels. Four-cornered terminations exhibited at the corners (§ 539, &c.) the four principal colours at once. But it was found possible to isolate and separate the colours more perfectly still at either pole. A two-pointed termination gave the two opposite colours of the meridian when the points lay in that line; of the parallels, when they lay in them (§ 561), just as a bar magnet showed them at its opposite ends. And a four-pointed termination yielded (§ 562,) all the four colours perfectly separated and unmixed, on either pole. Thus was demonstrated, at the same time, a double transversality, with the most beautiful distinctness and precision. A disc exhibited (§ 565,) on a horizontal surface all the four colours at once, and also all the mixed and transition tints. Lastly, a hollow sphere, uniting all these appearances (§ 571 et seq.), exhibited each distinctly and all collected on a single field of action, the surface of the sphere, on which a glance perceived them all in their mutual relations. To produce these effects, I sometimes used terrestrial magnetism, sometimes that of magnetic steel bars, at

other times that of an electro-magnet, but in the end they all come to the same principle, as is easily seen. We have seen, in numberless variations, in this section, that (to resume it briefly)—

590. *a.* The light, visible in the dark, sent out by magnets, is seen, by sensitives, of different colours at different distances, but is constant in colour for the same eye, at a certain fixed distance, which varies with the individual.

b. This light not only assumes different forms, but all known colours.

c. These colours embrace the whole spectrum, with its mixed colours and transition tints, as well as white and black combined in every shade of grey.

d. The colours often appear singly to the sensitive eye; generally blue at the northward or negative pole, red at the southward or positive pole, or grey at both.

e. But in most cases, and especially in cases of high intensity, several appear together; frequently all are seen.

f. When this is the case, and when the colours can arrange themselves freely, they always assume the order of the rainbow and of the spectrum.

g. When stratified horizontally one over the other, red is always lowest, blue highest.

h. Above the blue, and connected with it by the violet (which is an intermediate or transition colour thus produced), appears once more a pure red, so that the odylic spectrum, which begins with red, and passes through orange, yellow, green, blue, and violet, also ends in red.

i. These coloured luminous phenomena are produced, according to the same laws, by the magnetism of steel bars, by that of electro-magnets, and by that of the earth.

k. The last of these, relatively to us immoveable, impresses on the phenomena certain laws, which yield, for every point of the earth in which they are observed, different results.

l. Terrestrial magnetism produces them in any bar of soft iron.

m. They consist, in all observed cases, probably in all cases, of an Iris, except perhaps in certain directions, where grey alone occurs.

n. Within this Iris one colour usually, more rarely two, predominates in size and intensity. Often this prevailing colour is alone seen by the sensitive, while the feebler associated colours escape them.

o. In general, when directed downwards in the line of dip or inclination, they are loaded with grey. Towards north they are blue, yellow above, red towards the south. They are further yellow towards west, grey towards east. This applies to the vertical meridional circle, to the vertical circles in the parallels, and to the horizontal circle, with equal accuracy.

p. When the magnetism of bars, or of electro-magnets, comes, in virtue of their unconformable position, into conflict with that of the earth, the tints are enfeebled and discoloured. When the magnets lie conformably, the colours become stronger and more vivid. Intermediate positions yield intermediate tints and shades.

q. The odyle of crystals, of the animal body, and of every source of that influence, acts on the odylic light of magnets or of other bodies, when brought into conflict with it, just as the magnetism of the earth does.

r. A bar magnet, revolving on its axis, and with flames at both ends, does not exhibit, in any vertical or horizontal circle, nor in any other position, flames at its poles, the colours of which are complementary, although they stand in polar opposition.

s. The colours of the upper semicircle have more intensity of light than those of the lower. All the colours on the northward pole are brighter on the semicircle towards north, duller on the opposite side. This is reversed in the colours of the southward pole.

t. These coloured flames may be communicated from the magnet to other bodies.

u. Magnets, ending in several points, divide the colours among these, so that each has a colour corresponding to its position on the compass; and the Iris of every flame may thus be analysed into its elementary colours.

v. A four-cornered iron plate thus exhibits both magnetism and odyle transversely disposed, not only longitudinally in the meridian, but also in a line at right angles to this.

w. An iron disc, and, better still, an iron ball, with an electro-magnet passing through it, exhibits all these phenomena united, along with a number of new ones, giving to its lights all possible resemblance to the polar lights of the terrestrial sphere.

x. The odylic nature of the positive magnetic north pole of the earth, that of the east, and that of the ground (that below us), have a general character of agreement, in which they stand directly opposed to the negative magnetic south pole of the earth, to the west, and to the heavens (that above us).

ODYLIC LIGHT

IN THE MORE LIMITED SENSE OF THE TERM.

591. It is hardly necessary to explain, that we must accurately *distinguish odylic light* from odylic glow, odylic flame, odylic luminous fibre, odylic scintillations, and odylic smoke, and that these are all to be viewed as sources of odylic light. All this is included in the fact, that these phenomena are seen. For according to the structure of our eyes, we can only see that which gives out light; when we see any object, therefore, it emits luminous rays, or whatever name may be given to its emanations. I have already explained this clearly in § 20. The statement of Mlle. Reichel, there recorded, that the flame of the large nine-bar horse-shoe diffused light over the table twenty inches around, and the similar account of Mlle. Sturmann (§ 55.), have been in the mean time confirmed by numerous other observers. I shall here only name a few. Even the blind Bollmann observed that the flame of that magnet diffused light to the length of an arm round it, which he compared to a large luminous cloud. Mlle. Zinkel saw a bar magnet cast light on its armature lying near it. When she interposed her finger or any other body, the light was intercepted, and a shadow was seen on the corresponding part of

the armature. She also saw a light proceeding from the flames of two horse-shoes which I brought near each other, rendering the whole table visible to the distance of more than a foot. (§ 405. *c.*) Mlle. ATZMANNSDORFER saw not only the neighbouring objects, but in a certain degree the whole room, illuminated by the flame, five or six feet high, of the nine-bar horse-shoe, so that she could distinguish all large objects as in twilight.—It has been often mentioned that various other sensitives saw the ceiling so illuminated by the light from odylic flames and smoke, that they could distinguish the lines of the painting on it. Dr. NIED and M. DELHEZ saw on the ceiling over the nine-bar magnet a large round light spot, of the size of a common round table, and this at a height above the magnet of from thirteen and a half to nearly seventeen feet. Prof. ENDLICHER and Baroness VON AUGUSTIN saw the ceiling illuminated to a large extent over the same magnet, and still more over a strong electro-magnet. The latter lady saw the larger terrelle cast large lights both on the ceiling and floor, nearly three and a half feet in diameter. M. HOCHSTETTER saw a bright spot of the same size on the table, caused by the flame of the nine-bar magnet, when in the electrical atmosphere, as well as a bright round spot on the ceiling over the flame.—I must here mention a remarkable case, the more worthy of notice, as it stands hitherto isolated among my observations. The two terrelles, besides illuminating the ceiling vertically over them on a space about twenty-six or twenty-seven inches in diameter, *cast also a separate spot of light, of the form of the half moon*, laterally north-eastwards or perhaps northwards, on the ceiling; and this spot moved with the motions of the ball. This is a phenomenon, which must yield very important results, if followed up.

592. Mlle. ZINKEL once noticed, in the dark chamber, a small spot of light, on the opposite side of the room, and at a considerable height. When we went nearer, it appeared that the spot was on the ceiling, and that a feeble thread of light led from below to it. On searching below, we found a twenty-inch bar magnet, one square inch in section, fixed on a support, and having its southward pole upwards. From this, the

thread of light rose vertically, and formed the bright spot on the ceiling, which was as large as the bottom of an ordinary wine glass. On moving the magnet, the spot moved; on interposing the head, it disappeared. This observation is very remarkable, as proving that the cause which produced the spot of light *had proceeded to the distance of thirteen or fourteen feet, without dispersion;* for the section of the magnet and the spot of light were of nearly equal size.

593. I placed a seven-bar horse-shoe at the distance of two feet from the wall. There appeared two spots near each other. One, blackish, corresponded to the northward pole; the other, brighter and reddish, to the southward pole. Both, according to Mlle. ZINKEL, were as large as an egg. I cannot enter into the explanation of this interesting phenomenon here, but must reserve it for one of the following treatises. I only extract here the fact, that the flames of horse-shoes cause spots of light on side-walls, which remain close together.

594. Mlle. ZINKEL observed these phenomena in another form. When a copper wire, one-twelfth of an inch thick, was coiled round the northward pole of the nine-bar magnet, a beautiful flame flowed from the other end of the wire. It was eight inches long, and so luminous that the floor, at forty inches from it, was illuminated. The experiments in the next section will still further establish the results obtained in this one.

CONCENTRATION OF THE ODYLIC LIGHT.

595. The experiment described in § 18, in which I concentrated magnetic light by a large lens in presence of Mlle. REICHEL, has been since repeated with many others. I procured from Paris a large lens, which, with a diameter of twelve inches, had a focal length of 11.5 inches. This heavy glass was so framed, as to admit of motion in every direction. I now placed the nine-bar magnet at forty inches from the lens, towards which its poles were directed. I could not remove the magnet further off, because I should then have lost too much of its feeble light; but even at this distance,

since the flame was 10 or 12 inches wide, I might expect that a sufficient number of parallel rays would fall on the lens, to admit of concentration into one principal focus. I showed the whole, when thus arranged, at different times, to Mlle. ATZMANNSDORFER, Mlle. DORFER, Mme. KIENESBERGER, FRIEDRICH WEIDLICH, MM. KOTSCHY and TIRKA, JOHANN KLAIBER, the blind BOLLMANN, Mlle. ZINKEL, and Mlle. GLASER. Even BOLLMANN saw three lights in different places, and when I made him feel for the sources of them, his hand touched first the magnet, which he saw pale yellow, then the lens, which was reddish, and lastly the screen, where he saw the smallest, but the brightest and strongest light. All the others saw, when the screen was from twelve to sixteen inches distant from the lens, a round spot of light, from 0.96 of an inch to 3.2 inches in diameter. The most exact observers pointed out twelve inches as the distance at which the spot was smallest and brightest. (The focal length of the lens was 11.6 inches.) This was the opinion of Mme. KIENESBERGER, Mlle. ZINKEL, and Mlle. GLASER. They all saw the lens in red odylic glow, as the bell-jar of the air-pump was when a magnet was beneath it, but the focal image white. M. KOTSCHY and Mlle. ATZMANNSDORFER also described a distinct cone of light, the base of which was on the lens, the point in the focus, and which shone in the air. Mlle. ZINKEL and Mlle. GLASER saw the image enlarge, whenever the screen was moved either nearer to or further from the lens. Mme. KIENESBERGER added, that when the screen was moved further off, the enlarged image assumed *prismatic colours*, a dark red spot occupying the middle, a yellow zone round this, and externally a broader zone of blue. Mlle. GLASER, who moved the screen backwards and forwards for herself, saw sometimes a blue zone round the yellow, sometimes a blue spot on its middle. Mlle. ZINKEL also often observed the same prismatic colours (the three primary colours). Here then, also, an Iris had began to develope itself. Although I had entertained hopes of seeing the odylic light when concentrated, and of thus observing it for myself, I was unable to perceive a trace of it. When I placed a seven-bar magnet above the other, in order to strengthen the effect, Baroness VON AUGUSTIN saw on the screen a round spot of six inches diameter, in the middle of which was another brighter spot, less than an inch in diameter. This was obviously the focus of the parallel rays which fell on the lens. The Baroness had the kindness to make a painting of this in oils, as it appeared to her, and thus render it plain to others. I repeated this experiment with Mme. FENZL, who has the same degree of sensitiveness as Baroness VON AUGUSTIN. It was interesting to hear from her precisely the same description of what she saw. In order to test the observers, I made various changes which they could not be aware of nor understand, moving the screen in every direction, as well as the magnet, and turning the lens a little up or down. In all these cases, they uniformly described the position of the focus as it ought to have been, according to the known laws of dioptrics. Details are here unnecessary. By all these observations, made on four diseased and eight healthy persons, the former statements of Mlle. REICHEL were confirmed in every particular; and I can only hope that other conscientious observers may soon repeat them, and establish the fact indicated by them.*

* At the close of the proceedings of the self-styled Committee of Physicians, discord arose in its own bosom. Some of them, and especially Drs. WOTZELKA and STAINER, felt how much was insecure in their method of investigation, and the want of sufficiently trustworthy grounds for the conclusions drawn. The latter gentleman, before the publication of the proceedings, took the natural and sensible precaution of obtaining from some of the sensitive persons with whom I had experimented, some account of the method I pursued, and of the results they observed. As he obtained, from persons of the highest character, only confirmation of all I had stated as to the sensations and luminous phenomena, he expressed a wish, for his own conviction, to be permitted to assist at some of my experiments. With the greatest pleasure and readiness I invited him to a trial made with a new sensitive whom I had never before seen. Dr. STAINER was necessarily convinced by what he saw, of the care and caution with which my experiments were made; and it could not but happen, that serious misgivings should arise in his mind, lest the superficial labours of the Committee (?) might be refuted by me with proofs, and those who took part in them, sooner or later, exposed. All this induced him to deliver an address in the Society of Physicians of 16th November 1846, in which he earnestly opposed the publication of the Report of the Committee, directed against Mlle. REICHEL, and indirectly against me. In this address, which is printed in the protocol of

596. I must here mention an observation, made with Professor ENDLICHER. Being accustomed to use spectacles, he kept them on in the dark chamber. When I placed the magnet before him, and he did not see it so well as I expected from what I had seen of his sensitiveness, Dr. FENZL, who was present, suggested that possibly the spectacles interfered. M. ENDLICHER took them off. Immediately he saw the flames very distinctly, and just as I had supposed, from my knowledge of him, he would. The glasses had reflected one part, and absorbed another, of the light, and that which was transmitted was so insignificant a residue, that it only sufficed for a feeble perception, while the whole, as given out by the magnet, was quite easily seen, and formed a flame four inches high.

597. It having been thus shown, that the odylic light of magnets is transmitted through glass, and its passage suffers absorptions, reflections, and refractions, like those of ordinary light, it was necessary to test the former by a mirror, and compare its action on odylic light with its action on ordinary light. It was possible that odylic light might be absorbed, and thus be unfitted for producing catoptrical phenomena, inasmuch as the odyle from which it originates, is easily absorbed by all kinds of matter.

598. I tried this with ordinary glass mirrors. I placed one

that sitting, he said, among other things, "that from *such results* no absolute proof against the thing could be brought. For his part, he must expressly state, that he had had an opportunity of seeing Baron VON REICHENBACH experiment with persons for whose honour, love of truth, and impartiality in the matter, he could answer as for his own; and that he had been astonished at the coincidence of the statements of these persons with the assertions made by Baron VON REICHENBACH in his work. Other honourable persons, well known to him, had also assured him, that REICHENBACH's discoveries had been confirmed in their own persons." Consequently, Dr. STAINER expressed his opinion, that these proceedings could only compromise the Society, and that it would be best that they should either not be published at all, or at the utmost, published only within the Society, beyond whose limits they should, like a private matter, on no account be allowed to pass. On this there arose a violent debate in the general meeting of the Society, and it was decided by a large majority, that the report of the self-styled Committee *should not be published in the name of the Society*. How well this resolution was respected and fulfilled by the functionaries of the Society, any one may see, who looks at the title of the report in the Journal in which it appeared.—R.

before the bed of Mlle. ATZMANNSDORFER, and opposite to it, in the proper position, the nine-bar magnet, standing vertically. She saw in the mirror the image of the odylic glow of the metal, but not that of the flame, not even a trace of it, when she carefully looked for it in different very dark nights. I repeated this experiment a year later with Mlle. ZINKEL, who also saw the glow reflected, but could not perceive in the mirror the flame or any other emanation. The light of this flame was therefore so feeble, that after the usual absorption by the glass, there was not enough left to be reflected in a form visible to these two sensitives.

599. I had already become resigned to this, and had no hope of having the flame observed in the mirror, when, in a repetition with Mlle. GLASER, she firmly declared that she not only saw the glowing image of the magnet, but also an image of the flame, although a very feeble one. I now repeated the trial with Mlle. ZINKEL, during the catamenia. She also now saw a reflected image of the flame, which she could not perceive before. Lastly, I tried it with Mme. BAUER. She saw in the mirror many glowing objects quite easily; and she perceived the flames of the magnet so well, as to be able to distinguish the blue and red flames of the opposite poles. But she also said that the image was, beyond comparison, duller and feebler than the flame itself. In the latter, she could see, at both poles of a seven-bar magnet, traces of the Iris, yellow and violet, as well as smoke. These were indeed feeble, but on the image they had disappeared entirely.

600. In some cases, the odylic light can also pass *through semi-transparent bodies.* It can be seen, for example, *through the closed eyelids.* This was first observed by Mme. FENZL, who told me, that she could see the light in the dark chamber at a certain distance, even with closed eyes. Mme. BAUER and Mlles. ZINKEL and ATZMANNSDORFER confirmed this. They could not see the form, but they could tell, when a body which gave out odylic light was brought near to them, with certainty where the light was. They all saw, with closed lids, the light both of the flame and of the glow. They could point out the direction from which the light came, whether from one

or more points, and they could thus see both the glow and the flame, not by their peculiar character, but because both gave out light.

601. Now, an eyelid is strongly translucent. If a candle be brought near any one standing in the dark with closed eye, he becomes immediately aware of the arrival of light. This translucency is so great, that no one can endure to look at the mid-day sun through closed eye-lids for more than a very short time. *The odylic light, weak as it is, has yet force enough to pass through the closed eyelids, and become perceptible by sensitives.* We shall see hereafter that this circumstance, apparently insignificant, is not without importance, when we have to explain certain other rather striking phenomena observed in sensitive persons.

THE NORTHERN LIGHTS, OR AURORA BOREALIS.

602. I may now be permitted to return for a moment to the subject alluded to in § 2, and in § 21, in the very beginning of these researches. Even then, at a time when I knew far less concerning odylic light, I expressed the opinion that it was nothing else than the same phenomenon, which, on the great scale, appears as the aurora borealis, or rather as the polar light, south as well as north. This opinion, during the long period which I have since then devoted uninterruptedly to the subject of odylic phenomena, has not only not lost any of its probability, but has acquired additional strength and security from many parts of the investigation. The observations on the behaviour of odylic light, under the more or less exhausted receiver of the air-pump, the developement of colours in it so often seen in my later experiments, the motions which we can at pleasure communicate to it, the discovery that there are perfectly healthy persons in large numbers to whom it is visible; these and other facts of importance, not only do not oppose my original view, but all of them lend it new and powerful support. The theory of Sir HUMPHRY DAVY, that the aurora is not so much a magnetical as an electrical phenomenon, a slow tranquil equalization or neutralization of the atmospheric elec-

tricity, at heights where the atmosphere is highly rarefied, and supplanting the rapid process of neutralization of storms in our lower latitudes, and at lower atmospheric heights, meets with many difficulties on a full consideration of the subject. First, during winter, the meteorological phenomena of rain and snow occur without storms in the temperate, as well as in the arctic zones, while in the former, no appearance similar to the aurora is produced by the atmospheric electricity. Secondly, according to the multiplied observations made on the aurora since the time of DAVY, the aurora does not occur more frequently in winter than in summer. Thirdly, according to the knowledge we have acquired of the laws of electricity from other sources, such an accumulation of electricity as would be necessary to fill with light, as often happens, half a quadrant of the sky during perhaps more than half the night for nights together, cannot be well conceived to take place in our atmosphere, in the state of constant motion which belongs to it. Lastly, we can see no reason why such an accumulation of atmospheric electricity, even if it could exist, should occur exactly round the magnetic terrestrial poles.* DAVY'S suggestion has been admired as a very ingenious thought; but it has obtained no firm footing among the established laws of natural science, because it does not give a sufficient account of cause and effect in these remarkable phenomena. But now that we know, from the preceding researches, *that flaming lights exist over magnetic poles, larger than the magnets from which they flow;* when we learn *that these flaming appearances are moveable, undulating, often moving in serpentine windings, like those of a ribbon agitated by the wind, becoming at every moment larger or smaller, shooting out rays, scintillating, variegated in colour, and often nebulous, vaporous, and cloud-like;* when we find that *with our breath we can cause it to flicker backwards and forwards;* when we observe *that it increases in a rapid ratio, in size, intensity, and brilliancy, in rarefied air;* and lastly, when we see *it followed at every step by the play of rainbow colours, &c., &c.*

* It is not observed that those accumulations of electricity with which storms are certainly connected, have any such tendency towards the magnetic poles. —W. G.

—there remains hardly one essential mark of distinction between magnetic light and terrestrial polar light; unless we regard as such, the difference of intensity and amount of light, in virtue of which the polar light is visible to every ordinary eye, the magnetic light only to the sensitive eye.

The undulations and serpentine windings, which the aurora borealis often displays, are, on the supposition of the identity of the two lights, naturally and simply explained by the motion of the wind which causes the light of the earth to wave to and fro in more or less rarefied strata of air, precisely as our breath does with the odylic light of magnets. The constant alternations of greater and smaller size, in the aurora, correspond exactly to the unsteadiness of the magnetic light in our experiments. The powerful light from great and undetermined heights in the atmosphere, observed by some travellers to lie higher than the higher clouds, agrees beautifully with our observations on the magnetic light in the exhausted receiver, where the odylic light increased strikingly in size and brilliancy under half of the ordinary atmospheric pressure. But the equally well attested and even more numerous observations of other travellers, who have studied numerous polar lights with the most conscientious attention in the polar regions, to whom their height appeared very much less, and who often described them as luminous clouds, also harmonise perfectly with the nature of the odylic light of magnets. We have very frequently, in the course of these researches, met with the odylic phenomena, of luminous nebulæ or vapours, flame-like smoke, or whatever name may be given to the varieties of this appearance. It, also, increased in strength under diminished pressure. This is the cause of the appearance of luminous clouds constantly rising, which render complete the parallel between the odylic light of magnets and the polar light of the earth. Some have endeavoured to explain this phenomenon by the assumption of illuminated ordinary clouds. In many cases, such may have really occurred; but we have seen that they are not essential, and that odylic accumulations alone suffice to develope in the air luminous smoky appearances, that is, luminous clouds, which in my experiments rolled up in heavy masses, like cumuli, and sometimes illuminated the whole room up to the ceiling.—The higher the odylic emanations of the terrestrial magnetic poles rise in or beyond the atmosphere, the larger, more luminous, and more brilliant must their torrents of flame become, for the same reason that they increased in so striking a manner under the receiver of the air-pump, even with moderate rarefaction. Since healthy persons, if not sensitive, perceive no sensations caused by magnetism or by odyle, we can see why travellers, even at or near the magnetic poles of the earth, were not able to observe any unusual sensations. Even the observation frequently made by them, that in very high latitudes, the aurora borealis is seen to the south, is no longer enigmatical; for odylic light, wherever it is developed, shines from its locality with equal brightness in all directions.

603. If we now look at the rich combination of odylic luminous phenomena presented by the iron sphere, and described in § 571 et seq., we see in it *a kind of terrelle or miniature earth, exhibiting a northern and a southern aurora, or polar light, on the small scale.* As the earth is a magnet on the largest scale, so also is our terrelle, in virtue of the electro-magnet passing through its axis, *a magnet of the same form in little. The poles send forth, in that case on the larger, in this on the small scale, a delicate light, visible only in the darkness of night. At a certain height over both poles it bends round, and flows from all sides to the tropical zones, in the form of threads, or bands and rays.* All this takes place on the terrestrial sphere just as on the terrelle; and the light plays in all the glory of rainbow colours on the former as on the latter. If now we may venture to add that, in the two cases, the amounts and intensities of the odylic light will most probably be to each other in the ratio of the different mass in the two spheres (which, as is well known, increases as the cubes of the diameters, while the surface increases only as the squares of the diameters), and consequently the odylic light, sent out from the *mass*, but only spread over the *surface*, must be more concentrated and intense, in a ratio corresponding to the difference of size, over the terrestrial poles than over those of the

terrelle,—if we may suppose this; then the vastly greater luminous power of the polar light, as compared with that of the terrelle under the ordinary pressure, becomes in some degree conceivable.—*A second cause of difference in intensity between the large and the small polar lights, is to be sought for in the odylic action of the sun and moon on the earth.* I have noticed generally, in § 95 et seq., and in § 118 et seq., some experiments, from which it appears that both these heavenly bodies communicate a powerful odylic charge to all objects on which their rays fall. The surface of the earth, on which they, in some part of the globe, constantly shine, receives, therefore, uninterruptedly, a supply of odyle from the sun, just as it receives its supply of solar light and heat. It is therefore not alone the odyle inherent in the earth, considered as a magnet, which appears in a concentrated form at its poles, but in addition to this, there is the copious supply derived from the sun and moon. According to the laws which regulate it, the odyle, when it reaches the earth's surface, not only flows towards those parts where, at the moment, the odylic charge is weaker, but it immediately becomes polarised; that is, the current sets towards the poles, and there increases the intensity of the phenomena caused by the odyle inherent in the mass of the earth. The inequalities in the state of the surface, according as it is covered with clouds or not, the odylic rays being in the former case arrested in the atmosphere, in the latter falling on the solid mass of our planet, and the different positions of sun and moon in regard to the earth, according as they act in combination with, or in opposition to one another;—all these things necessarily cause fluctuations and irregularities in the occurrence of the polar light, just as they do in the weather, or in meteorological phenomena generally. The differences in the intensity of the odylic light between the poles of the terrelle and those of the earth are thus explained, and to a certain extent accounted for.

604. Should it be objected, that no magnet passes through the earth's axis, as through that of my terrelle, I think such an objection can be easily and satisfactorily met. For it cannot annihilate the proved and established fact, that the earth

actually has magnetic poles, whether its magnetism proceed alone from the surface and the sun's rays falling on it, as modern philosophers think they are entitled to suppose, or be also derived from internal conditions of the whole planet. But in reference to this last opinion, there exists a circumstance favourable to it, which, so far as I know, has not yet been taken into account in physics. I shall here venture to explain it. It is now generally admitted that meteorites are of planetary origin, and are small stars, deriving their genealogy from the same source, and revolving in their paths round the sun in the same way as the larger and smaller planets. And indeed the distance is not greater, from the mighty Jupiter to the small Vesta or Astræa, hardly larger than the island of Ceylon, than from these asteroids to the huge meteorites of the Senegal, which are said to form small hills of iron. From such meteorites, through those of Bahia, Durango, or Zacatecas, which are masses of iron of many tons in weight, to that at Aix-la-Chapelle, of a few tons, and to our ordinary meteorites of a few pounds in weight, the transition links are everywhere to be found.* From Jupiter, therefore, to the smallest meteorite, is an unbroken series. There is no difference between a planet and a meteorite, except that of size. If this be admitted, and indeed we can hardly now avoid admitting it, then the structure of meteorites gives us a highly valuable key to open up considerations and doctrines concerning the true internal condition of our earth. The law which presided at the formation of meteorites, and fixed their structure and composition, presided also, on this theory, at the formation of our globe. Now, meteorites consist, chiefly, either of metallic masses, most commonly iron, with a portion of nickel (cobalt, &c.), or of stony masses, almost always abundantly mixed with the same kind of iron. This is well known; but it is not well known, and it is an essential point in the enquiry, that in the large majority of meteorites, the abundant iron is not in the form of irregular, accidental, and scattered portions, *but in that*

* Baron von Reichenbach has long paid much attention to the subject of meteorites, and possesses a magnificent collection of them, from all parts of the world, superior, probably, to any collection in Europe.—W. G.

of a coherent net-work of cells pervading their mass. This can be seen with the naked eye in the meteorites of Krasnojarsk, of Atacama, in the original mass of iron of Bittburg, in the meteorites analogous to these in the museum at Gotha, of unknown origin, and in many others. But even those more minutely mixed, such as that of Smolensk, of Seres, of Blansko, of Taber, of Barbotan, &c. &c. have the same structure. We can remove the stony parts, and obtain a coherent net-work of iron cells. The specific gravity of these stones varies from 3 to 4 and 5. That of the earth is about 4.7, and it is therefore probable that it has an internal constitution like that of a meteorite with a coherent mass of iron cells; that it contains, throughout its stony mass, a pervading web of metallic iron, like the large majority of meteoric stones. *But it is in all probability this iron in which the magnetism of the earth resides.* The inequality of its distribution, as seen in every meteorite, and as we must also suppose to exist in the earth, renders the fourfold polarity of the earth easily conceivable.

605. This view of the constitution of our globe, of which geology has not as yet made any use, but which, as we see, rests on a foundation of fact in natural history, is not in contradiction to the generally admitted supposition, that liquid strata of melted matter may yet exist under our lowest geological formations. Such liquid strata exist as certainly as that meteorites, when they reach the earth, are still hot, and have *a crust* of slag, recently melted. The remarkable meteorite of Clairborn in Alabama was, when I received it, encrusted all round with a slag, from one-fifth to one-third of an inch thick. The meteoric mass of iron of Caryfort, Decalb county, North America, came to me with a crust of about one-sixth of an inch, (0.016 inch) thick. I possess the aerolite of Nanjemoy in Maryland, which is covered, in parts, with a porous slag about one-seventh of an inch (0.014 inch). But from such a covering *to the igneous fusion of the mass throughout* is a long way. The earth has certainly a red-hot or still fluid covering, several leagues thick, analogous to the coating of slag on the meteorites. There are too many concordant and convincing

proofs of this, to permit us to doubt it; but we are far from being thence entitled to conclude, that it must therefore be in a state of fusion through its whole mass, as is generally, but rather hastily assumed. It is beyond all comparison more probable, that it has internally the structure of a meteorite, and indeed of one containing iron, possibly of several combined, as it daily unites under our eyes with new meteorites.

606. Some philosophers, and among them Dr. FARADAY, have asserted, that the earth could not be magnetic through its mass, because it is internally in a state of igneous fusion, and because we know that high temperatures are not reconcileable with the presence of magnetism in bodies. This argument has weight, only so long as we regard *the whole* mass as a lava. But not only is there no strict proof of this, but many considerations are strongly opposed to it, of which I shall here only refer to that above derived from the analogy of meteorites. All of them are melted only on the surface, but internally they are crystalline formations, proceeding from a force which can be shewn to have operated entirely at a low temperature. A moment has come for each meteorite, in which an excessive heat has acted on it externally, has melted the surface, and formed lavas on it. But this has only been a moment, and that a very short one, which has only been able to melt the surface to a very limited depth. Such a moment there has once been for the earth, but could melt down only a thin crust of its mass, which now forms the red hot liquid stratum below its congealed surface, from which molten fountain Geologists suppose our volcanoes, our basalts, trachytes, and porphyries to have flowed, and the existence of which, if admitted, satisfactorily explains the fact, that in every shaft and pit the heat increases as we descend. The hypothesis of the existence of this fluid stratum, has been generally received with favour, as being founded on sufficient data. But such a mere crust of glowing or fused matter is, even if several miles in thickness, altogether insufficient to destroy, or even materially to diminish the effects of the magnetism of the vast iron-pervaded cold sphere, 8000 miles in diameter.

According to what has here been stated, then, there does

actually exist a magnet, passing through the earth, exactly as I passed one through my terrelle (although of a different form), in imitation of it, and thus endeavoured to place the small sphere in conditions analogous, in a certain degree, to those of the terrestrial globe.

607. But this iron, which I hold we must assume to be interwoven with the whole internal mass of the earth, as of meteorites, *is also uniformly crystallised.* When meteoric stones are cut, polished, and very slowly etched by diluted nitric acid, crystalline figures appear on every portion of iron, however small, and may be distinctly recognised under the microscope. These have been shown to be the same appearances, which, on masses of meteoric iron, are called the figures of Widtmannstetten, and they are nothing else than crystalline forms of the metallic constituents. I have examined many meteorites in my own collection, and have uniformly found the reguline iron crystallise, exactly in the same way, and according to the same laws, as the large masses of meteoric iron. I have picked out of the Blansko meteorite* small portions of iron, which, when polished and etched, notwithstanding their small size, exhibited not only planes of cleavage, but crystallised pyrites enclosed in the iron, exactly as in the large masses. Meteorites, therefore, include a twofold source of magnetic and odylic polarity; *they consist in part of cellular but yet coherent masses of iron;* and *these metallic masses are crystallised*, and perhaps form, in each case, a single large crystal, externally indeed irregular, but, internally, cohering according to the laws of crystallization. But we know, from § 55, that all crystals permanently pour out odylic light from their poles. Here, therefore, the presence of iron, and the crystalline form, act together in producing magnetico-odylic poles in the earth, and these are the reasons which to me render it probable, that the magnetism and the odyle of the earth are to be ascribed, not only to external causes, such as the solar rays, but also in great part, perhaps for the most part, to internal causes. In following these out further, the facts,

* BARON VON REICHENBACH long lived at Blansko, in Moravia, and, if I am not mistaken, saw this meteorite fall.—W. G.

otherwise so strange, that the astronomical and magnetic poles do not coincide, that the earth has not two, but four magnetic poles, and many others, will no longer appear unaccountable, but will admit of easy explanation. Thus, the earth may, on good grounds, deduced from the laws of magnetism and of odyle, be supposed to draw its polar light in great part from internal sources, as my terrelle did its odylic light. We may consider both spheres as being diametrically penetrated by bodies in which reside active magnetism and odyle. And thus the analogy between our planet and the imitation of it by artificial means on the small scale, exists also in this view of the phenomena as fully as in the former.

608. In connection with this part of the subject, an old remark made by the Swedish philosopher WILKE deserves to be recalled to memory; namely, *that disturbances of the magnetic needle always* PRECEDE *the appearance and the motions of the aurora borealis.** This, as we have seen, agrees most exactly with the phenomena of odylic light; for these always occur later and more slowly than the associated magnetic or electric effects, which are only followed by the odylic effects after an observable pause. The same facts which I have ascertained in my dark chamber were therefore, many years ago, noticed in the wide expanse of heaven by other observers.

609. Lastly, I may here adduce the observation of WARGENTIN, *that the polar lights render the needle less acute and delicate in its indications, or blunt it.* This is now naturally explained. The aurora borealis is an odylo-magnetic emanation, ponderable or imponderable, material or immaterial. Where it issues from the north pole of the earth, it is positively magnetic. But the *northward* pole of the needle is negative. When + M passes over the needle, therefore, from north to south, as the aurora does, the effect is the same as if a very feeble southward pole of a magnet were drawn uninterruptedly along the needle, from north to south, during the whole continuance of the aurora. The effect of the aurora must be the same as that of the ordinary reversed stroke of a weak magnet. And every one knows, that a needle thus

* GEHLER's Physical Dictionary, I. 161.

stroked against its polarity, must lose power; that is, as WARGENTIN says, must be blunted. The aurora exerts on the needle the influence of a real, though feeble reverse stroke, the tendency of which is to concentrate at the southward pole, the south or negative polarity of the northward pole of the needle; in other words, to reverse its polarity. Although the intensity of the aurora is not sufficient to reverse the polarity of the needle, it yet weakens or blunts it, and consequently, as WARGENTIN adds, renders it less sensible to future polar lights.

610. All the facts that have been ascertained, in regard to the properties of odyle, harmonise, therefore, with constantly increasing consistency, in the conclusion that the polar light is to be regarded as a vast manifestation of magnetic odylic flame, odylic vapour, and odylic light. The terrelle exhibits the phenomena of the polar light very perfectly in miniature.

END OF PART II.

APPENDIX.

APPENDIX.

A.

When I became acquainted, in 1845, with the researches of the Author, by the publication of Part I., I was naturally anxious to have an opportunity of observing, for myself, the phenomena presented by sensitive persons. I was deeply convinced of the truth and exactitude of Baron von Reichenbach, but it was impossible not to feel a desire personally to become acquainted with these interesting facts. I therefore enquired for such persons, affected with nervous disease, as were likely to be sensitive in a high degree, and I very soon found one who would certainly have stood very high in the Author's list. This was a young woman, recently delivered of a healthy child at one of the charitable Institutions in this city. Her appearance was that of perfect and vigorous health, and she was nursing a thriving child. But I was informed, that she was so excessively nervous, that if the hand of another person were placed on her head, she was very soon thrown into convulsions, on the cessation of which, she was found to be in a state of somnambulism. I was also told, by eye-witnesses of intelligence, education, and veracity, that she exhibited, in this state, the phenomena of transference of one of the senses. As it was obviously not prudent to excite convulsions, in order to produce the somnambulistic state, I did not see her in that state at all. Neither was it found practicable to make observations in the dark, so that I was here confined to her sensations. I tried on her passes with various weak magnets, the only ones at the time in my possession, and, in every particular, she described the same sensations, without any questions being put, as the sensitives examined by the Author; namely, different apparent temperature from the same end on the upward and downward pass, so that she could instantly tell which pole or which pass was employed. These sensations were very vivid and strong, the cooler being pleasant, the warmer unpleasant. I obtained precisely similar results with several crystals of moderate size. Both magnets and crystals caused twitchings and tendency to spasm. I then tried the effect of holding her left hand in my right, adding afterwards her right in my left. Here also the sensation of a current, as described by the Author in § 85, was very strongly and vividly felt, and was not disagreeable to her, although tending after a time also to excite spasm. A few minutes later, while she was describing some of her sensations, which she did very clearly and intelligently, I quietly crossed my hands, and took hold of both of hers, unconformably, that is, right in right, left in left. I had not thus held them for half a minute, when she stopped speaking, tore away her hands, and gave, word for word, the same description of the dreadful sensations she experienced, as is given in § 85, to which I refer. She added that, had it continued a little longer, she would have had a fit; of which, from the above history of her case, I entertain no doubt, but of course I did not attempt to verify it by experiment. As in the Author's cases, no persuasions could induce her to allow this experiment to be repeated. This took place in the spring of 1846, while I was preparing my Abstract. There can, I think, be no doubt, that this patient would have seen the odylic light. Indeed, her sensitiveness was so extreme, and her case in all respects so interesting, that I had the intention of pursuing the investigation under more favourable circumstances, when she left the Institution; but she only did so to join her husband abroad, and I have never since heard of her.

The next case that occurred was that of a lady, since dead, who had long laboured under a complication of disorders, chiefly of the nervous system. My learned colleague, Sir W. Hamilton, showed to her and to her sisters my Abstract; when they all immediately declared, that in that work they found the explanation of the fact, observed for years, that the patient could never feel tolerably comfortable, or obtain sleep, except when lying in the position from north to south. As this proved her to be sensitive, besides confirming the Author by observations made long before his work existed, I was desirous to test her on other points. Sir W. Hamilton introduced me to the family, and the lady most obligingly consented to make some trials. She felt strongly the action of magnets and of crystals, precisely as the Author describes them. In an extemporised and imperfect darkness, she was yet able, being highly sensitive, to see light from crystals; (I did not try magnets, as mine were but weak); and was able, in the dark, when no one of five or six other persons could see any thing, instantly to discover a rock crystal wherever it might be placed, by the light it emitted.

Another lady, a member of my family, subject, occasionally, to very severe headaches, and who felt very distinctly the action of magnets and crystals, even of small size and feeble intensity, was also able to discover a small rock crystal by its light in the same way.

A boy, nine or ten years of age, the son of a gentleman residing near Ambleside, and whom I saw playing on the lawn, in perfect health, was called in, and shut up in a short passage between two drawing rooms, which could be tolerably, but not perfectly darkened. He held one end of a copper wire ten or twelve feet long, which passed through the key-hole. When the other end was placed on one pole of a weak horse-shoe magnet, he soon described a warm sensation passing up his arm from the wire. When I immersed the end of the wire in weak nitric acid, the sensation was different. The wire soon felt hot, and after a pause, the boy called out, "I see a fire." On further enquiry, he described this in his own language, but found a difficulty in comparing it to any thing he knew; but at last he said it was *a shining green cloud* or smoke. Of course I had carefully avoided telling him any thing whatever of the phenomena. It is most interesting to see that this little boy, who probably never knew what kind of wire he held, and was certainly, if he had known it, not aware that copper and its compounds tinge ordinary flame green, described a green flame from the point of the wire, or surrounding the wire; a fact always observed by the Author in the odylic light when copper was used.

The same boy, in various crystals of rock crystal, rock salt, gypsum, heavy spar, galena, and Iceland spar, instantly detected the warm and cool poles in the lines of the crystallographic axes, pointing out, in some crystals, three pairs of such poles. I marked these, unknown to him, and tried him again in the dark; he always fixed on the same poles as warm or cold.

I was not able to devote the necessary time and labour to a further study of such cases. I had seen enough to confirm my previous knowledge of the Author's minute accuracy. But I have, at different times, and including the above four cases, tried twelve persons with magnets and crystals. More than half perceived sensations, more or less distinct, from passes made with these, and, as we have seen, four were highly sensitive. This agrees with the Author's estimate of the frequency of sensitive persons; and, as far as they go, my imperfect, but carefully performed experiments, confirm what he has told us on several important points. I hear almost daily of persons exhibiting some of the phenomena of sensitiveness, so that those who wish to investigate the subject, can have no difficulty in finding the means of doing so.

I have lately been informed of the following case. A lady, during her last illness, and when, I understand, speechless (probably from some cerebral affection), was observed by her family invariably, after having been, as they supposed, comfortably arranged and settled in bed, to have moved in a short time so as to lie *across* the bed. After they had long in vain tried to counteract this tendency, her relatives, very sensibly, turned the bed at right angles to its former position. From that time till she died, although in precisely the same state, the patient never changed her position once, as she had done before, and gave signs of satisfaction at the change. Her relatives had not the slightest idea of the cause of this, and it occurred some years before the Abstract was published. But on lately seeing that work, they at once recognised the facts they had observed, and obtained the natural explanation of them.

WILLIAM GREGORY.

B.

The experiments to be here briefly noticed, were made by a gentleman, well versed in science, on a lady, when in the mesmeric sleep. This lady exhibited various remarkable phenomena, including clairvoyance in different forms; but into these I do not here enter, as being foreign to the special object of this work. All somnambulists, spontaneous or otherwise, are in a high degree sensitive, and we are, therefore, here to consider this lady simply in reference to her odylic sensitiveness when in the mesmeric sleep. It will be seen that her descriptions coincide perfectly with those of others examined by the Author in the ordinary state (a point of much interest); which again, in all points, agree with those given by Mlle. ATZMANNSDORFER in the spontaneous somnambulistic state, as recorded in § 480. It is clear that the only difference, in regard to these phenomena, between the ordinary state and that of somnambulism, is, that in the latter the sensitiveness is more intense.

April 9, 1846.—Mrs. K., aged twenty-eight, in the mesmeric sleep, produced in the course of a minute by Mr. F. Crystals of quartz and fluor spar, caused the hand to follow them. When placed in the hand, it was spasmodically contracted, and became rigid.

April 13.—A rock crystal, two inches long and half an inch thick, was convulsively grasped in the hand, and the arm rose, and became rigid. A bilateral crystal of selenite (gypsum) four inches by one and a half inches, exhibited, in the diffused daylight, a "soft and dimmish light, but the summit very bright, and with rays, yellow and red." The light soon became too intense, and "made the eyes water; she could not bear it." A short time after, the flame above the crystal was from one to two inches higher, the light as intense as before. She saw five points of light (still in the ordinary light of a room, with the blinds drawn) on the points of Mr. F.'s fingers.

April 17.—A large translucent double crystal of quartz, each crystal five inches by one and a half, held at three feet distance, gave out a light too brilliant to be borne. At six feet, it was "pleasant and very beautiful, light, yellow, and white, rising two inches above the crystal, flickering. A glow of light all over." Water was charged by these crystals, and much relished by the patient. It was somewhat different from water charged by the hand; it tasted warm and like peppermint, without flavour. A crystal of fluor spar charged water less powerfully, but distinctly.

April 20.—The light of the gypsum crystal, at less than six feet, was too intense to be borne. At six feet, the colours were "beautiful, varied, white, yellow, green, and blue." It soon became too powerful even there. On interposing the hand, the crystal was not seen, but only a strong light diffused all round it. Even three or four feet further off, in front of an unclosed window, it became too powerful, and was removed on account of the over-excitement it caused. Passes with the quartz crystal were felt very vividly.

A notice of these experiments appeared in the Critic of May 16, 1846, from which, with the permission of the highly qualified observer, I have extracted them.

W. G.

Plate I

Fig. 1.

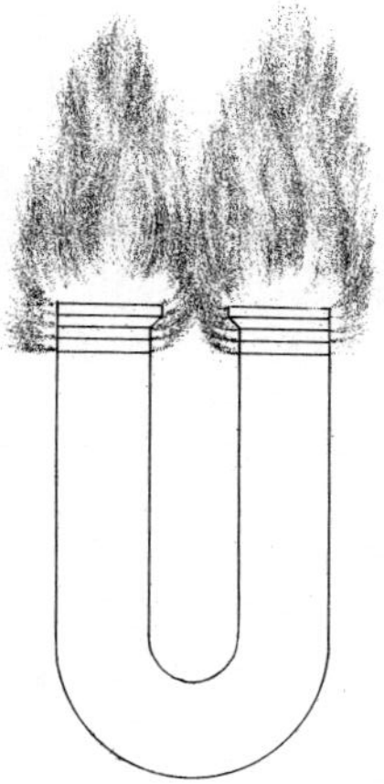

Fig. 2.

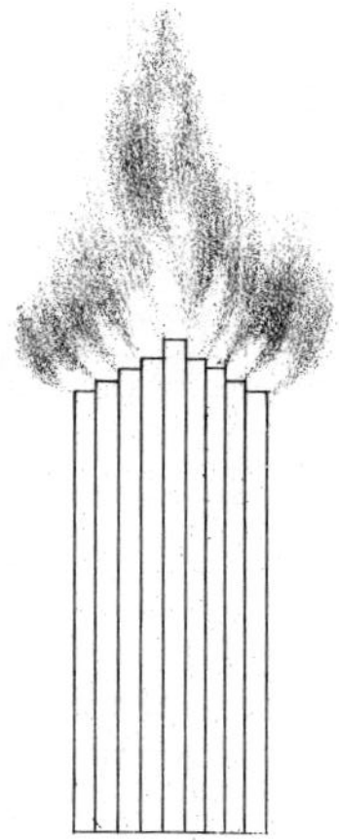

Fig. 3.

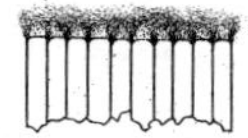

Fig. 4.

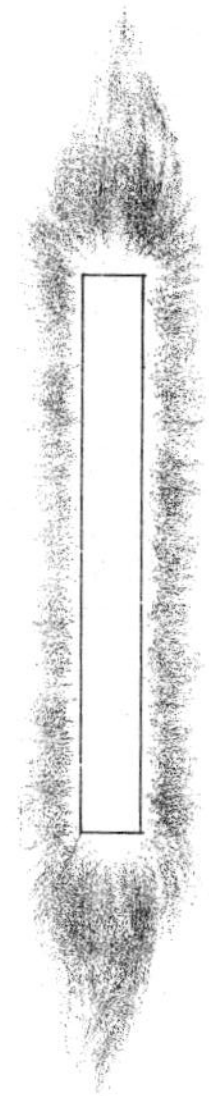

Fig. 5.

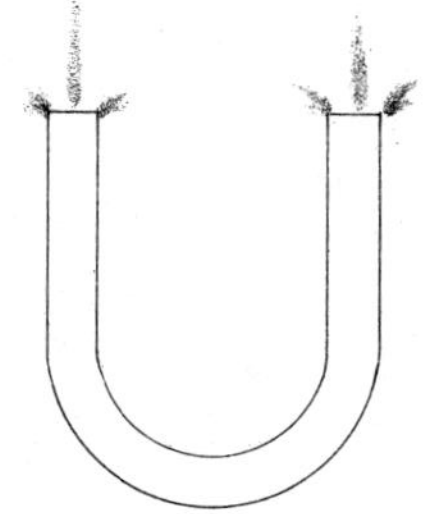

Fig. 6.

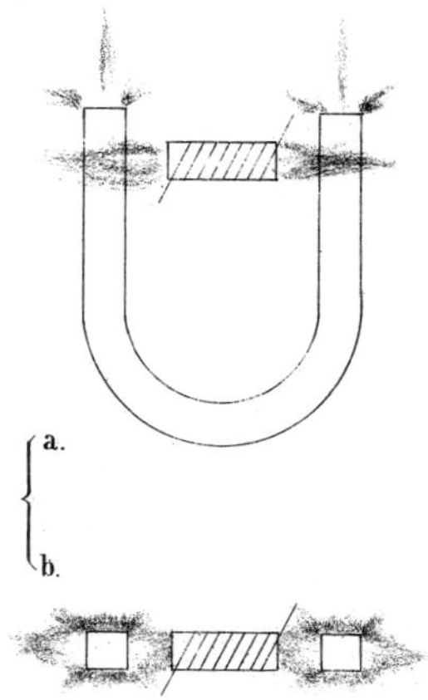

Fig. 7.

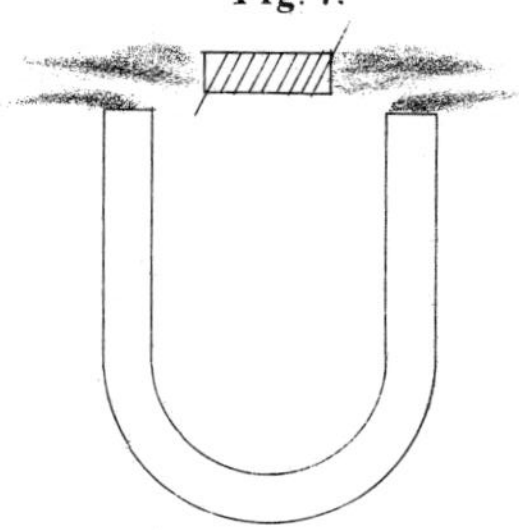

Fig. 8.

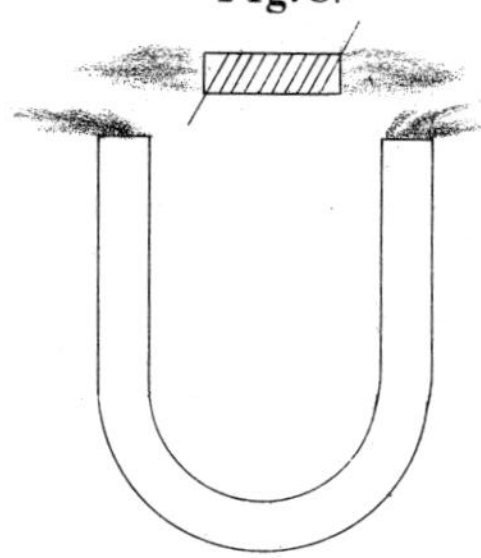

Fig. 9.

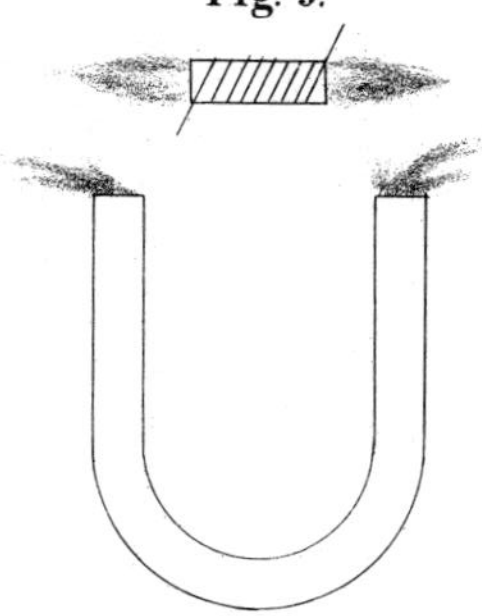

Fig. 10.

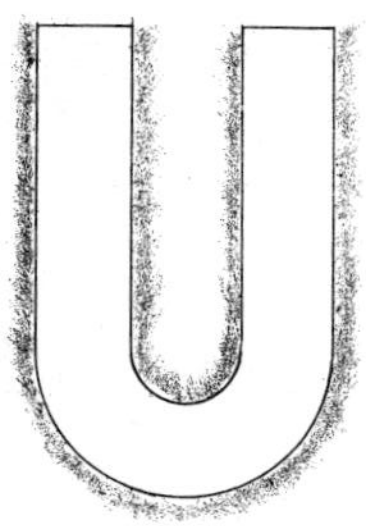

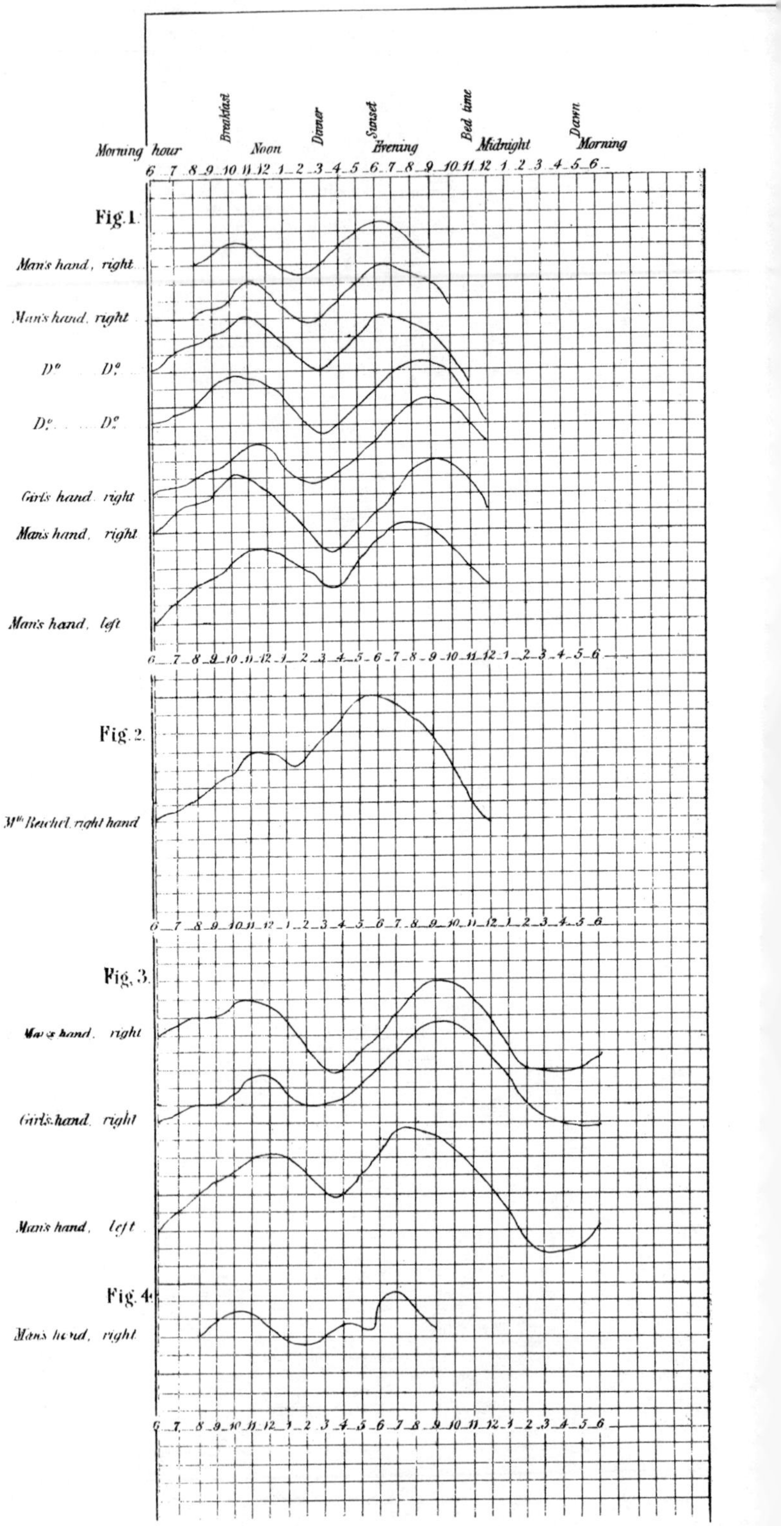
Morning hour
Breakfast
Noon
Dinner
Sunset
Evening
Bed time
Midnight
Dawn
Morning
6 7 8 9 10 11 12 1 2 3 4 5 6 7 8 9 10 11 12 1 2 3 4 5 6
Fig. 1.
Man's hand, right
Man's hand, right
D° D°
D° D°
Girl's hand, right
Man's hand, right
Man's hand, left
Fig. 2.
Mr. Reichel, right hand
Fig. 3.
Man's hand, right
Girl's hand, right
Man's hand, left
Fig. 4.
Man's hand, right

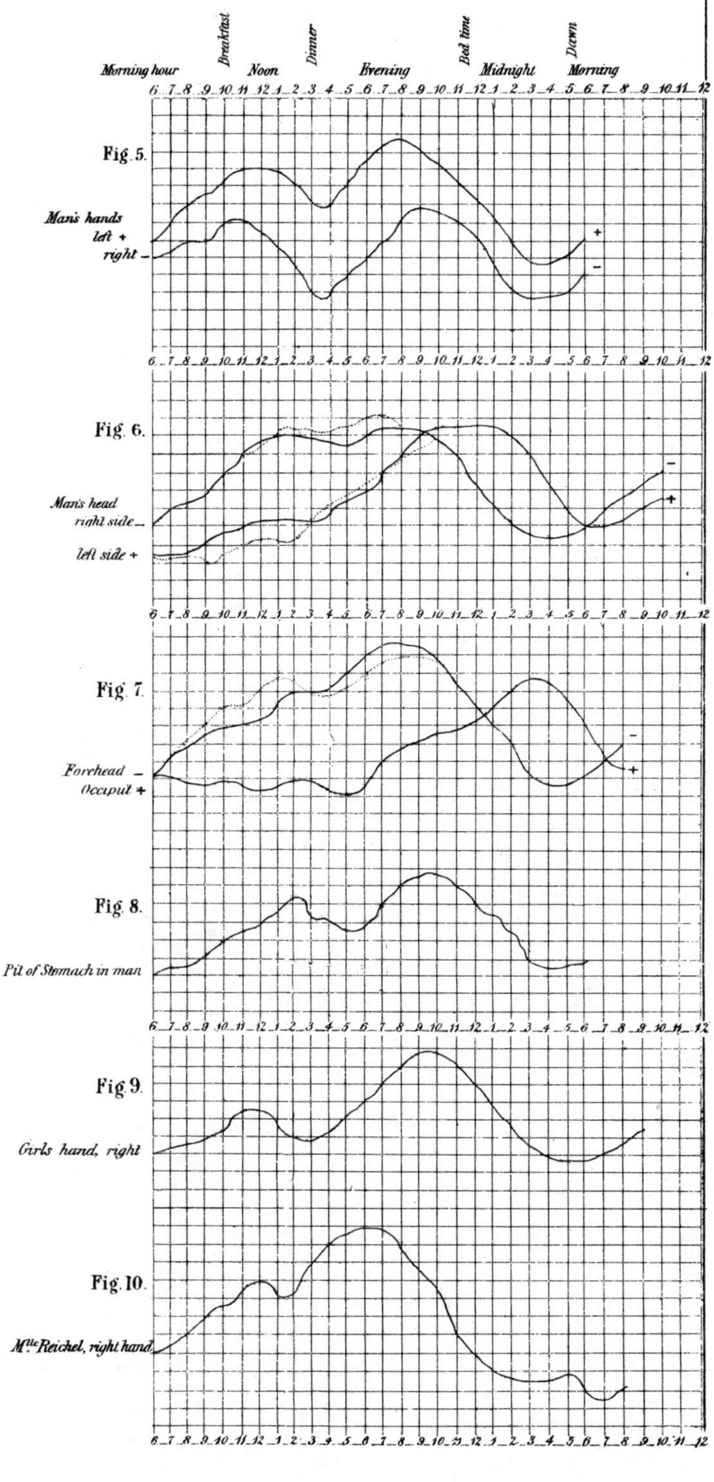
Breakfast
Dinner
Bed time
Dawn
Morning hour
Noon
Evening
Midnight
Morning
Fig. 5.
Man's hands
left +
right –
Fig. 6.
Man's head
right side –
left side +
Fig. 7.
Forehead –
Occiput +
Fig. 8.
Pit of Stomach in man
Fig. 9.
Girls hand, right
Fig. 10.
Mlle Reichel, right hand

a
b
c
d
d
e
f
g
h